# 4th International Conference on Structure, Processing and Properties of Materials (SPPM 2018)

AA006976

IOP Conference Series: Materials Science and Engineering Volume 438

Dhaka, Bangladesh
1 – 3 March 2018

ISBN: 978-1-5108-7656-9
ISSN: 1757-8981

**Printed from e-media with permission by:**

Curran Associates, Inc.
57 Morehouse Lane
Red Hook, NY 12571

CURRAN ASSOCIATES INC. proceedings.com

**Some format issues inherent in the e-media version may also appear in this print version.**



Printed by Curran Associates, Inc. (2019)



Institute of Physics
Dirac House, Temple Back
Bristol BS1 6BE UK

Phone: 44 1 17 929 7481
Fax: 44 1 17 920 0979

techtracking@iop.org

**Additional copies of this publication are available from:**

Curran Associates, Inc.
57 Morehouse Lane
Red Hook, NY 12571 USA
Phone: 845-758-0400
Fax: 845-758-2633
Email: curran@proceedings.com
Web: www.proceedings.com

# TABLE OF CONTENTS

## TECHNICAL SESSION 3A FATIGUE, FRACTURE, AND PERFORMANCE



## TECHNICAL SESSION 3B POLYMERS AND COMPOSITES 2



## TECHNICAL SESSION 4A NANOMATERIALS AND THIN FILM 2

IOP Conf. Series: Materials Science and Engineering **438** (2018) 011001 doi:10.1088/1757-899X/438/1/011001

# Souvenir

4th International Conference On

## Structure, Processing and Properties of Materials

SPPM2018

1 – 3 March 2018
BUET, Dhaka

Organised By:

Department of Materials and Metallurgical Engineering
Bangladesh University of Engineering and Technology
Dhaka-1000, Bangladesh

IOP Conf. Series: Materials Science and Engineering **438** (2018) 011001 doi:10.1088/1757-899X/438/1/011001

4th Intl. Conf. on Structure, Processing and Properties of Materials, SPPM2018
1 – 3 March 2018, BUET, Dhaka

IOP Conf. Series: Materials Science and Engineering **438** (2018) 011001 doi:10.1088/1757-899X/438/1/011001

4th Intl. Conf. on Structure, Processing and Properties of Materials, SPPM2018
1 – 3 March 2018, BUET, Dhaka

# List of Sponsors

University Grants Commission of Bangladesh

IOP Conf. Series: Materials Science and Engineering **438** (2018) 011001 doi:10.1088/1757-899X/438/1/011001

4th Intl. Conf. on Structure, Processing and Properties of Materials, SPPM2018
1 – 3 March 2018, BUET, Dhaka

# International Advisory Committee

IOP Conf. Series: Materials Science and Engineering **438** (2018) 011001 doi:10.1088/1757-899X/438/1/011001

4th Intl. Conf. on Structure, Processing and Properties of Materials, SPPM2018
1 – 3 March 2018, BUET, Dhaka

# Organising Committee

Chair
Prof. M. Moniruzzaman

Organising Secretary
Prof A. K. M. Bazlur Rashid

Members
Prof. A. A. M. R. Haque
Prof. M. N. Haque
Prof. M. A. Islam
Prof. A. Sharif
Prof. F. Gulshan
Prof. K. M. Shorowordi
Dr. M. Hasan
Mr. M. M. Rahman
Dr. H. M. M. A. Rashed
Dr. M. M. Billah
Mr. M. H. Kabir
Mr. S. A. Ayon
Ms. S. Majumder
Mr. S. Mondal

# Sub-Committees

Welcome Sub-Committee
Prof. A. A. M. R. Haque
Prof. M. N. Haque
Prof. M. Moniruzzaman
Prof. A. K. M. B. Rashid
Prof. M. A. Islam

Invitation, Publicity and Public Relation Sub-Committee
Prof. M. Moniruzzaman
Prof. A. K. M. B. Rashid
Prof. M. A. Islam
Mr. S. A. Ayon
Ms. S. Majumder

Fund-Raising and Finance Sub-Committee
Prof. M. Moniruzzaman
Prof. M. A. Islam
Prof. A. K. M. B. Rashid
Mr. M. M. Rahman

Registration, Accommodation and Transportation Sub-Committee
Dr. M. Hasan
Dr. M. M. Billah
Mr. S. A. Ayon
Ms. S. Majumder

Editorial Sub-Committee
Prof. A. K. M. Bazlur Rashid
Prof. M. A. Islam
Dr. H. M. M. A. Rashed

Publication and Website Management Sub-Committee
Prof. A. K. M. Bazlur Rashid
Prof. M. A. Islam
Dr. H. M. M. A. Rashed

Venue Management Sub-Committee
Prof. A. Sharif
Prof. K. M. Shorowordi
Dr. M. M. Billah
Mr. M. M. Rahman
Engr. M. F. Hossain
Engr. A. A. Maksud
Engr. M. N. Haque

Food and Entertainment Sub-Committee
Dr. H. M. M. A. Rashed
Mr. M. H. Kabir
Mr. S. Mondal

IOP Conf. Series: Materials Science and Engineering **438** (2018) 011001 doi:10.1088/1757-899X/438/1/011001



# Conference Secretariat

Prof. A. K. M. Bazlur Rashid
Department of Materials and Metallurgical Engineering
Bangladesh University of Engineering and Technology
Dhaka-1000, Bangladesh
Tel. and Fax: (+8802) 966 5618
Email: sppm2018@mme.buet.ac.bd

IOP Conf. Series: Materials Science and Engineering **438** (2018) 011001 doi:10.1088/1757-899X/438/1/011001



# Welcome Message from the Head

The Department of Materials and Metallurgical Engineering (MME) of Bangladesh University of Engineering and Technology (BUET) feels extremely happy to welcome you all to the 4th International Conference on "Structure, Processing and Properties of Materials," SPPM2018. The department started in 1952 as the Department of Metallurgical Engineering, and subsequently changed to its present name in 1997. The Department of MME at BUET is the leading seat of comprehensive education and research in the field of materials science and metallurgical engineering in the country. In Bangladesh, this Department has made significant contributions in the fields of steel, ceramics, polymer and composite materials. MME Department is equipped with a substantial body of expertise and experience in different branches of materials science and engineering. Through the concerted efforts over the last years this department now frequently providing expert services in different national activities. The department has developed a network of industrial contacts within Bangladesh with large companies and with small and medium sized enterprises (SMEs) and performs joint research and development projects where the department of MME plays the role of a knowledge centre.

On this occasion, the department of MME feels highly honoured of having Professor Abdul Mannan, Chairman, University Grants Commission of Bangladesh as Chief Guest to inaugurate the programme. We thank Professor Dr. Saiful Islam, Vice-Chancellor of BUET and Dr. Gauranga Chandra Mohanta ndc, Project Director (Additional Secretary), Higher Education Quality Enhancement Project (HEQEP) to grace the occasion as Special Guests.

This is the fourth in a row that this department is involved in organising the Conference SPPM. We hope that the conference will provide a useful forum for the materials scientists and engineers, both from home and abroad, to share their experience and research results. The participants of this conference will go back to their workplaces with updated knowledge and contribute more effectively to their respective professions.

We appreciate the financial assistance from HEQEP CPSF 3117 sub-project. The contribution of different organisations and institutions, reflecting the growing realization of role of materials to all engineering development, is highly commendable. We believe the success of the conference will greatly depend on the volunteer efforts of all members of Department of MME. We thank all those who helped make this conference possible.

Professor Dr. Md. Moniruzzaman

IOP Conf. Series: Materials Science and Engineering **438** (2018) 011001 doi:10.1088/1757-899X/438/1/011001



# Programme Summary

| 01 March 2018; Thursday | | |
|---|---|---|
| 03:00 am – 06:00 pm | **EARLY REGISTRATION**<br>Venue: Professor Ibrahim Library, Department of MME | |
| **02 March 2018; Friday** | | |
| 09:00 am – 10:00 am | **LATE REGISTRATION**<br>Venue: Council Bhabhan, BUET | |
| 10:00 am – 11:00 am | **INAUGURAL CEREMONY**<br>Venue: Council Bhabhan, BUET | |
| 11:00 am – 11:30 am | Tea / Coffee Break | |
| 11:30 am – 12:15pm | **PLENARY SESSION 1**<br>Venue: Council Bhabhan, BUET | |
| 12:15 pm – 02:30 pm | Prayer / Lunch Break | |
| 02:30 pm – 04:00 pm | **TECHNICAL SESSION 1A:**<br>Nanomaterials & Thin Films 1<br>Venue: Seminar Room, DMME | **TECHNICAL SESSION 1B:**<br>Polymers & Composites 1<br>Venue: PI Library, DMME |
| 04:00 pm – 04:30 pm | Tea / Coffee Break | |
| 04:30 pm – 06:00 pm | **TECHNICAL SESSION 2A:**<br>Processing, Fabrication & Properties<br>Venue: Seminar Room, DMME | **TECHNICAL SESSION 2B:**<br>Ceramic & Biomaterials<br>Venue: PI Library, DMME |
| **03 March 2018; Saturday** | | |
| 09:30 am – 10:45 am | **PLENARY SESSION 2**<br>Venue: Seminar Room, Department of MME | |
| 10:45 am – 11:00 am | Tea / Coffee Break | |
| 11:00 am – 12:30 pm | **TECHNICAL SESSION 3A:**<br>Fatigue, Fracture & Performance<br>Venue: Seminar Room, DMME | **TECHNICAL SESSION 3B:**<br>Polymers & Composites 2<br>Venue: PI Library, DMME |
| 12:30 pm – 01:00 pm | **POSTER SESSION**<br>Venue: Department of MME | |
| 01:00 pm – 02:30 pm | Prayer / Lunch Break | |
| 02:30 pm – 04:30 pm | **TECHNICAL SESSION 4A:**<br>Nano-materials & Thin Films 2<br>Venue: Seminar Room, DMME | **TECHNICAL SESSION 4B:**<br>Miscellaneous<br>Venue: PI Library, DMME |
| 04:30 pm – 05:00 pm | Tea / Coffee Break | |
| 05:00 pm – 05:30 pm | **CONCLUDING SESSION**<br>Venue: Seminar Room, DMME | |

IOP Conf. Series: Materials Science and Engineering **438** (2018) 011001 doi:10.1088/1757-899X/438/1/011001

4th Intl. Conf. on Structure, Processing and Properties of Materials, SPPM2018
1 – 3 March 2018, BUET, Dhaka

# Inaugural Ceremony

**02 March 2018, Friday Council Bhabhan, BUET 10:00 am – 11:00 am**

| | |
|---|---|
| 09:50 am | Guests take their seats |
| 10:00 – 10:05 am | Welcome address by the Conference Secretary<br>**Dr. A. K. M. Bazlur Rashid**<br>Professor, Department of MME, BUET |
| 10:05 – 10:20 am | Speech by the Special Guest<br>**Dr. Gauranga Chandra Mohanta** *ndc*<br>Project Director (Additional Secretary), HEQEP, UGC of Bangladesh |
| 10:20 – 10:35 am | Speech by the Special Guest<br>**Prof. Dr. Saiful Islam**<br>Vice-Chancellor, BUET |
| 10:35 – 10:50 am | Speech by the Chief Guest<br>**Prof. Abdul Mannan**<br>Chairman, University Grants Commission of Bangladesh |
| 10:50 – 11:00 am | Vote of thanks by the Chair<br>**Dr. Md. Moniruzzaman**<br>Professor and Head, Department of MME, BUET |
| 11:00 am | Refreshment |

IOP Conf. Series: Materials Science and Engineering **438** (2018) 011001 doi:10.1088/1757-899X/438/1/011001



# Technical Session Details

## Plenary Session 1
02 March 2018, Friday, 11:30 am – 12:15 pm, Council Bhabhan, BUET
**Chair:** Prof. Dr. A. S. M. A. Haseeb, University of Malaya, Malaysia

**[A01]**
Full-Scale Fire Testing of CFRP-Strengthened Reinforced Concrete Column
*Sobia Anwar Qazi, Hamidah Mohd. Saman, Azmi Ibrahim, Mahmood Tahir, Shek Poi Ngian*

## Technical Session 1A – Nanomaterials and Thin Film 1
02 March 2018, Friday, 02:30 pm – 04:00 pm, Seminar Room, Department of MME
**Chair:** Dr. Mohammed Abdul Basith, Asso. Prof., Department of Physics, BUET

**[K03]**
$Bi_{0.9}Ho_{0.1}FeO_3$/ $TiO_2$ Composite Thin Films: Synthesis and Study of Structural and Optical Properties
*Md. Rafiqul Islam, M. A. Zubair, M. S. Bashar and A. K. M. B. Rashid*

**[K20]**
Effect of Activation Site Density on Copper Encapsulation of MWCNTs by Electroless Deposition
*Md Muktadir Billah and Dr. Quanfang Chen*

**[K10]**
Fabrication of Silver Nanoparticles Decorated Graphene-Magnetic Carbon Nanotube with High Catalytic Efficacy
*Md. Ferdous, Md. Rafiqul Islam and Md. Shafiul Azam*

**[K14]**
Synthesis and Characterization of Boron Doped Bismuth Oxide Thin Films via a Simple Chemical Route for Fuel Cell Applications
*B. C. Dev, M. H. Babu and J. Podder*

**[K15]**
Study of Structural and Optical Properties of Spray Deposited Undoped and Zinc (Zn) Doped Bismuth Oxide Thin Films
*F. Basharat, J. Podder and Ishtiaque M. Syed*

IOP Conf. Series: Materials Science and Engineering **438** (2018) 011001 doi:10.1088/1757-899X/438/1/011001



**[K05]**
Effect of Process Variables on Deposited Copper Oxide Thin Film by Sol-gel Spin Coating Technique
*Dipika Rani Bhowmik, Aninda Nafis Ahmed, M. A. Gafur, Muhammed Yusuf Miah and Dipa Islam*

**[K16]**
Electroless Deposition of p-type Cuprous Oxide Thin Film for Solar Application
*Asif Mahmud, Shadman Sharar Mahmud and Md. Moniruzzaman*

## Technical Session 1B – Polymers and Composites 1

02 March 2018, Friday, 02:30 pm – 04:00 pm, Professor Ibrahim Library, Department of MME
**Chair:** Dr. Mubarak Ahmad Khan, Previous DG, AERE, Savar

**[G23]**
Synthesis and Characterization of Reduced Graphene Oxide (rGO) Reinforced Polymer Matrix Composite
*V. C. Saha, M. Alam, P. Shams, S. Rahman, M. R. Qadir, M. R. Islam and F. Gulshan*

**[G01]**
Post-stress Induced Tensile Property Modification of Jute Polymer Composite
*M. K. Nahian, S. Z. Ali and M. A. Islam*

**[G24]**
Fabrication and Characterization of Sugarcane Bagasse Reinforced Polypropylene (PP) Composites
*Mohammad B. Hoque, Asiqun Nobi, Anich Mia, Mamun Hossain, Md. Sabbir Azam, Abu H. M. Kamal, Md. Razzak, Muhammad B. Uddin and Ruhul A. Khan*

**[G04]**
Effect of Fiber Ratio and Chemical Treatment on Properties of Banana and Betel Nut Fiber Reinforced Hybrid Polypropylene Composites
*B. Sutradhar, M. B. Mesbah and M. Hasan*

**[G02]**
Flexural Property Enhancement of Jute Polyester Composite
*S. Z. Ali, M. K. Nahian and M. A. Islam*

**[G08]**
Effect of Age of Bamboo Fibers on Mechanical Properties of Bamboo Fiber Reinforced Composites: An Experimental Investigation.
*Md. Kharshiduzzaman, Raju Ahmed, Abul Kalam Azad and Aminul Ahasan*

**[G21]**
Mechanical Properties of Silk and Glass Fiber Reinforced Hybrid Polypropylene Composites
*Md. R. H. Mazumder, F. Numera, A. Al-Asif and M. Hasan*

IOP Conf. Series: Materials Science and Engineering **438** (2018) 011001 doi:10.1088/1757-899X/438/1/011001

4th Intl. Conf. on Structure, Processing and Properties of Materials, SPPM2018
1 – 3 March 2018, BUET, Dhaka

## Technical Session 2A – Processing, Fabrication and Properties

02 March 2018, Friday, 04:30 pm – 06:00 pm, Seminar Room, Department of MME
**Chair:** Dr. Md. Abdul Gafur, Principal Scientific Officer, IFRD, BCSIR, Dhaka

**[B03]**
Effects of Ca Addition on Microstructure and Hot Deformation of AZ61 Magnesium Alloy
*I. A. F. Rabbi, M. M. A. Shawon and H. M. M. A. Rashed*

**[B05]**
Study of Electromigration Failure of Indium-Silver Alloy
*S. A. Chowdhury, K. M. Shorowordi and M. A. Matin*

**[B07]**
Effect of $Mg_2Sn$ Particles on Microstructure and Mechanical Properties of Aged AZ Series Magnesium Alloy
*M. E. Hossain, H. M. M. A. Rashed and M. A. Zubair*

**[C12]**
Thermal Modification of Ni Content Al-Si-Cu Automobile Engine Block Alloys
*A. Hossain, F. Gulshan and A. S. W. Kurny*

**[C01]**
Effect of Copper/Magnesium Ratio on Mechanical Properties of an AA7xxx Aluminium Alloy
*Md. Jasim Uddin and H. M. M. A. Rashed*

**[C03]**
Development of a High Strength Al–$Mg_2Si$–Mg–Zn Based Alloy
*Mahbuba Akhter Liza and A. K.M. B. Rashid*

**[C02]**
Effect of Cu on Mechanical Properties of an Al-Si Alloy
*E. A. Ashrafi, A. Hasan and H. M. M. A. Rashed*

**[C19]**
Synthesis and Micro Hardness Properties of Ball Milled rGO Reinforced Aluminium Matrix Composite
*P. Shams, S. Rahman, V. C. Saha, M. Alam, S. Ahmed, Md. Hasanuzzaman and F. Gulshan*

IOP Conf. Series: Materials Science and Engineering **438** (2018) 011001 doi:10.1088/1757-899X/438/1/011001

4th Intl. Conf. on Structure, Processing and Properties of Materials, SPPM2018
1 – 3 March 2018, BUET, Dhaka

## Technical Session 2B – Ceramics and Biomaterials

02 March 2018, Friday, 04:30 am – 06:00 pm, Professor Ibrahim Library, Department of MME
**Chair:** Dr. Md. Fakhrul Islam, Professor, Department of Glass and Ceramic Engineering, BUET

**[I01]**
Theoretical and Experimental Evidence of Modified Structure, Magnetism and Optical Properties in Ba and Mn Co-Substituted $BiFeO_3$
*Tilottoma Saha, Fabia Farlin Athena, M. A. Matin and Ahmed Sharif*

**[I02]**
Study of the Structural and Magnetic Properties of Li Substituted Cu-Mn Mixed Ferrites
*M. A. M. Quraishi, M. S. Ullah and A. K. M. Akther Hossain*

**[I03]**
Effect of Sr and Mn Co-doping on Multiferroic Properties of $BiFeO_3$ via Sol-gel Chemical Process
*Sumaiya Jannat, Md. Rafiqul Islam, M. A. Matin and Ahmed Sharif*

**[J01]**
A Comparative Analysis of Conventional Biomaterials as Bone Substitutes for Frontal Human Skull
*M. M. Billah, A. Mazumder and A. Bhattacharjee*

**[I04]**
Processing of Ultralight Porous $Al_2O_3$-Ceramics by Biotemplating of Bamboo
*M. A. Rahman and K. M. Shorowordi*

**[L04]**
Utilization of Hard Rock Dust in Ceramic Glaze Formulation
*Mst. Shanjida Sultana, Aninda Nafis Ahmed, Mohammad Nazim Zaman, Md. Aminur Rahman*

**[H05]**
Review Paper on Investigation of a Possibility of Electronic Battery
*Azrin Akhter Chowdhury*

**[I05]**
Effects of Gd and Cr Co-doping on Multiferroic Properties of $Bi_{0.9}Gd_{0.1}Fe_{(1-x)}Cr_xO_3$ (x=0-0.08)
*M. N. Hossain, M. A. Matin, M. A. Hakim, M. F. Islam*

**[I06]**
Effects of $CoFe_2O_4$ Substitution on Magnetic Properties of $NiFe_2O_4$ Spinel Ferrite
*S. A. Ali, M. A. Matin, M. A. Hakim, M. F. Islam*

IOP Conf. Series: Materials Science and Engineering **438** (2018) 011001 doi:10.1088/1757-899X/438/1/011001



## Plenary Session 2

03 March 2018, Saturday, 09:30 am – 10:45 pm, Seminar Room, Department of MME
**Chair:** Prof. Dr. Hamidah Mohd. Saman, Universiti Teknologi MARA, Malaysia

**[A02]**
Multi-Scaled Heterogeneities in Microstructures and Deformation Behaviour in Steels
*Y. Tomota*

**[A03]**
Reactive Nanoparticle Doped Flux for High Reliability Electronic Micro-Joints
*A. S. M. A. Haseeb, M. Nasir Bashir, G. K. Sujan*

## Technical Session 3A – Fatigue, Fracture, and Performance

03 March 2018, Saturday, 11:00 am – 12:30 pm, Seminar Room, Department of MME
**Chair:** Dr. Md. Mohar Ali, Previous Professor, Department of MME, BUET

**[B06]**
Fatigue Properties of Overaged Cast Aluminium-7Silicon-0.3Magnesium Alloy
*R. Islam, A. Rahman and A. K. M. B. Rashid*

**[C11]**
Effect of Boron Addition on the Wear Properties of As-Cast AZ91 Magnesium Alloy
*N. Akram, K. M. Shorowordi, S. Roy and A. Mahmud*

**[D05]**
Fatigue Life Dependency on the Cleanliness and Ductility of Structural Steels
*M. A. Islam, R. A. Mizan and M. A. S. R. Saadi*

**[F02]**
Finite Element Analysis of an Aluminum Alloy Using Progressive Failure Algorithm
*M. F. Hossain, K. M. Shorowordi and M. S. Islam*

**[F04]**
Stress, Deformation and Failure Analysis of Parabolic Leaf Spring by Finite Element Analysis with Material Optimization
*Tasmeen Inteeser Khan, M. I. Washif Rahman Omi, Md. Jubayer Hossain, Md. Merajul Haque and Fazlar Rahman*

**[F05]**
Investigation of Elastic Properties of Fiber-Reinforced Composites by Micromechanical Modeling Method
*Anik Das Anto, Sobahan Mia, Md Abdul Hasib, Mahedy Hasan and Tazdiqul Alam Chowdhury*

IOP Conf. Series: Materials Science and Engineering **438** (2018) 011001 doi:10.1088/1757-899X/438/1/011001



**[F07]**
Studies the Tensile Fracture at Various Strain Rates and Impact Fracture Behaviour of Heat Treated Cu Content Al-Si-Ni Alloys
*A. Hossain, F. Gulshan and A. S. W. Kurny*

## Technical Session 3B – Polymers and Composites 2

03 March 2018, Saturday, 11:00 am – 12:30 pm, Professor Ibrahim Library, Department of MME
**Chair:** Dr. Md. Nasrul Haque, Professor, Department of MME, BUET

**[G03]**
Effect of Fiber Content and Post Stress on Moisture Absorption of Jute Polyester Composite
*S. Z. Ali, M. K. Nahian and M. A. Islam*

**[G06]**
Mechanical Properties of Chemically Treated Coir and Betel Nut Fiber Reinforced Hybrid Polypropylene Composites
*M. M. Rahman, M. Mondol and M. Hasan*

**[G07]**
Thermal Conductivity Enhancement of Polypropylene Based Composites Filled by Organic and Inorganic Particles
*R. A. Mizan and M. A. Islam*

**[G09]**
Mechanical, Thermal and Morphological Properties of Pineapple and Betel Nut Husk Fiber Reinforced Hybrid Polypropylene Composites
*M. Akter, E. Jahan and M. Hasan*

**[G16]**
Physico-Mechanical Properties of Pineapple Leaf and Banana Fiber Reinforced Hybrid Polypropylene Composites: Effect of Fiber Ratio and Sodium Hydroxide Treatment
*Sangita Das, Maliha Rahman and Mahbub Hasan*

**[G18]**
Mechanical, Morphological and Thermal Properties of Alkali Treated Ladies Finger Fiber Reinforced Polypropylene Composites
*Md. I. Hossain and M. Hasan*

**[G22]**
Properties of Jute and Sheep Wool Fiber Reinforced Hybrid Polypropylene Composites
*J. Tusnim, N. S. Jenifar and M. Hasan*

IOP Conf. Series: Materials Science and Engineering **438** (2018) 011001 doi:10.1088/1757-899X/438/1/011001



## Technical Session 4A – Nanomaterials and Thin Film 2

03 March 2018, Saturday, 02:30 pm – 04:30 pm, Seminar Room, Department of MME
**Chair:** Dr. Engr. Sheikh Manjura Hoque, Chief Scientific Officer & Head, Materials Science Division, AEC, Dhaka

**[K07]**
Effect of Cu Doping on Morphological, Structural, Optical and Electrical Properties of $MnO_2$ Thin Films Deposited by Spray Pyrolysis Method
*M. Zahan, M. Sharmin and J. Podder*

**[K04]**
Synthesis and Characterization of $Ba^{2+}$, $La^{3+}$ and $Zr^{4+}$ Co-Modified Multiferroic $BiFeO_3$ Nanoparticles
*Fahim Khan, Sudipta Mondal, Hasan Muhammad Usama and Ahmed Sharif*

**[K02]**
Size Dependent Magnetic and Optical Properties of Mn Doped $Bi_{0.9}Ho_{0.1}FeO_3$ Nanoparticles
*Fabia Farlin Athena, Md. Rafiqul Islam, Tilottoma Shaha and Ahmed Sharif*

**[K12]**
Structural and Optical Properties of Undoped and Zinc Doped Titanium Dioxide Thin Films by Spray Pyrolysis Technique
*F. Anjum, M. S. Ullah, J. Podder, M. A. Matin, M. Shahjahan and M. M. Rahman*

**[K13]**
Effect of Fe Doping on the Surface Morphological, Structural, Optical and Electrical Properties of Nano-Crystalline $SnO_2$ Thin Films Deposited by Thermal Spray Pyrolysis
*M. H. Babu, J. Podder, B. C. Dev and M. A. Zubair*

**[K18]**
Al- and Al-Ca-doped ZnO Thin Films by Sol-Gel Method for Solar Cell Application
*Shubhrodev Bhowmik, Md. Ibrahim Raiyan and Md Moniruzzaman*

**[K11]**
Effect of Concentration of Precursor Solution on Surface Morphology and Optical Properties of Titania Thin Film
*N. Barua, J. F. Moushumi and A. K. M. B. Rashid*

**[L10]**
Binding and Release of Ciporofloxacin Hydrochloride on Nanocrystalline Cellulose and its Modified Nanocrystalline Form
*Mohammad Motiur Rahman, Khalide Hasan Parvej, M Mozammal Hosen and Shakhawat H. Firoz*

**[L05]**
Fabrication of Biodegradable Graphene Oxide Based 2D-crosslinked Polylactic Acid (PLA) with Enhanced Mechanical Strength and Thermal Stability
*Majharul Islam Sujan and Md. Shafiul Azam*

**[J02]**
In- vitro Anthelmintic activity of Zinc Oxide nanoparticles against Pheretima Posthuma
*Faruk Hosain, Abdullah-Al-Ragib, Md. Shafiul Islam and H. B. M. Zukaul Islam*

## Technical Session 4B – Miscellaneous

03 March 2018, Saturday, 02:30 pm – 04:30 pm, Professor Ibrahim Library, Department of MME
**Chair:** Dr. A. A. Md. Rezaul Haque, Professor, Department of MME, BUET

**[L02]**
Quantification and Qualification of Silica Sand Extracted from Padma River Sand
*A. S. M. M. Rafi, U. F. Tasnim and M. S. Rahman*

**[C04]**
Electroless Deposition of p-type Cuprous Oxide Thin Film for Solar Application
*Asif Mahmud, Shadman Sharar Mahmud and Md. Moniruzzaman*

**[L08]**
Comparison of Mechanical Properties of Induction Furnace Steel Slag and Electric Arc Furnace Steel Slag as a Replacement of Fine Aggregate in Construction Materials
*R. Atahar, R. M. Rezaul, T. B. Tasnim, A. S. W. Kurny, R. Momtaz and F. Gulshan*

**[L09]**
Characterization and Upgradation of Silica Sand of Sylhet
*M. H. Tania, A. S. W. Kurny and F. Gulshan*

**[L11]**
Effect of Surface Coating on the Bonding Strength of Reinforcing Steel Bars in Concrete Structures
*M. N. Uddin and M. A. Islam*

**[C07]**
Cheap and environment-friendly solar cell by electrodeposition
*M. Habib, A. Roy and M. Moniruzzaman*

**[L03]**
Property Test of Micro-Concrete Reinforced with GI Wire and Polymer Fiber
*Aditi Saha*

**[C17]**
Effects of Quality of Coarse Aggregates and Compaction Mode on the Compressive Strength of Concrete
*J. Roy and M. A. Islam*

**[D04]**
Microstructural Changes Under Fire in TMT 500W Structural Steel Rebars
*M. S. Shahriar, M. R. Nawaz and M. A. Islam*

**[D01]**
Fire Resistance of Locally Produced 500W Structural Steel Bars
*M. R. Nawaz, M. S. Shahriar and M. A. Islam*

**[F03]**
Stress Analysis of a Cylinder Subjected to Thermo-mechanical Loads by Using FEM
*Somnath Somadder and Md. Shahidul Islam*

**[F06]**
Structural Analysis of a FML Wing based on FE Method
*S. B. Rayhan, S. Jian and Y. Chunjin*

## Poster Session

03 March 2018, Saturday, 12:30 pm – 01:00 pm, Department of MME

**[P01]**
Fabrication of Jute Fabrics Reinforced Polyethylene Matrix Composites: Effect of Colors
*Nasrin A. Kona, Shahirin Shahida, Md. Naimul Islam, Arwah Ahmed, Anich Mia, Mamun Hossain, Abu H. M. Kamal, Md. B. Uddin, Ruhul A. Khan*

**[P02]**
Comparative Studies of Mechanical Properties between Jute fiber and E-glass Fiber-Reinforced Unsaturated Polyester-based Composites
*Mamun Hossain, Arwah Ahmed, Anich Mia, Md. SabbirAzam, Muhammad B. Uddin, Md. Razzak, A. H. M. Kamal, Md. Zahirul I. Mollah, Ruhul A. Khan**

**[P03]**
Fabrication and Characterization of Heavy Mineral Reinforced Polymer-based Concrete Composite for Gamma Radiation Shielding
*Md. Muminul Haque, Md. Sabbir Azam, M. Shamsuzzaman, Abid Imtiaz, Abu Z. M. Salahuddin, Ruhul A. Khan*

**[P05]**
Investigation of Structural, Morphological, Optical and Electrical Properties of Spray Synthesized $Fe_2O_3$ Thin Films for Optoelectronic Applications
*M. Sharmin, M. Zahan, J. Podder*

**[P08]**
Synthesis and Characterization of Hydrogels by Gamma Radiation for Biomedical Applications
*Fariha Afrose, Sadia Afroz, Abu H. M. Kamal, Md. Razzak, Mamun Sarkar, Md. Ashraful Alam and Ruhul A. Khan*

IOP Conf. Series: Materials Science and Engineering **438** (2018) 011001 doi:10.1088/1757-899X/438/1/011001



**[P09]**
Mechanical and Spectroscopic Characterization of Polyethylene Oxide/Dimethylacrylamide Based Hydrogels for Wound Healing Applications
*Sadia Afroz, Fariha Afrose, Abu H. M. Kamal, Md. Razzak, Muhammed Y. Miah, Md. Ashraful Alam and Ruhul A. Khan*

**[P11]**
Fabrication and Characterization of Natural and Synthetic Fiber Reinforced Polypropylene-based Composites
*Shahirin Shahida, Md. Naimul Islam, Nasrin A. Kona, Anich Mia, Ruhul A. Khan*

**[P12]**
Mechanical Characterization of Electronic-glass Fiber Mat Reinforced Unsaturated Polyester Resin-based Composites
*Md. Naimul Islam, Shahirin Shahida, Nasrin A. Kona, Anich Mia, Mamun Hossain, Ruhul A. Khan*

**[P15]**
Mechanical and Interfacial Characterization of Jute Fabrics Reinforced Unsaturated Polyester Resin Composites
*Anich Mia, Muhammad B. Uddin, Arwah Ahmed, Mamun Hossain, Md. Razzak, Abu H. M. Kamal, Ruhul A. Khan*

IOP Conf. Series: Materials Science and Engineering **438** (2018) 011001 doi:10.1088/1757-899X/438/1/011001

4th Intl. Conf. on Structure, Processing and Properties of Materials, SPPM2018
1 – 3 March 2018, BUET, Dhaka

# Concluding Ceremony

**03 March 2018, Saturday, 05:00 pm – 05:30 pm, Seminar Room, Department of MME**

| | |
|---|---|
| 04:55 pm | Guests take their seats |
| 05:00 – 04:05 pm | Welcome address by the Conference Secretary<br>**Dr. A. K. M. Bazlur Rashid**<br>Professor, Department of MME, BUET |
| 05:05 – 05:10 pm | Best Paper and Best Poster Paper Award Ceremony |
| 05:10 – 05:20 pm | Speech by the Chief Guest<br>**Prof. Dr. Ehsanul Haque**<br>Previous Member, UGC, Bangladesh<br>and Previous Vice-Chancellor, KUET |
| 05:20 – 05:25 pm | Vote of thanks by the Conference Chair<br>**Dr. Md. Moniruzzaman**<br>Professor and Head, Department of MME, BUET |
| 05:25 – 05:30 pm | Concluding Remarks by the Conference Secretary<br>**Dr. A. K. M. Bazlur Rashid**<br>Professor, Department of MME, BUET |

IOP Conf. Series: Materials Science and Engineering **438** (2018) 011001 doi:10.1088/1757-899X/438/1/011001



# List of Participating Authors

Afrose, F., Institute of Radiation and Polymer Technology, BAEC, Dhaka, Bangladesh

Afroz, S., Institute of Radiation and Polymer Technology, BAEC, Dhaka, Bangladesh

Ahasan, A., Department of Mechanical and Production Engineering, AUST, Dhaka

Ahmed, A. N., Pilot Plant and Process Development Centre, BCSIR, Dhaka

Ahmed, A., Institute of Radiation and Polymer Technology, BAEC, Dhaka, Bangladesh

Ahmed, R., Department of Mechanical and Production Engineering, AUST, Dhaka

Ahmed, S., Department of Materials and Metallurgical Engineering, BUET, Dhaka

Akram, N., Department of Materials and Metallurgical Engineering, BUET, Dhaka

Akter, M., Department of Materials and Metallurgical Engineering, BUET, Dhaka

Alam, M. A., Department of Applied Chemistry and Chemical Engineering, NSTU, Bangladesh

Alam, M., Department of Materials and Metallurgical Engineering, BUET, Dhaka

Al-Asif, A., Department of Materials and Metallurgical Engineering, BUET, Dhaka

Ali, S. A., Department of Glass and Ceramic Engineering, BUET, Dhaka

Ali, S. Z., Department of Materials and Metallurgical Engineering, BUET, Dhaka

Anjum, F., Department of Physics, BUET, Dhaka

Anto, A. D., Department of Mechanical Engineering, KUET, Khulna

Ashrafi, E. A., Department of Materials and Metallurgical Engineering, BUET, Dhaka

Atahar, R., Department of Materials and Metallurgical Engineering, BUET, Dhaka

Athena, F. F., Department of Materials and Metallurgical Engineering, BUET, Dhaka

Azad, A. K., Department of Mechanical and Production Engineering, AUST, Dhaka

Azam, M. S., Department of Chemistry, BUET, Dhaka

Azam, M. S., Institute of Radiation and Polymer Technology, AERE, Dhaka, Bangladesh

Babu, M. H., Department of Physics, BUET, Dhaka

Barua, N., Department of Materials and Metallurgical Engineering, BUET, Dhaka

Bashar, M. S., Institute of Fuel Research and Development, BCSIR, Dhaka

Basharat, F., Department of Physics, BUET, Dhaka

Bashir, M. N., Centre of Advanced Materials, University of Malaya, Malaysia

Bhattacharjee, A., Department of Mechanical Engineering, BUET, Dhaka

Bhowmik, D. R., Department of Applied Chemistry and Chemical Engineering, NSTU, Noakhali

Bhowmik, S., Department of Materials and Metallurgical Engineering, BUET, Dhaka

Billah, M. M., Department of Mechanical Engineering, BUET, Dhaka

Billah, M. M., MEMS and Nanomaterials Lab, University of Central Florida, USA

Chen, Q., MEMS and Nanomaterials Lab, University of Central Florida, USA

Chowdhury, A, A, University of Wollongong, Australia

Chowdhury, S. A., Department of Materials and Metallurgical Engineering, BUET, Dhaka

Chowdhury, T. A., Department of Mechanical Engineering, KUET, Khulna

Chunjin, Y., Nanchang Hangkong University, China

Das, S., Department of Materials and Metallurgical Engineering, BUET, Dhaka

Dev, B. C., Department of Physics, BUET, Dhaka

Ferdous, M., Department of Chemistry, BUET, Dhaka

Firoz, S. H., Department of Chemistry, BUET, Dhaka

Gafur, M. A., Pilot Plant and Process Development Centre, BCSIR, Dhaka

Gulshan, F., Department of Materials and Metallurgical Engineering, BUET, Dhaka

Habib, M., Department of Materials and Metallurgical Engineering, BUET, Dhaka

Hakim, M. A., Department of Glass and Ceramic Engineering, BUET, Dhaka

Hamidah, M. S, Faculty of Civil Engineering, Universiti Teknologi MARA, Malaysia

Haque, M. M., Department of Mechanical and Production Engineering, AUST, Dhaka

Haque, M. M., Institute of Radiation and Polymer Technology, BAEC, Dhaka, Bangladesh

Hasan A., Department of Materials and Metallurgical Engineering, BUET, Dhaka

Hasan, M., Department of Materials and Metallurgical Engineering, BUET, Dhaka

Hasan, M., Department of Mechanical Engineering, KUET, Khulna

Hasanuzzaman, M., Department of Glass and Ceramic Engineering, BUET, Dhaka

Haseeb, A. S. M. A., Centre of Advanced Materials, University of Malaya, Malaysia

Hasib, M. A., Department of Mechanical Engineering, KUET, Khulna

Hoque, M. B., Institute of Radiation and Polymer Technology, AERE, Dhaka, Bangladesh

Hosain, F., Department of Applied Chemistry and Chemical Engineering, NSTU, Noakhali

Hosen, M. M., Department of Chemistry, BUET, Dhaka

Hossain, A. K. M. A., Department of Physics, BUET, Dhaka

Hossain, A., Sylhet Gas Field Limited, Bangladesh

Hossain, M. E., Department of Materials and Metallurgical Engineering, BUET, Dhaka

Hossain, M. F., Department of Materials and Metallurgical Engineering, BUET, Dhaka

Hossain, M. I., Department of Materials and Metallurgical Engineering, BUET, Dhaka

IOP Conf. Series: Materials Science and Engineering **438** (2018) 011001 doi:10.1088/1757-899X/438/1/011001



Hossain, M. J., Department of Mechanical and Production Engineering, AUST, Dhaka

Hossain, M. N., Department of Glass and Ceramic Engineering, BUET, Dhaka

Hossain, M., Institute of Radiation and Polymer Technology, BAEC, Dhaka, Bangladesh

Ibrahim, A, Faculty of Civil Engineering, Universiti Teknologi Malaysia

Islam, D., Institute of Food, Science and Technology, BCSIR, Dhaka

Islam, H. B. M. Z., Department of Chemistry, BUET, Dhaka

Islam, M. A., Department of Materials and Metallurgical Engineering, BUET, Dhaka

Islam, M. F., Department of Glass and Ceramic Engineering, BUET, Dhaka

Islam, M. N., Institute of Radiation and Polymer Technology, BAEC, Dhaka, Bangladesh

Islam, M. R., Department of Materials and Metallurgical Engineering, BUET, Dhaka

Islam, M. R., Department of Physics, BUET, Dhaka

Islam, M. S., Department of Applied Chemistry and Chemical Engineering, NSTU, Noakhali

Islam, M. S., Department of Mechanical Engineering, KUET, Khulna

Islam, M. S., Department of Naval Architecture and Marine Engineering, BUET, Dhaka

Islam, R., Department of Materials and Metallurgical Engineering, BUET, Dhaka

Jahan, E., Department of Materials and Metallurgical Engineering, BUET, Dhaka

Jannat, S., Department of Materials and Metallurgical Engineering, BUET, Dhaka

Jenifar, N. S., Department of Materials and Metallurgical Engineering, BUET, Dhaka

Jian, S., Nanchang Hangkong University, China

Kamal, A. H. M., Institute of Radiation and Polymer Technology, BAEC, Dhaka, Bangladesh

Khan, F., Department of Materials and Metallurgical Engineering, BUET, Dhaka

Khan, R. A., Institute of Radiation and Polymer Technology, BAEC, Dhaka, Bangladesh

Khan, T. I., Department of Mechanical and Production Engineering, AUST, Dhaka

Kharshiduzzaman, M., Department of Mechanical and Production Engineering, AUST, Dhaka

Kona, N. A., Institute of Radiation and Polymer Technology, BAEC, Dhaka, Bangladesh

Kurny, A. S. W., Department of Materials and Metallurgical Engineering, BUET, Dhaka

Liza, M. A., Department of Materials and Metallurgical Engineering, BUET, Dhaka

Mahmud, A., Department of Materials and Metallurgical Engineering, BUET, Dhaka

Mahmud, S. S., Department of Materials and Metallurgical Engineering, BUET, Dhaka

Matin, M. A., Department of Glass and Ceramic Engineering, BUET, Dhaka

Mazumder, A., Department of Chemical Engineering, BUET, Dhaka

Mazumder, M. R. H., Department of Materials and Metallurgical Engineering, BUET, Dhaka

Mesbah, M. B., Department of Materials and Metallurgical Engineering, BUET, Dhaka

Mia, A., Institute of Radiation and Polymer Technology, BAEC, Dhaka, Bangladesh

Mia, S., Department of Mechanical Engineering, KUET, Khulna

Miah, M. Y., Department of Applied Chemistry and Chemical Engineering, NSTU, Bangladesh

Miah, M. Y., Institute of Fuel Research and Development, BCSIR, Dhaka

Mizan, R. A., Department of Materials and Metallurgical Engineering, BUET, Dhaka

Mollah, M. Z. I., Institute of Radiation and Polymer Technology, BAEC, Dhaka, Bangladesh

Momtaz, R., Department of Civil Engineering, BUET, Dhaka

Mondal, S., Department of Materials and Metallurgical Engineering, BUET, Dhaka

Mondol, M., Department of Materials and Metallurgical Engineering, BUET, Dhaka

Moniruzzaman, M., Department of Materials and Metallurgical Engineering, BUET, Dhaka

Moushumi, J. F., Department of Materials and Metallurgical Engineering, BUET, Dhaka

Nahian, M. K., Department of Materials and Metallurgical Engineering, BUET, Dhaka

Nawaz, M. R., Department of Materials and Metallurgical Engineering, BUET, Dhaka

Nobi, A., Institute of Radiation and Polymer Technology, BAEC, Dhaka, Bangladesh

Numera, F., Department of Materials and Metallurgical Engineering, BUET, Dhaka

Omi, M. I. W. R., Department of Mechanical and Production Engineering, AUST, Dhaka

Parvej, K. H., Department of Chemistry, BUET, Dhaka

Podder, J., Department of Physics, BUET, Dhaka

Qadir, M. R., Pilot Plant and Process Development Centre, BCSIR, Dhaka

Qazi, S. A., Faculty of Civil Engineering, Universiti Teknologi MARA, Malaysia

Quraishi, M. A. M., Department of Physics, BUET, Dhaka

Rabbi, I. A. F., Department of Materials and Metallurgical Engineering, BUET, Dhaka

Rafi, A. S. M. M., Department of Glass and Ceramic Engineering, RUET, Rajshahi

Ragib, A.-Al, Department of Applied Chemistry and Chemical Engineering, NSTU, Noakhali

Rahman, F., Department of Mechanical and Production Engineering, AUST, Dhaka

Rahman, M. A., Department of Materials and Metallurgical Engineering, BUET, Dhaka

Rahman, M. A., Institute of Mining, Mineralogy and Metallurgy, BCSIR, Joypurhat

Rahman, M. M., Department of Chemistry, BUET, Dhaka

Rahman, M. M., Department of Materials and Metallurgical Engineering, BUET, Dhaka

Rahman, M. M., Department of Physics, BUET, Dhaka

Rahman, M. S., Department of Glass and Ceramic Engineering, RUET, Rajshahi

Rahman, M., Department of Materials and Metallurgical Engineering, BUET, Dhaka

Rahman, S., Department of Materials and Metallurgical Engineering, BUET, Dhaka

IOP Conf. Series: Materials Science and Engineering **438** (2018) 011001 doi:10.1088/1757-899X/438/1/011001



Raiyan, M. I., Department of Materials and Metallurgical Engineering, BUET, Dhaka

Rashed, H. M. M. A., Department of Materials and Metallurgical Engineering, BUET, Dhaka

Rashid, A. K. M. B., Department of Materials and Metallurgical Engineering, BUET, Dhaka

Rayhan, S. B., Nanchang Hangkong University, China

Razzak, M., Institute of Radiation and Polymer Technology, BAEC, Dhaka, Bangladesh

Rezaul, R. M., Department of Materials and Metallurgical Engineering, BUET, Dhaka

Roy, A., Department of Materials and Metallurgical Engineering, BUET, Dhaka

Roy, J., Department of Materials and Metallurgical Engineering, BUET, Dhaka

Roy, S., Department of Materials and Metallurgical Engineering, BUET, Dhaka

Saadi, M. A. S. R., Department of Mechanical and Aerospace Engineering, George Washington University, USA

Saha, A., Department of Civil Engineering, BUET, Dhaka

Saha, T., Department of Materials and Metallurgical Engineering, BUET, Dhaka

Saha, V. C., Department of Materials and Metallurgical Engineering, BUET, Dhaka

Salahuddin, A. Z. M., Department of Nuclear Science and Engineering, MIST, Dhaka, Bangladesh

Sarkar, M., Department of Applied Chemistry and Chemical Engineering, NSTU, Bangladesh

Shaha, T., Department of Materials and Metallurgical Engineering, BUET, Dhaka

Shahida, S., Institute of Radiation and Polymer Technology, BAEC, Dhaka, Bangladesh

Shahjahan, M., Department of Physics, BUET, Dhaka

Shahriar, M. S., Department of Materials and Metallurgical Engineering, BUET, Dhaka

Shams, P., Department of Materials and Metallurgical Engineering, BUET, Dhaka

Shamsuzzaman, M., Institute of Nuclear Science and Technology, BAEC, Dhaka, Bangladesh

Sharif, A., Department of Materials and Metallurgical Engineering, BUET, Dhaka

Sharmin, M, Department of Physics, BUET, Dhaka

Shawon, M. M. A., Department of Materials and Metallurgical Engineering, BUET, Dhaka

Shorowordi, K. M., Department of Materials and Metallurgical Engineering, BUET, Dhaka

Somadder, S., Department of Mechanical Engineering, KUET, Khulna

Sujan, G. K., Centre of Advanced Materials, University of Malaya, Malaysia

Sujan, M. I., Department of Chemistry, BUET, Dhaka

Sultana, M. S., Institute of Mining, Mineralogy and Metallurgy, BCSIR, Joypurhat

Sutradhar, B., Department of Materials and Metallurgical Engineering, BUET, Dhaka

Syed, I. M., Department of Physics, BUET, Dhaka

Tania, M. H., Department of Materials and Metallurgical Engineering, BUET, Dhaka

IOP Conf. Series: Materials Science and Engineering **438** (2018) 011001 doi:10.1088/1757-899X/438/1/011001



Tasnim, T. B., Department of Materials and Metallurgical Engineering, BUET, Dhaka

Tasnim, U. F., Department of Glass and Ceramic Engineering, RUET, Rajshahi

Tomota, Y., Advanced Industrial Science and Technology, Japan

Tusnim, J., Department of Materials and Metallurgical Engineering, BUET, Dhaka

Uddin, M. B., Institute of Radiation and Polymer Technology, BAEC, Dhaka, Bangladesh

Uddin, M. J., Department of Materials and Metallurgical Engineering, BUET, Dhaka

Uddin, M. N., Department of Materials and Metallurgical Engineering, BUET, Dhaka

Ullah, M. S., Department of Physics, BUET, Dhaka

Usama, H. M., Department of Materials and Metallurgical Engineering, BUET, Dhaka

Zahan, M., Department of Physics, BUET, Dhaka

Zaman, M. N., Institute of Mining, Mineralogy and Metallurgy, BCSIR, Joypurhat

Zubair, M. A., Department of Glass and Ceramic Engineering, BUET, Dhaka

# Peer review statement

All papers published in this volume of *IOP Conference Series: Materials Science and Engineering* have been peer reviewed through processes administered by the proceedings Editors. Reviews were conducted by expert referees to the professional and scientific standards expected of a proceedings journal published by IOP Publishing.

IOP Conf. Series: Materials Science and Engineering **438** (2018) 012001 doi:10.1088/1757-899X/438/1/012001

# Effect of Process Variables on Deposited Cupric Oxide Thin Film by Sol-Gel Spin Coating Technique

**Dipika Rani Bhowmik[1], Aninda Nafis Ahmed[2], M A Gafur[3], Muhammed Yusuf Miah[4] and Dipa Islam[5]**

[1,4] Department of Applied Chemistry and Chemical Engineering (ACCE), Noakhali Science and Technology University (NSTU), Noakhali, Bangladesh
[2] Pilot Plant and Process Development Centre (PP&PDC), Bangladesh Council of Scientific and Industrial Research (BCSIR), Dhaka, Bangladesh
[3] Institute of Fuel Research and Development (IFRD), Bangladesh Council of Scientific and Industrial Research (BCSIR), Dhaka, Bangladesh
[5] Institute of Food, Science and Technology (IFST), Bangladesh Council of Scientific and Industrial Research (BCSIR), Dhaka, Bangladesh
E-mail: [2]adhi88bd@yahoo.com

**Abstract.** Cupric oxide were prepared by sol-gel technique and deposited onto glass substrates as thin films using spin coating method. The aim of this research was to study the effects of different spin coating speeds and solution concentrations of cupric oxide thin films on the structural, optical, electrical and physical properties of thin films. Thin films were deposited with concentration variation of the solution ranging from 0.5 M-1 M and speed variation from 1200 rpm-2400 rpm. X-ray diffraction (XRD) and UV-Vis spectrophotometer were used to characterize the structural and optical properties of the deposited films, respectively. The as-synthesized CuO thin films were also characterized using SEM. The dc electrical properties were characterized using nano-electrometer. Based on the results obtained, it was found that the electrical resistivity of the cupric oxide thin films increases as spin coating speeds increase. The optical band gap values of CuO thin films for concentration variation were found 1.47-1.27 eV and for speed variations, it was observed 1.45-1.95 eV. Decreased values of the optical energy gap were observed with increasing concentration while reverse effect was found with increasing speed of spin coating.

## 1. Introduction

Cupric oxide (CuO) is a black color compound with slight translucence. CuO is a promising candidate for many applications owing to the abundance of its components in nature, low-cost production, good thermal stability, and electrochemical properties. This combined property enables CuO thin films to be an excellent semiconductor for several applications such as high-temperature superconductors [1], solar cells [1], gas sensors [2], magnetic storage media [3], catalyst etc. It has direct optical band gap energy in the range of 1.2 to 1.9 eV [4]. CuO has a monoclinic structure with the lattice parameters a = 4.684A, b = 3.425A, c = 5.129A, and β = 99.28A. The transparent conductive cupric oxide thin films had been developed by several techniques such as electrodeposition, thermal evaporation, sol-gel technique [5] and wet chemical method [6] which could be deposited on various substrates [7]. However, the most prepared technique used is sol-gel method due to its simplicity, lower cost, reproducibility, ease of composition control and large area deposition [8]. Sol-gel method easily to deposit on different types of substrates that could be performed under non-vacuum environment. Therefore in this research, sol-gel method was used to form the solution of cupric oxide which then deposited on glass substrates using spin coating technique. The main goal of this paper was to study the effects of different spin coating speeds and solution concentrations of the cupric oxide thin films on the optical, structural, electrical and physical properties on glass substrates which were characterized using UV-Vis spectrophotometer, X-ray diffraction pattern, Nano-electrometer and Scanning Electron Microscope (SEM), respectively.

## 2. Experimental details

The procedures for preparing cupric oxide thin films were summarized in the following.

### *2.1. Preparation of substrates*

In this work, the substrates were initially cleaned with detergent soap in tap water. Then they were cleaned using methanol, acetone and deionized water and allowed to dry with dryer.

### *2.2. Preparation of cupric oxide solution using sol-gel method*

In the sol-gel process, the materials used were copper acetate powder [9], diethanolamine, isopropanol and ethylene glycol without any further purification. 0.5 M, 0.75 M and 1 M of solution were formed by dissolving 0.5 g, 0.75 g and 1 g of copper acetate powder in 9 ml isopropyl alcohol and 0.5 ml diethanolamine which act as precursor, solvent and complexing agent, respectively. Then, 0.5 ml ethylene glycol was added into the solution. Next, the cupric oxide solutions were stirred at room temperature until a transparent solution were obtained. The solutions with concentration of 0.5, 0.75 and 1 M were a clear and dark blue without any suspension of particles.

### *2.3. Preparation of cupric oxide thin films deposition by spin coating technique*

Prior to the thin films deposition process, quarter area of the substrate was covered with capton tape while the other area of the substrate was left uncovered for the layer thickness measurements. After stirring process, a double step spinning program was applied to obtain the homogenous precursor films. In this step, the solutions were spread onto the substrates at 100 rpm for 10 seconds. The solutions were spin coated with concentration of 0.5 M, 0.75 M and 1 M with spinning speed of 1200 rpm and with spinning speed of 1200, 1800 and 2400 rpm of 0.75 M for 3 minutes each. Immediately after coating process, the sample was dried for 5 minutes at 250°C for each layer. Five layers of cupric oxide thin film were coated one over the above. After the drying of the last layer, sample was annealed for 30 minutes at 600°C.

### *2.4.Samples Characterization*

UV-Vis spectrophotometer was used to observe the absorbance of cupric oxide thin films. To determine the crystallinity, X-ray diffraction (XRD) was used. Current-voltage measurements for concentration variation and speed variation were performed by using 6517B Electrometer with applied voltage -5 to 5 V. Prior to the measurements, the thin films were coated with silver (Ag) as the metal contacts. The width of metal contacts deposited onto cupric oxide thin films were 2.3 mm. The morphology of the deposited CuO thin films was observed by SEM (Scanning Electron Microscope).

## 3. Results and discussion

### *3.1. Optical properties of cupric oxide thin films*

Figures 1 and 2 show the optical properties of deposited CuO thin films at various solution concentrations and spinning speeds on glass substrates have been determined by using absorbance spectra in the region of (300-1100) nm. It was observed that film's higher absorption on the shorter wavelength side (ultraviolet region), and lower absorption on the higher wavelength side (visible region) [10].

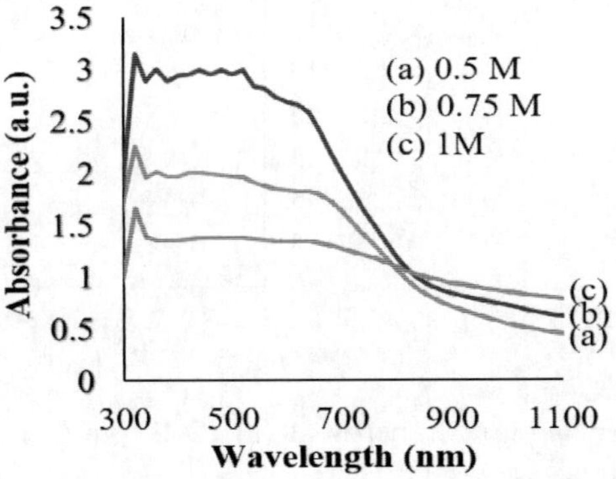


**Figure 1.** Optical absorbance vs. wavelength for various concentration of CuO at 1200 rpm.

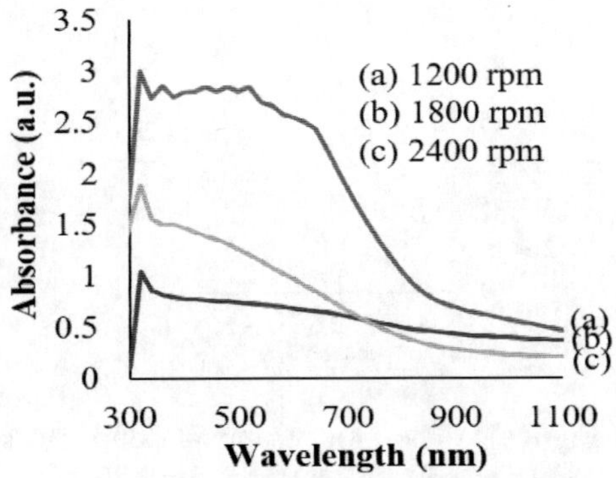


**Figure 2.** Optical absorbance vs. wavelength for 0.75 M of CuO at various spinning speed.

From figures 1 and 2, the highest absorbance was observed for 0.75 M in case of concentration variation at 1200 rpm and for speed variation of 0.75 M, highest absorbance was found for 1200 rpm. Because, the absorption was occurred when electron made the transition from a lower energy state in valence band to high energy state in conduction band while absorbing the excess energy as photon. It was observed that almost absorption edges were found around 319-321 nm present in the spectra indicating the absorption properties of cupric oxide in the UV region. The optical absorption at absorption edge related to the transition from valence band to conduction band and was closed with band gap energy [11]. The absorption coefficient (α) for the prepared thin films was calculated by the following equation.

$$\alpha = \{ln\ (100/(\%T))\}/d \qquad (1)$$

Where T is the transmittance spectrum of the thin film and d is the thin film thickness. The absorption coefficient, α of the cupric oxide thin films was related to the photon energy, hv [12], where hv is the photon energy (eV). These optical band gap were obtained by tangent line intersection with energy axis on the photon energy (hv) of the $(\alpha hv)^2$ versus photon energy plot [13]. The photon energy (hv) was calculated using equation 2:

$$hv = hc/\lambda \qquad (2)$$

Where h is the Plank constant which is $6.626 \times 10^{-34}$, c is the light constant that is $3 \times 10^8$ and λ is wavelength. Optical band gap was calculated and showed in figures 3 and 4.

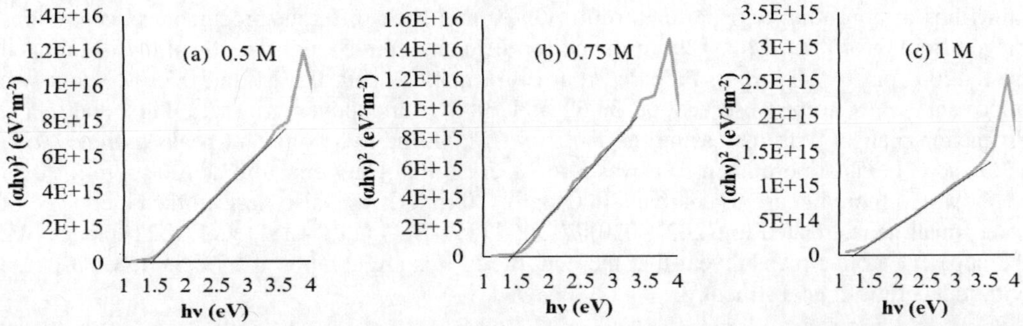


**Figure 3.** The plot of optical band gap energy versus photon energy at 1200 rpm for (a) 0.5 M, (b) 0.75 M and (c) 1 M CuO thin films.

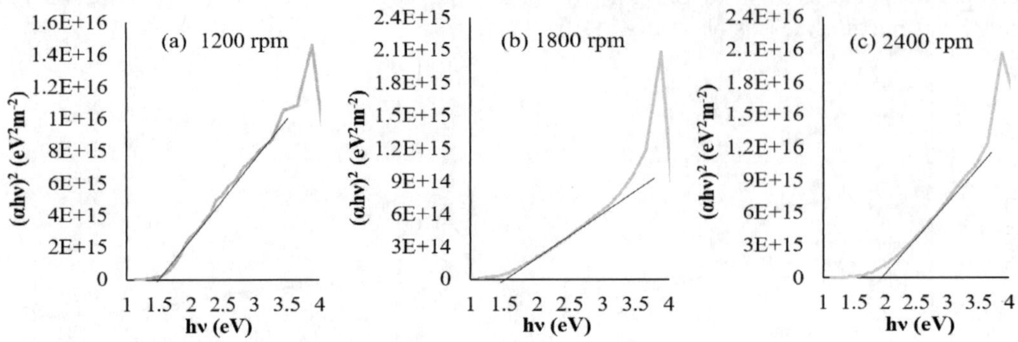


**Figure 4.** The plot of optical band gap energy versus photon energy at (a) 1200 rpm, (b) 1800 rpm and (c) 2400 rpm for 0.75 M CuO thin films.

From figures 3 and 4, the values of the optical band gap of CuO thin films were tabulated in tables 1 and 2. The optical band gap energy for concentration variation were found to be in the range of 1.47 to 1.27 eV which were decreased as the solution concentration increased, while for speed variation were found to be in the range of 1.45 to 1.95 eV which were increased as the spin coating speeds increased and also were found to be affected by film thickness. It was reported that the optical band gap decreased with the increased of the grain size [13, 14] and crystallinity.

**Table 1.** Thickness of thin films and optical band gap at different solution concentrations.

| Solution concentration (M) | Thickness (nm) | Optical band gap (eV) |
|---|---|---|
| 0.5 | 184.694 | 1.47 |
| 0.75 | 220.99 | 1.45 |
| 1 | 293.77 | 1.27 |

**Table 2.** Thickness of thin films and optical band gap at different spin coating speeds for 0.75 M concentration.

| Spin speed (rpm) | Thickness (nm) | Optical band gap (eV) |
|---|---|---|
| 1200 | 220.99 | 1.45 |
| 1800 | 183.35 | 1.53 |
| 2400 | 117.24 | 1.95 |

*3.2. Structural properties of CuO thin films*

In figures 5 and 6, it was reported that the X-ray diffraction patterns (XRD) of prepared CuO thin films at annealing temperatures (600)°C (30 min.) for different precursor concentrations ranged from 0.5 M to 1 M at 1200 rpm and for different spin coating speeds of 0.75 M ranged from 1200 rpm to 2400 rpm. Prepared thin film samples using 0.5 M and 0.75 M show two important peaks were observed at 35.5° and 38.6° corresponds to the (-111) and (111) diffraction planes. With increasing the molarity to 1 M, the two dominant peaks assigned to (-111) and (111) and more intense. It was also observed that CuO thin film at 1200 rpm for 0.75 M show two more intense peaks than 1800 and 2400 rpm. It was also noticed the emergence of some small peaks related to (-202) (020)(202) (-113) (-311) (113) (311) and (-222) planes. All the apparent atomic plans reveal that the deposited films are composed of a single CuO phase with monoclinic crystal structure.

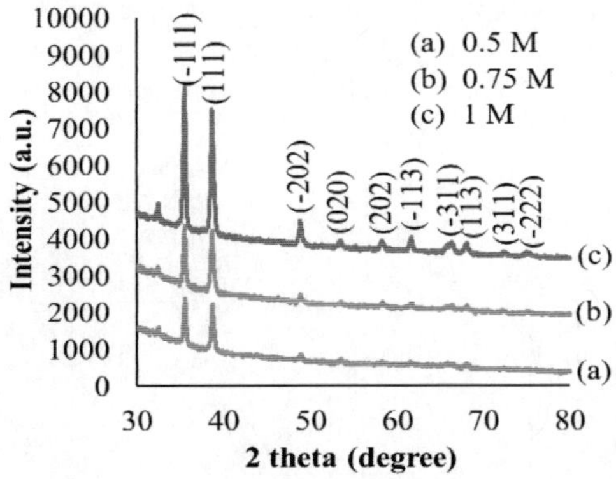


**Figure 5.** X-ray diffraction of CuO thin films elavorated with concentration variations at 1200 rpm.

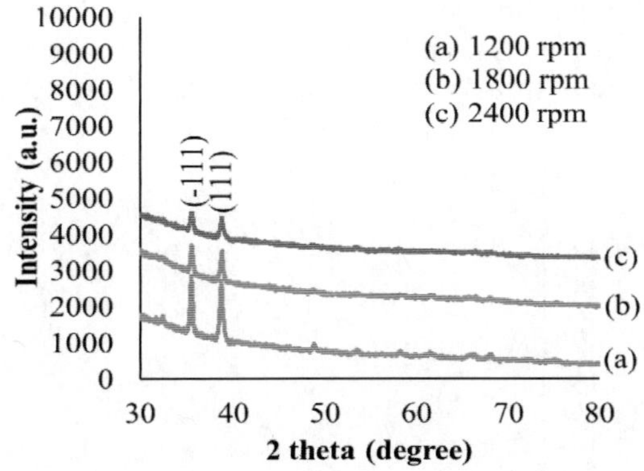


**Figure 6.** X-ray diffraction of CuO thin films elaborated with spin coating speed variations for 0.75 M.

*3.3. Electrical properties of cupric oxide thin films*

Figure 7 shows three curves for different concentrations and figure 8 shows another three curves for different spin coating speeds of CuO thin films on soda lime glass deposited by spin coating method which shown that all films indicated the Ohmic behavior with linear curve.

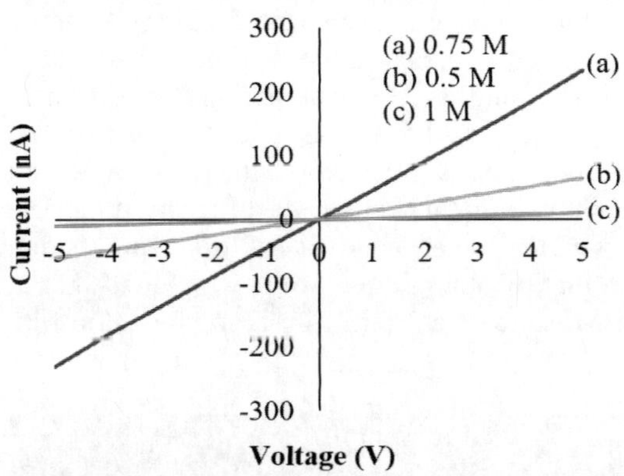


**Figure 7.** The plot of curves for concentration variation at 1200 rpm.

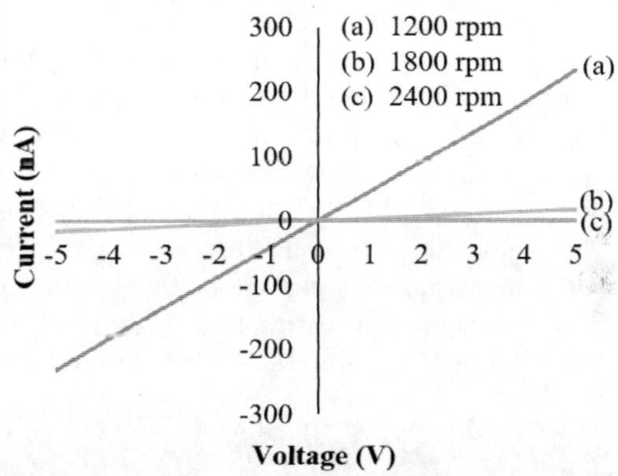


**Figure 8.** The plot of curves for 0.75 M at speed variation.

It was observed that slope of curves increased with decreasing of speed and the curves revealed that the current at 2400 rpm was the lowest while the current at 1200 rpm was the highest among the other curves. On the other hand, higher slope of curves was found at 0.75 M in case of concentration variation and it was found that the current for 1 M was the lowest while the current for 0.75 M was the highest at 1200 rpm. CuO thin film for 0.75 M at 1200 rpm have active layer uniformly on the glass substrates [15]. As a result, the recombination process took place effectively as more holes and electrons were transported to the active area, which translated into better electrical characteristics [15].

The calculated resistivity obtained in this work to be 3.34, 1.09 and 31.82 Ωm for thin films prepared at 1200 rpm for 0.5, 0.75 and 1M, respectively, while 1.09, 13.03 and 134.68 Ωm were obtained as the calculated resistivity for 0.75 M at 1200,1800 and 2400 rpm, respectively. The lowest resistivity was about 1.09 Ωm for 0.75 M, while the highest resistivity was about 31.82 Ωm for 1 M at concentration variation. Again, the lowest resistivity was about 1.09 Ωm for 0.75 M at 1200 rpm while the highest resistivity was about 134.68 Ωm for 0.75 M at 2400 rpm. This was due to increase of electron and hole as carrier concentration and hence decreased the resistivity [9]. Therefore, it was observed that CuO thin film with highest concentration and spin coating speed with the decreased carrier concentration produced the highest resistivity.

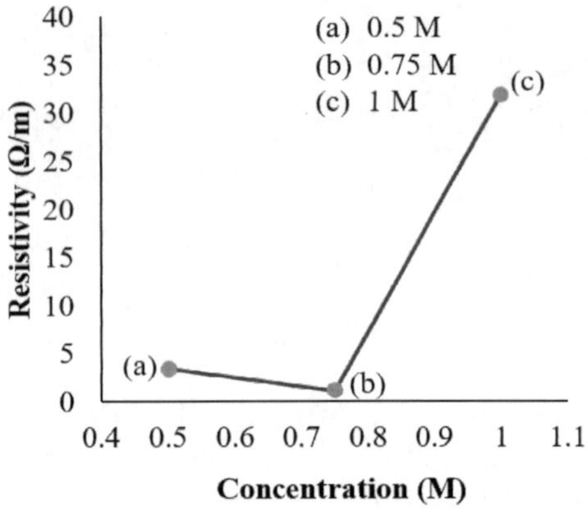


**Figure 9.** The plot of resistivity of CuO thin films vs. concentrations at 1200 rpm.

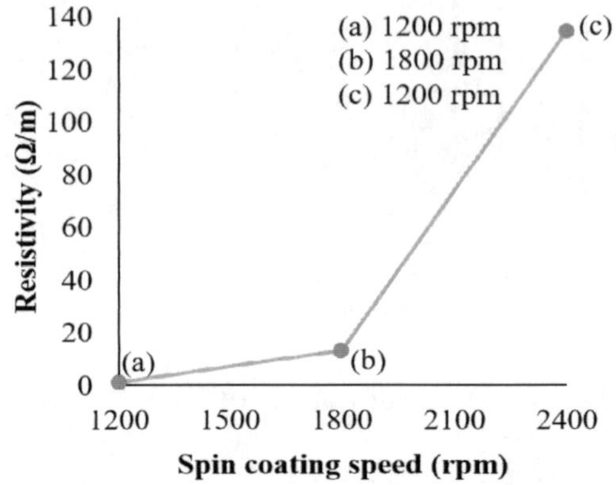


**Figure 10.** The plot of resistivity of CuO thin films vs. spin coating speeds for 0.75.

*3.4. Physical properties of CuO thin films*

The surface morphology of as-synthesized CuO thin films was investigated using Scanning Electron Microscope (SEM). The SEM photographs of concentration variation and speed variation of CuO thin films are shown in figures 11 and 12. On comparison of the three concentration photographs at 1200 rpm, it was observed that 0.5 M coated CuO thin film has more pore than that of 0.75 M thin film and it's particle size is greater than that of 0.75 M thin film. Again, 0.75 M CuO thin film coating show highly homogeneous layer of reduced particle size, improved texture, lowest void and pore. On the other hand, 1 M thin film is not homogeneous and it's particle size is greater than that of other films. So, it was found that in case of concentration variation at 1200 rpm, 0.75 M coated CuO thin film is the best thin film among 0.5 M, 0.75 M and 1 M coated thin films.

**Figure 11.** The FE-SEM images for CuO films deposited with: (a) 0.5 M, (b) 0.75 M and (c) 1 M at 1200 rpm.

On comparison of speed variations for 0.75 M, CuO thin film coated at 1200 rpm shows good homogeneous layer, besides it has good crystallinity according to XRD and highest absorbance. Again, thin film prepared at 1800 rpm shows more homogeneity but according to XRD, it's crystallinity is less than that of film at 1200 rpm and it's absorbance is lower than that of thin film at 1200 rpm and 2400 rpm. On the other hand, CuO coated film at 2400 rpm has more pore and it's pore is greater than that of prepared films at 1200 rpm and 1800 rpm, besides it's crystallinity is less than that of other two films. Thus it was concluded that CuO coated thin film at 1200 rpm for 0.75M is an excellent semiconductor and would be more suitable for photovoltaic applications.

**Figure 12.** The FE-SEM images for CuO films deposited with: (a) 1200 rpm (b) 1800 rpm and (c) 2400 rpm for 0.75 M.

## 4. Conclusion

In summary, thin film of CuO was deposited using spin coating method on glass substrate with different thickness by varying solution concentrations and spin coating speeds. The optical, structural, electrical and physical properties of thin films were measured by UV-Visible spectrophotometer, X-ray diffraction, Electrometer and FE-SEM. Film with 0.75 M at 1200 rpm showed the highest absorbance. The decrease of the optical band gap with the increased of the thickness might attributed by the increased of crystallinity. CuO thin film with 0.75 M at 1200 rpm showed the lowest of electrical resistivity. The XRD analysis indicates that the obtained films have a monoclinic structure. Crystallinity for films prepared increased with increasing of concentration and decreasing of spinning speed. The SEM micrographs show that deposited films are homogeneous except 1M sample. From SEM micrographs, it was concluded that film elaborated at precursor concentration and spin coating speed equal to 0.75 M at 1200 rpm shows good quality. Finally from our research, it was observed that 0.75 M at 1200 rpm CuO thin film would be more suitable for solar cells conversion. Future investigation should be done on the optical, structural, electrical and physical properties of CuO thin films using doping technique.

## 5. Reference

[1] Shockley W, Queisser HJ. *Detailed balance limit of efficiency of p-n junction solar cells.* Journal of applied physics. 1961 Mar;32(3):510-9

[2] Tamaki J, Maekawa T, Miura N, Yamazoe N. *CuO-SnO2 element for highly sensitive and selective detection of H2S.* Sensors and Actuators B: Chemical. 1992 Oct 1;9(3):197-203

[3] Huang LS, Yang SG, Li T, Gu BX, Du YW, Lu YN, Shi SZ. *Preparation of large-scale cupric oxide nanowires by thermal evaporation method.* Journal of Crystal Growth. 2004 Jan 2;260(1-2):130-5

[4] Kumar DA, Xavier FP, Shyla JM. *Investigation on the variation of conductivity and photoconductivity of CuO thin films as a function of layers of coating.* Archives of Applied Science Research. 2012;4(5):2174-83

[5] Armelao L, Barreca D, Bertapelle M, Bottaro G, Sada C, Tondello E. *A sol–gel approach to nanophasic copper oxide thin films.* Thin solid films. 2003 Oct 1;442(1-2):48-52

[6] Qu Y, Li X, Chen G, Zhang H, Chen Y. *Synthesis of Cu2O nano-whiskers by a novel wet-chemical route.* Materials letters. 2008 Mar 15;62(6-7):886-8

[7] Al-Kuhaili MF. *Characterization of copper oxide thin films deposited by the thermal evaporation of cuprous oxide (Cu2O)*. Vacuum. 2008 Feb 19;82(6):623-9

[8] Roy B, Perkins JD, Kaydanova T, Young DL, Taylor M, Miedaner A, Curtis C, Kleebe HJ, Readey DW, Ginley DS. *Preparation and characterization of sol–gel derived copper–strontium–oxide thin films*. Thin Solid Films. 2008 Apr 30;516(12):4093-101

[9] Hashim H, Shariffudin SS, Saad PS, Ridah HA. *Electrical and Optical Properties of Copper Oxide Thin Films by Sol-Gel Technique*. InIOP Conference Series: Materials Science and Engineering 2015 (Vol. 99, No. 1, p. 012032). IOP Publishing

[10] Kadhim RG, Kzar K. *Structural and Optical Properties of CuO Doped (Li) Thin Films Prepared by Sol-Gel Technique*. World Scientific News. 2016;56:56-66

[11] Rodzi AS, Musa MZ, Mamat MH, Mohamad Z, Rusop M, Berhad MN. *Optical properties of annealed nanostructured Zinc Oxide thin film exposed at different temperatures*. InElectronic Devices, Systems and Applications (ICEDSA), 2010 Intl Conf on 2010 Apr 11 (pp. 401-403). IEEE

[12] Ray SC. *Preparation of copper oxide thin film by the sol–gel-like dip technique and study of their structural and optical properties*. Solar energy materials and solar cells. 2001 Jun 1;68(3-4):307-12

[13] Shariffudin SS, Khalid SS, Sahat NM, Sarah MS, Hashim H. *Preparation and Characterization of Nanostructured CuO Thin Films using Sol-gel Dip Coating*. InIOP Conference Series: Materials Science and Engineering 2015 (Vol. 99, No. 1, p. 012007). IOP Publishing

[14] Hong R, Huang J, He H, Fan Z, Shao J. *Influence of different post-treatments on the structure and optical properties of zinc oxide thin films*. Applied Surface Science. 2005 Apr 15;242(3-4):346-52

[15] Reshak AH, Shahimin MM, Juhari N, Suppiah S. *Electrical behaviour of MEH-PPV based diode and transistor*. Progress in biophysics and molecular biology. 2013 Nov 1;113(2):289-94

**Acknowledgements**
The authors would like to thank to the Pilot Plant and Process Development Centre (PP&PDC), Bangladesh Council of Scientific and Industrial Research (BCSIR), Dhaka, Bangladesh for the laboratory facilities and for funding this project.

IOP Conf. Series: Materials Science and Engineering **438** (2018) 012002 doi:10.1088/1757-899X/438/1/012002

# Electroless Deposition of p-type Cuprous Oxide Thin Film for Solar Application

**Asif Mahmud, Shadman Sharar Mahmud and Md. Moniruzzaman**

Department of Materials and Metallurgical Engineering, Bangladesh University of Engineering and Technology, Dhaka 1000, Bangladesh

**Abstract.** Solar cells hold a promising prospect in the foreseeable future. Presently cuprous oxides ($Cu_2O$), due to their excellent photovoltaic properties, are being considered as p-type semiconductor materials for solar cell. This work aims at electroless deposition of p-type $Cu_2O$ thin film on copper substrate and studying its various properties through characterizations. A bath containing varying amounts of $CuCl_2.2H_20$, $CoCl_2.6H_20$, $NaH_2PO_2.6H_20$ and $KNaTar.4H_2O$ (Potassium sodium tartrate tetrahydrate) was used for the process. NaOH pellets were used in order to get the desired pH level ranging from 11 to 14. The bath temperature was varied between 30°C to 70°C. The deposited films were characterized by Scanning Electron Microscope (SEM) equipped with Energy Dispersive Spectroscopy (EDS), X-ray diffractometer (XRD) and UV spectroscopy. It was observed that thin films were deposited at different conditions and uniform deposition of $Cu_2O$ thin film was visible at pH 14. The coating thickness was found to be in the range of 0.48 to 4.6 μm. The bandgap for the deposited $Cu_2O$ was found to be in the range of 1.3-1.4 eV. Some variation in results at different pH were observed and the reasons were identified.

## 1. Introduction

Cuprous oxide is a material that has been investigated for years with regard to solar cells due to its high theoretical efficiency and low potential cost [1-4]. It has direct bandgap of ($E_g$ =1.85-2.2 eV) and a high absorption coefficient .It also absorbs visible wavelength upto 650nm. In addition, $Cu_2O$ has several advantages such as non-toxicity and low cost, can be prepared with simple and cheap methods on large scale. Moreover, this material is known to have sufficient mobility and a relatively large minority carrier diffusion length. Its outstanding exciton properties including a large exciton binding energy (~140 meV) have been the target of much research efforts during the past decades [5]. Due to these specific features, fabrication of solar cells based on a $Cu_2O$ photo absorber has been studied for a long time.

The goal of this work is to deposit cuprous oxide on copper plate by electroless technique as the absorber layer of a solar cell where copper plate acted as the back contact and to measure the bandgap of the deposited film. The structural and morphological properties of the as-deposited films were also evaluated.

## 2. Experimental

Electroless deposition technique was used to deposit cuprous oxide thin film on the copper substrate. The setup for this deposition consisted of hotplate, solution of Copper (II) chloride, Cobalt chloride, sodium hypophosphite and Sodium potassium tartrate as the electrolyte. The deposition parameters were chiefly bath composition and pH value. We prepared 24 samples altogether. The first set of samples were prepared keeping the bath composition fixed but changing the pH value. This helped us to find the best deposition

IOP Conf. Series: Materials Science and Engineering **438** (2018) 012002 doi:10.1088/1757-899X/438/1/012002

based on physical appearance. Further, samples were prepared by keeping the pH value fixed but with varied compositions. Finally, 7 samples were selected based on physical appearance of the deposition which were used for different characterizations. Table 1 denotes the first set of samples in which the bath composition was fixed but pH varied. Table 2 denotes the second set of samples in which the pH was fixed but bath composition varied.

**Table 1.** Different Experimental Conditions (pH altered)

| Sample | pH | Bath Compositions | Time (mins) | Temperature (degree) |
|---|---|---|---|---|
| A | 11 | $CuCl_2.2H_2O$=0.024M<br>$CoCl_2 6H_2O$=.0005M<br>$NaH_2.PO_2.H_2O$=0.2M<br>$KNaTar.4H_2O$=0.037M | 60 | 60 |
| B | 12 | | | |
| C | 13 | | | |
| D | 14 | | | |

**Table 2.** Different Experimental Conditions (bath composition altered)

| Sample | pH | Bath Compositions | Time (mins) | Temperature (degree) |
|---|---|---|---|---|
| D | 14 | $CuCl_2.2H_2O$=0.024M<br>$CoCl_2 6H_2O$=.0005M<br>$NaH_2.PO_2.H_2O$=0.2M<br>$KNaTar.4H_2O$=0.037M | 60 | 60 |
| E | | $CuCl_2.2H_2O$=0.036M<br>$CoCl_2 6H_2O$=.00075M<br>$NaH_2.PO_2.H_2O$=0.3M<br>$KNaTar.4H_2O$=0.052M | | |
| F | | $CuCl_2.2H_2O$=0.038M<br>$CoCl_2 6H_2O$=.0006M<br>$NaH_2.PO_2.H_2O$=0.3M<br>$KNaTar.4H_2O$=0.037M | | |
| G | | $CuCl_2.2H_2O$=0.04M<br>$CoCl_2 6H_2O$=.0009M<br>$NaH_2.PO_2.H_2O$=0.3M<br>$KNaTar.4H_2O$=0.037M | | |

## 3. Result and discussion

We used SEM for the morphological study of the deposited thin film, XRD for the structural analysis, EDS for the compositional analysis and UV-VIS Spectroscopy for bandgap measurement.

### *3.1. SURFACE Morphology- SEM studies for deposited film*

It is evident from Figure 1 that non uniform deposition was dispersed throughout the plate. The SEM image of sample C at pH 13 with concentration of copper being 0.024M showed minute granules with tiny porosity. The SEM image of sample D at pH 14 with the same conc. of copper revealed deposition dispersed throughout the plate. The SEM analysis of sample E at pH 14 with the concentration of copper being 0.038M showed comparatively thick deposition as well as granules dispersed with some porosity around. Even though the morphological study of all the samples gave an abstruse view of the images, keeping the solution at a pH of 14 will give a more uniform deposition.

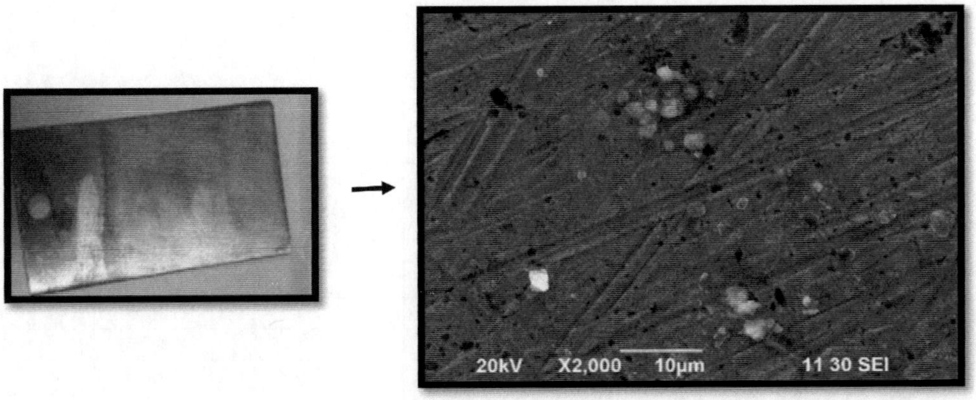


(a)

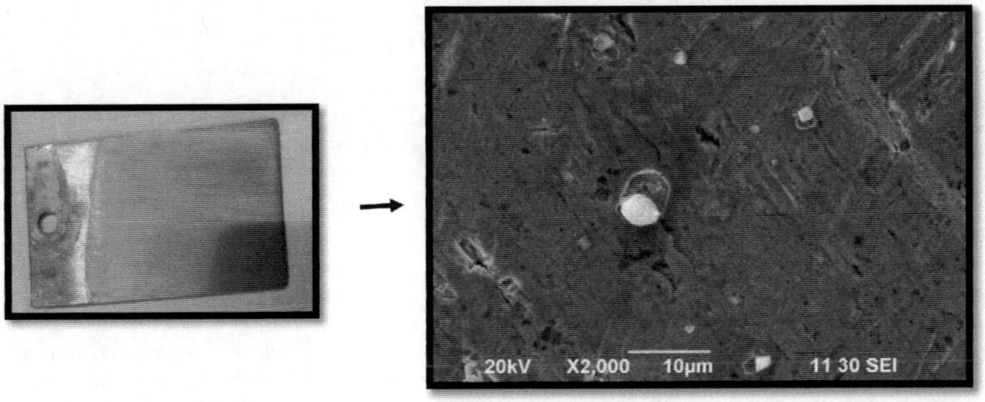


(b)

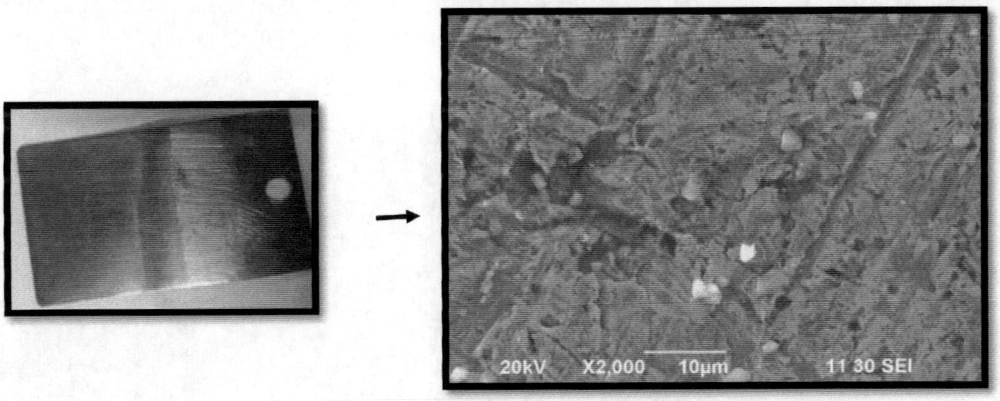


(c)

**Figure 1:** Deposited thin films and its corresponding SEM images (a) sample C (pH 13, conc. of Cu 0.024M) , (b) sample D ((pH 14, conc. of Cu 0.024M) (c) Sample E (pH 14, conc. of Cu 0.036M)

IOP Conf. Series: Materials Science and Engineering **438** (2018) 012002 doi:10.1088/1757-899X/438/1/012002

*3.2. Coating thickness calculation:*

**Table 3.** Thickness of the deposited films

| Sample | Coating thickness (μm) | | | | Average thickness (μm) | Standard deviation |
|---|---|---|---|---|---|---|
| A | 1.1 | 1.3 | 0.8 | 1.2 | 1.1 | 0.216025 |
| B | 3 | 3.3 | 2.1 | 2 | 2.6 | 0.648074 |
| C | 3.9 | 3.7 | 4.1 | 3.4 | 3.78 | 0.298608 |
| D | 0.79 | 0.39 | 0.24 | 0.5 | 0.48 | 0.232522 |
| E | 3.3 | 2.7 | 3.8 | 3 | 3.2 | 0.469042 |
| F | 1.9 | 1.5 | 2.6 | 2.8 | 2.2 | 0.60553 |
| G | 4 | 4.5 | 4.8 | 5.1 | 4.6 | 0.469042 |

Table 3 denotes the multiple readings of the coating thickness and average thickness of all the 7 samples chosen for our experiment. The thickness of the coating was calculated by coating thickness meter named as Fischer Dualscope. The thickness increased as solution gets more basic but dropped at maximum pH value while the concertation was kept fixed. When the pH was kept fixed thickness increased with increase in conc. of copper then dropped & then rose again.

*3.3. EDS analysis- for deposited film*

EDS analysis was performed for the analysis of mass percentage of copper and oxygen in the deposited films. To become cuprous oxide 11.18% oxygen and 88.82% copper by mass should be present whereas in cupric oxide 20.13% oxygen and 79.87% copper by mass should be present [6]. EDS was performed on sample (B) having pH 12 and concentration of copper being 0.024M. From Figure 2 we found that the mass % of copper was 91.23% and surprisingly the mass % of oxygen was close to 2%. Though the mass % of oxygen suggests that the deposition is not of cuprous oxide rather of copper alone and this deviation may be due to the non-uniform thin film deposition and presence of large percentage of copper is due to copper substrate itself.

IOP Conf. Series: Materials Science and Engineering **438** (2018) 012002 doi:10.1088/1757-899X/438/1/012002

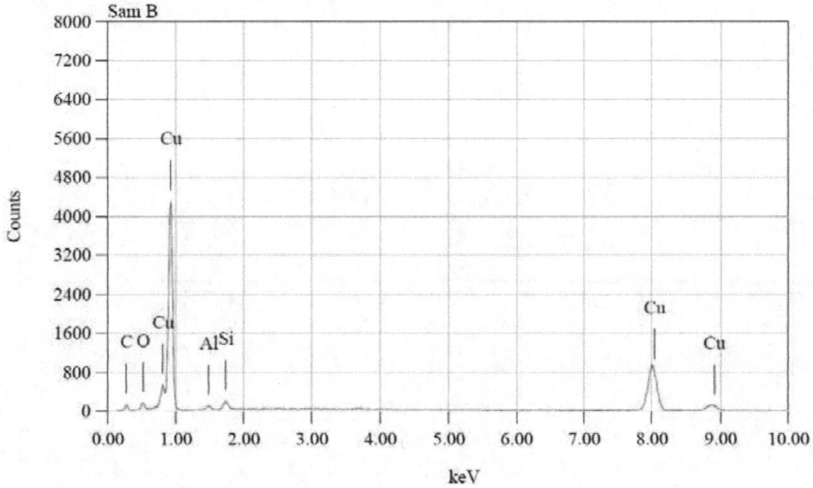


Acquisition Parameter
Instrument : 6490(LA)
Acc. Voltage : 20.0 kV
Probe Current: 1.00000 nA
PHA mode : T3
Real Time : 50.58 sec
Live Time : 50.00 sec
Dead Time : 1 %
Counting Rate: 1781 cps
Energy Range : 0 - 20 keV

Thin Film Standardless Standard Quantitative Analysis
Fitting Coefficient : 0.3803

| Element | (keV) | Mass% | Counts | Error% | Atom% | Compound | Mass% | Cation | K |
|---|---|---|---|---|---|---|---|---|---|
| C K | 0.277 | 3.70 | 539.04 | 0.04 | 15.61 | | | | 0.7025 |
| O K | 0.525 | 1.89 | 721.32 | 0.10 | 5.96 | | | | 0.2673 |
| Al K | 1.486 | 1.02 | 517.80 | 0.34 | 1.92 | | | | 0.2022 |
| Si K | 1.739 | 2.15 | 1070.95 | 0.18 | 3.87 | | | | 0.2052 |
| Cu K (Ref.) | 8.040 | 91.23 | 9325.37 | 0.03 | 72.64 | | | | 1.0000 |
| Total | | 100.00 | | | 100.00 | | | | |

**Figure 2:** EDS analysis of sample (B) with intensity peak

### *3.4. X-ray Diffraction (XRD) Analysis*

XRD analysis was performed on sample F (0.038M, pH 14) and sample G (0.04M, pH14) to identify the phases present in the deposited films. In case of sample (F) the XRD spectra showed particularly two peaks- one at 43.158 ° (d spacing 2.09348 Å and relative intensity 100%) and the other at 50.272° (d spacing 1.8134 Å and relative intensity 73.42%) whereas in case of (G) two peaks were found at 43.325° (d spacing 2.08671 Å and relative intensity 100%) and 50.446° (d spacing 1.8134 Å and relative intensity 82.96%) respectively. According to JCPDS card no. 4-836 both of these peaks is the evidence of copper. The reason for obtaining sharp copper peaks and no cuprous oxide peaks as seen in Figure 3 may be due to the fact that the deposited films were very thin and X-ray penetration was large; hence, the larger copper peaks were those of the copper of the substrate.

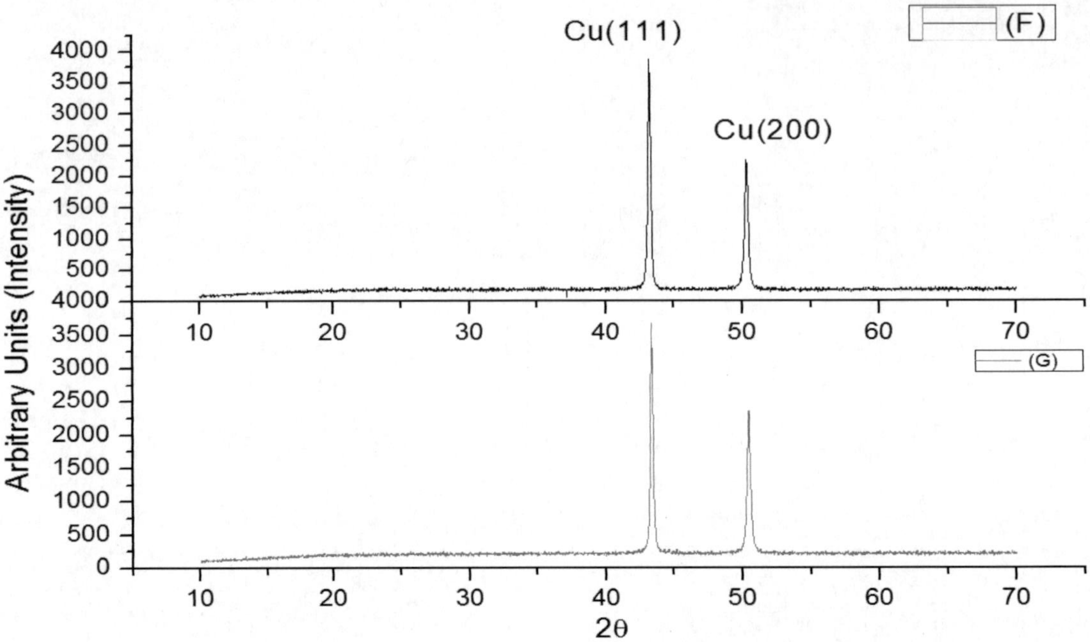


**Figure 3:** Overlay XRD spectra of sample (F) and (G)

*3.5. UV-VIS SPECTROMETRY*

To measure the optical absorption profile of the deposited films UV-Vis spectrometry was used. From Figure 4 for all the 7 samples it is observed that as the solution gets more basic (Table 1 and 2) absorption intensity gradually increased when the concentration was fixed. On the other hand, when pH was kept fixed at 14, on increase in concentration of copper absorption intensity increased but not significantly.

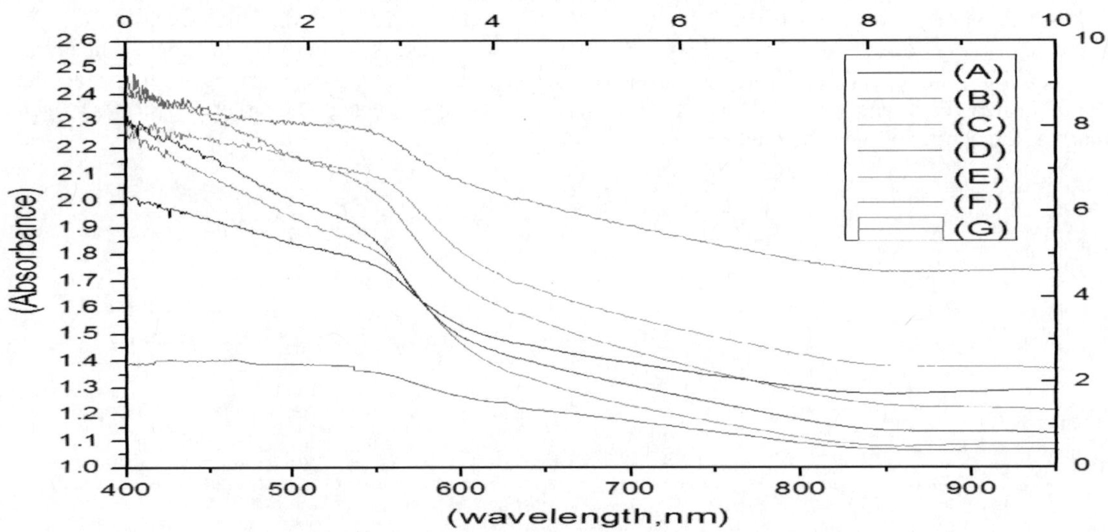


**Figure 4:** Overlay absorbance spectra of all the samples

### *3.6. BANDGAP MEASUREMENT*

From the UV-Vis spectra, optical bandgap of the deposited films were calculated. Absorption coefficient α was calculated first and from this value $(\alpha h\nu)^2$ vs hν curve was plotted where hν is the photon energy. Cuprous oxide has a bandgap value of about 1.85-2.2 eV but the average bandgap value of the deposited films in this work was about 1.4 eV as seen in Figure 5 which is close to the bandgap value of metallic copper which might be due to the fact that lower film thickness may have caused greater penetration of the UV-Vis radiation and substrate spectrum may have overlapped with that of the film.

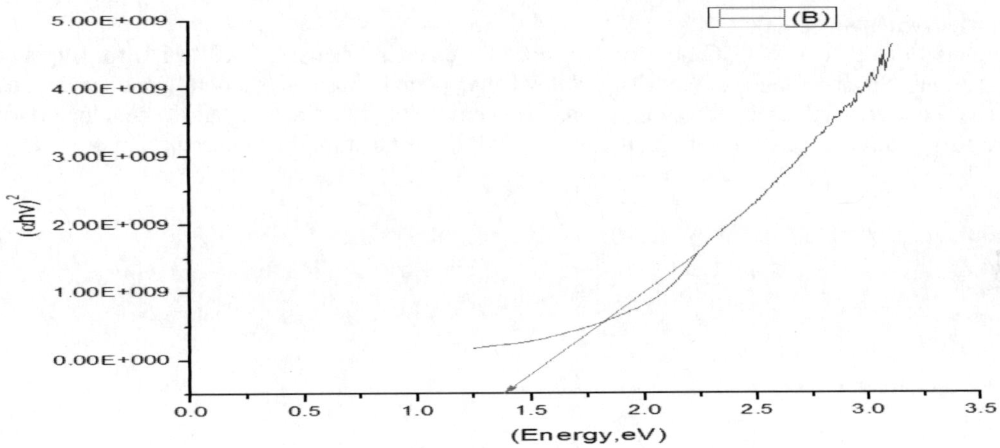


**Figure 5:** Bandgap measurement of sample (B).

## 4. Conclusion

The best parameters were pH 14 and concentrations of copper being 0.04M.The XRD pattern reveals the presence of Cu peak and EDS analysis confirms the presence of Copper and oxygen. The absorption intensity increased as the solution gets more basic and the bandgap value for the parameter (pH 14 and Conc. of Cu 0.04M) was about 1.4 eV. The thickness of the deposited film increased with increased pH value.

## 5. References

[1] Abdu, Y.* and Musa, *A.O*, "Copper(I) Oxide $Cu_20$ Based Solar Cells – A review*" Bayero Journal of Pure and Applied Sciences*, **2(2): 8 – 12**

[2] Charles Yianni Cummings," Electrodeposition and Characterization of Semiconductor Materials"

[3] Pim Reith & Gerben Hopman," Investigating electrodeposition to grow CZTS thin films for solar cell applications"

[4] Lytovchenko Oleksiyi, "Study of Thin Film Solar Cell based on Copper Oxide Substrate by DC Reactive Magnetron Sputtering" V. S. Arunach

[5] Abdelhamid El-Shaer , Abeer Ramadan Abdelwahed , AbdElraouf Tawfik , Mohsen Mossad , Dalal Hemada ," Effect of Deposition Parameters on Electrodeposited Cuprous Oxide Thin Films*," International Journal of Emerging Technology and Advanced Engineering*, Volume **4,** Issue **12.**

[6] Abu Sadat Md. Sayem Rahman, Md. Anisul Islam, Kazi Md. Shorowordi. Electrodeposition and Characterization of Copper Oxide Thin Films for Solar Cell Applications. *International Journal of Advanced Engineering and Nano Technology (IJAENT)*, ISSN: 2347-6389, Volume-2 Issue-2, January 2015

[7] Charles B. Greenberg1 and J. Shannon Breininger, "Electroless Deposition of Cuprous Oxide." *J. Electrochem. Soc.* 1977 volume **124**, issue **11**, 1681-1685.
[8] Direct electroless deposition of cuprous oxide films. US 4315055 A. Feb 9, 1982.
[9] G Papadimitropoulos, N Vourdas, V Em Vamvakas and D Davazoglou. "Deposition and characterization of copper oxide thin films" © 2005 IOP Publishing Ltd *Journal of Physics: Conference Series*, Volume **10**, Volume **10.**
[10] R.Neskovska, M.Ristova, J.Velevska, M.Ristov "Electrochromism of the electroless deposited cuprous oxide films" Volume **515**, Issue **11**, 9 April 2007, Pages 4717-4721

**Acknowledgements**
The authors would like to thank Centre for Advanced Research in Sciences (CARS), DU for helping us in doing various characterizations. We also acknowledge partial financial support from Department of Materials and Metallurgical Engineering, Bangladesh University of Engineering and Technology (BUET). We are also grateful to the laboratory assistants of MME department for their constant support.

# Effect of Activation Site Density on Copper Encapsulation of MWCNTs by Electroless Deposition

**Dr. Md. Muktadir Billah[1] and Dr. Quanfang Chen[2]**

[1,2]MEMS and Nanomaterials Lab, University of Central Florida, Orlando, FL 32816, USA

E-mail: [1]mdmuktadirbillah@knights.ucf.edu

**Abstract.** Importance of the activation site density on final encapsulation of multi-walled carbon nanotubes (MWCNTs) with copper was investigated in this study. MWCNTs, sensitized with tin and activated with silver, were used as precursors to prepare copper-decorated nanotubes. The effect of activation bath pH was found to be critical to control the density of silver-activation sites and to produce uniform encapsulation of carbon nanotubes (CNTs) with copper in the subsequent electroless deposition method. The morphology of the copper decorated nanotubes was studied using Field Emission Gun Scanning Electron Microscope (FEG SEM) and Energy Dispersive Spectroscopy (EDS).

## 1. Introduction

Since its discovery by Iijima [Iijima, S., 1991], multi-walled carbon nanotubes (MWCNTs) have been attracted by the researchers and academicians worldwide because of their extraordinary combination of superlative mechanical properties, excellent thermal conductivity (3000 W/m-K) with superior thermal stability and high electric conductivity [Endo et. al., 2002; Berber et. al., 2000; Salvetat et. al., 1999]. The intrinsically strong in-plane atomic bonding of carbon nanotubes (CNTs) offer high young's modulus (1.0–1.8 TPa), high tensile strength (30–200 GPa) and high elongation at break (10–30%) [Lau and Hui, 2000; Wong et. al., 1977; Salvetat-Delmotte and Rubio, 2002; Poncharal et. al, 1999; Qian et. al, 2002]. However, there is still challenge to exploit CNTs especially in metal matrix composites because of issues stemmed from the weak CNT-metal interface. Interest in metallization of CNTs has been growing over the years to overcome these issues. Such metallization, which offers excellent potential for CNTs for their reinforcement in the metal matrix, can be achieved by controlled electroless deposition [Chand, 2000; Li et. al., 1999; Arai and Endo, 2003; Arai and Endo, 2004]. For interconnecting materials, metal coated CNTs are the potential reinforcing agent. Among all the coating metals, copper got immense interest among the researchers for ages [Peng and Chen, 2012; Arai and Fujii, 2011; Wang et. al., 2014].

Numerous studies have been accompanied for uniform encapsulation of CNTs with metals. As reported widely, uniform and smooth metal coating is determined by the activation site density and their distribution [Qunqing Li et. al., 1997; Serp et. al., 2003; M. A. Fraga, 2002]. Extensive studies have been reported on the effect of activation site density to decorate CNTs with Nickel. However, for copper reported studies are still limited. After replacing Pb, silver has been used as catalytic seed for electroless copper plating in two-step activation process for long [Uzunlar et. al., 2013; Billah and Chen, 2015; Wei and Roper, 2014]. This study focused on the effect of activation site density of silver on copper encapsulation of CNTs. Since, silver reduction rate is controlled by the pH of the electroless solution; here authors studied silver activation of CNTs as a function of pH value of the silver nitrate solution to optimize pH for uniform copper coating on CNTs.

## 2. Experimental procedure

The multi-walled carbon nanotubes of diameter 30-50 nm and axial dimension 10-20 μm were procured from Cheap Tubes Inc. The purified nanotubes had purity greater than 95%. For the initial surface treatment as received nanotubes were ultra-sonicated for 2 hours at 60°C in an aqueous solution of $HNO_3$ (70%). The nanotubes were then rinsed with de-ionized (DI) water until pH 7 was reached. Then 50 mg CNTs were ultrasonically dispersed in 50 ml 0.1 M $SnCl_2$/0.1 M HCl for 30 minutes for the pre-activation (surface catalysts) [Y. Wang et. al. 2006]. After the pre-activation the $Sn^{2+}$- sensitized CNTs were separated and washed with DI water thoroughly. The activation of the sensitized CNTs were accomplished by stirring 50 mg sensitized CNTs for 2 hours in 50 ml aqueous solution of $AgNO_3$ (9 g/L) and $NH_3$ solution. The pH of the activation solution was varied over a range of 7-12 with $NH_3$ solution to study the effect of pH on silver deposition. The activated CNTs were then washed with DI water again and filtered off. Ultra-sonication was employed periodically to prevent agglomeration.

Finally the activated CNTs were transferred to the copper electroless bath for decorating with copper. The composition of the electroless copper bath was 0.03 M $CuSO_4{\cdot}5H_2O$, 0.25M $EDTANa_2$, $CHOCOOH{\cdot}H_2O$ and 10 ppm 2, 2'-Bipyridine ($(C_5H_4N)_2$). $CuSO_4{\cdot}5H_2O$ was used as copper ion source, $EDTANa_2$ as complexing agent, $CHOCOOH{\cdot}H_2O$ as reducing agent and 2, 2'-Bipyridine ($(C_5H_4N)_2$) as stabilizer [Mallory and Hajdu, 1990; Caturla, 1995]. CNTs were added to the electroless bath at 60°C and temperature was held constant during the deposition period. The electroless bath was placed into an ultrasonic cell bath where the frequency was fixed at 42 kHz and power at 100 W. The deposition was carried out in short time for 2 minutes after bubbles came out. The pH of the solution was adjusted at 12.2 with NaOH. Finally the copper decorated CNTs were rinsed with DI water, filtered off and dried in vacuum. The Sn-sensitized, Ag-activated and copper coated nanotubes were investigated with FESEM and EDS.

## 3. Results and discussion

The as received MWCNTs were first oxidized using $HNO_3$ (70%) since the oxidation allows better dispersion of the CNTs in the electrolytes [Peng et. al. 2007]. Also because of the oxidation metal nano particles form strong covalent bond with CNTs. Covalent bonds are much stronger than the weak van der Waals bond that is formed in absence of oxidation [Zhong and Lukes, 2006]. The sensitized CNTs were observed under FEG SEM (Fig. 1a) and the EDS spectrum confirmed the presence of Sn (Fig. 1b). The sensitization temperature was selected at higher than room temperature to produce highly tin-rich catalyst sites. It is evident that either heating of the catalytic solution or increasing pH of the solution promotes crystallization of tin-rich catalytic sites during the sensitization process [Froment et. al., 1995].

Silver was deposited on the sensitized CNTs by electroless deposition method using silver nitrate solution. The pH of the solution was varied from 7-12. All the CNTs were then studied under FEG SEM and presence of silver was confirmed with the EDS spectra. In Fig. 2 wt. percentage of the deposited silver is plotted against pH value of silver nitrate solution. The weight percentage of deposited silver was used as a measure of the plating rate. It is evident here that silver was deposited over a wide range of pH value with quite different rate. Silver deposition was apparent for pH value 7. For further increase of pH of the solution, deposition rate was gradually accelerated exhibiting a maxima in the deposition rate at pH value 10. Later deposition rate was drastically decreased for higher pH values.

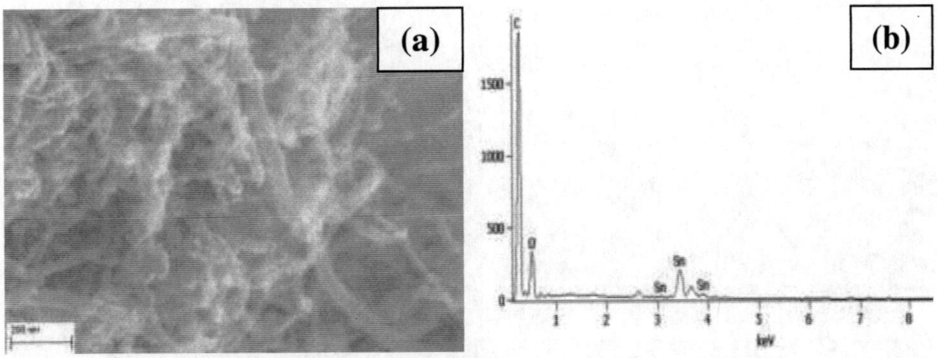


**Figure 1.** SEM image (a) and EDS (b) of sensitized MWCNTs..

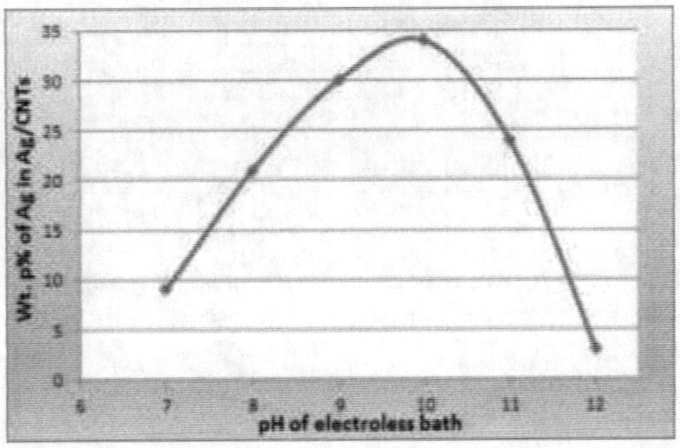


**Figure 2.** Rate of silver deposition as a function of pH.

SEM shows the plating morphology of the Ag/CNTs in Fig. 3. Though the presence of silver was identified with EDS for pH value 7-8, CNTs were not coated uniformly with silver (Fig. 3a & b). Fig. 3c shows that CNTs are completely encapsulated with silver. A large magnification is shown in Fig. 4 with the EDS spectrum. The Ag/CNTs composite was fibrous without agglomeration as ultra-sonication was employed periodically during silver deposition. But for higher pH values coated CNTs showed agglomerating tendency due to excessive silver deposition with large silver crystals identified on CNT surfaces (Fig. 3d & e). For pH 12 silver deposition was of trace amount (Fig. 3f).

Fig. 5 shows the Cu encapsulated MWCNTs. Cu was deposited on CNTs activated with silver as particles at pH 8 (Fig. 5a) and as the CNTs were fully covered with silver at pH 9 (Fig. 5b). It appeared that as the CNTs were entirely coated with silver, excessive normal growth rate of copper caused large copper crystals and some copper particles were detached from CNT surfaces. However, CNTs were better coated with copper as silver was present as high density activation sites rather than complete encapsulating the CNTs. High density activation sites allowed significant lateral growth of the subsequent coating material to get uniform coating without excessive crystal growth [Qunqing Li et. al., 1997, Jagannatham M. et. al., 2015].

IOP Conf. Series: Materials Science and Engineering **438** (2018) 012003 doi:10.1088/1757-899X/438/1/012003

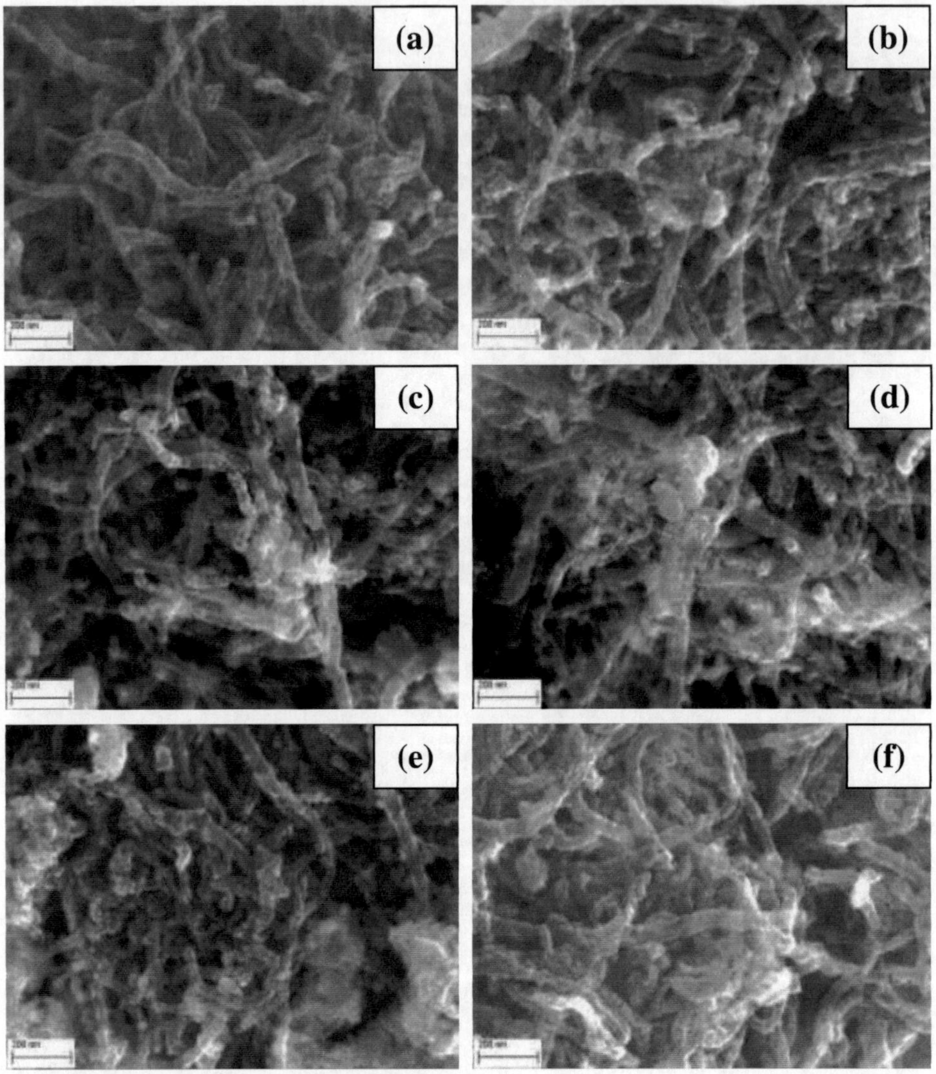


**Figure 3.** Ag/MWCNTs for different pH value of silver nitrate solution, pH=7 (a), pH=8 (b), pH=9 (c), pH=10 (d), pH=11 (e), pH=12 (f).

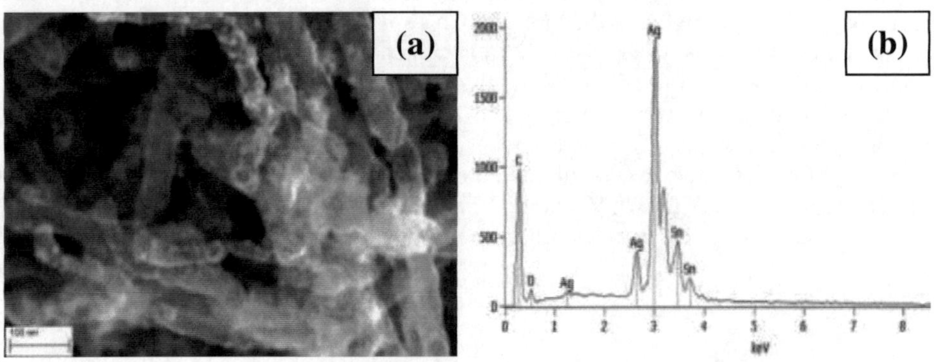


**Figure 4.** Ag encapsulated MWCNTs (a) and EDS spectrum (b).

IOP Conf. Series: Materials Science and Engineering **438** (2018) 012003 doi:10.1088/1757-899X/438/1/012003

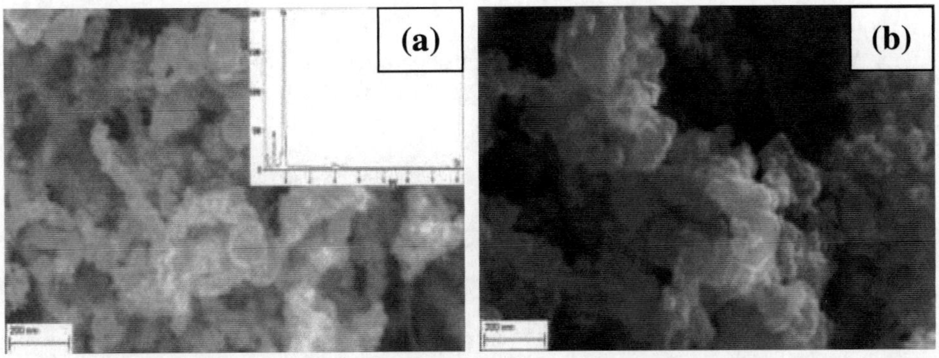


**Figure 5.** Cu/MWCNTs for Ag activation at pH=8 (a) and pH=9 (b).

## 4. Conclusion

In this study the deposition of silver and copper on MWCNTs were investigated using electroless deposition method. The acid treated CNTs were successfully coated with silver and copper. Silver was found to be deposited for a wide range of pH of silver nitrate solution. The pH was optimized for complete encapsulation of CNTs with silver. However for uniform copper deposition high density silver activation sites were found to be more effective than uniform silver coating on CNT surfaces.

## 5. Reference

[1] S. Arai and M. Endo, Electrochemical Common, 5 (2003) 797
[2] S. Arai, M. Endo, N. Kaneko, Carbon, 42 (2004) 641
[3] S. Berber, Y.K. Kwon and Tomanek D, Physics Rev Letter, 84 (2000) 4613
[4] M.M. Billah and Q. Chen, Journal of Electronic Materials, 45 (2016) 1
[5] F. Caturla, F. Molina, M. Sabia, F. Rodriguez-Reinoso and A. Esteban, Journal of Electrochemical Society, 142 (1995) 4084
[6] S. Chand, Journal of Material Science, 35 (2000) 1303
[7] M. Endo, Y.A. Kim, T. Fukai, T. Hayashi, K. Oshisa, M. Terrones, T. Yanagisawa, S. Higaki and M.S. Dresselhaus, Applied Physics Letter, 80 (2002) 1267
[8] M.A. Fraga, E. Jord˜ao, M.J. Mendes, M.M.A. Freitas, J.L. Faria, and J.L. Figueiredo, Journal of Catalysis, 209 (2002) 355
[9] M. Froment, E. Queau, J. R. Martin, and G. Stremsdoerfer, Journal of Electrochemical Society, 142 (1995) 3373
[10] Iijima, S. Nature, 56 (1991) 354
[11] M. Jagannatham, S. Sankaran, H. Prathap, Applied Surface Science, 324 (2015) 475
[12] K.T. Lau and D. Hui, Journal of Electronic Materials, 33 (2000) 263–77
[13] Y.B. Li, B.Q. Wei, C.L. Xu, J. Jiang and D.H. Wu, Journal of Materials Science, 34 (1999) 5281
[14] Y.T. Peng, Y.Z. Hu and H.Wang, Tribological Letter, 3 (2007) 247
[15] Y. Peng and Q. Chen, Journal of Electrochemical Society, 159 (2012) 72
[16] D. Qian, Appl. Mechanics, 55 (2002) 495
[17] L. Qunqingi, S. Fan, W. Han, C. Sun and W. Liang, Journal of Applied Physics, 36 (2011) 501
[18] J.P. Salvetat, J.M. Bonard, N.H. Thomson, A.J. Kulik, L. Forro, W. Benoit and L. Zuppiroli, Applied Physics A, 69 (1999) 255
[19] P. Serp, M. Corrias, and P. Kalck, Applied Catalysis A, 253 (2003) 337
[20] S. Arai and J. Fujii, Journal of Electrochemical Society, 158 (2011) 506
[21] E. Uzunlar, Z.Wilson and P.A. Kohl, Journal of Electrochemical Society, 160 (12) 3237
[22] F. Wang, S. Arai and M. Endo, Electrochemistry Communications, 6 (2004) 1042
[23] Y. Wang, X. Xu, Z. Tian, Y. Zong, H. Cheng, and C. Lin, Chemical European Journal,

(2006) 2542
[24] X. Wei and D.K. Roper, Journal of Electrochemical Society, 161 (2014) 235
[25] H. Zhong and J.R. Lukes, Physics Rev. B, 74 (2006)125

# Post-stress Induced Tensile Property Modification of Jute Polymer Composite

**M.K. Nahian, S.Z. Ali and M.A. Islam**
Department of Materials and Metallurgical Engineering, Bangladesh University of Engineering and Technology, Dhaka 1000, Bangladesh

**Abstract.** Jute fiber reinforced composites is an emerging area in composite science because of their low cost, low density and high specific properties. These composites are biodegradable and non-abrasive. In this research work, 10 and 15 wt% raw jute fiber reinforced polyester composite were made in hand lay-up method. Then each type of composite was again fabricated but this time, different levels of post-stress of 2.5, 5 and 10 kPa were applied to observe their effect. Later on, the developed composites were subjected to the tensile test. All composites had higher tensile properties (tensile strength and tensile modulus) than pure polyester. Tensile properties were observed to increase with increasing post-stress for same weight percentage of jute fiber.

*Keywords*: Jute fiber composite, Polyester, Post-stress, Tensile strength

## 1. Introduction

Over the past few decades, polymers have replaced many of the conventional materials in various applications. This is due to the advantages polymers offer over conventional materials. The most important advantages of using polymers are the ease of processing, productivity, lightweight and cost reduction. Property of polymers can be modified and enhanced by reinforcing fillers. Today synthetic polymers are combined with various reinforcing fillers in order to improve the mechanical properties and obtain the characteristics demanded in actual applications [1-3]. Fibers are most commonly used as fillers. Natural fibers have recently attracted the attention of scientists and technologists because of the advantages that these fibers provide over conventional reinforcement materials, and the development of natural fiber composites has been a subject of interest for the past few years [4].

Growing attention nowadays is being paid to jute fiber. Jute fiber is one of the most common natural fiber since it is produced on a large scale in the Indian sub-continent, especially Bangladesh, and has a minimal effect on the environment because of its biodegradable properties [5,6]. Jute is the cheapest lignocellulosic long vegetable bast fiber and abundantly available in South Asia. But Jute has an affinity towards moisture. Due to its hygroscopic nature, dimensions of the filament, as well as its mechanical and electrical properties, are changed [7]. So proper wetting of the two phases is naturally poor which does not conduct the desired properties in composite materials.

Various types of polymers are used to make composites that are normally used in engineering applications. Normally they are classified as Thermoplastic and Thermoset polymers. Thermoset polymers outperform thermoplastics in a number of areas, including mechanical properties, chemical resistance, thermal stability, and low raw material cost.

In this research work thermoset composites were made under different amount of pressure. This was done by applying certain amount of weight when the resin started to coagulate during the composite fabrication process which is called post-stress. Mechanical properties, including the tensile strength, flexural strength, tensile modulus and flexural modulus of the composites, also declined with the increase of the void content which is created by air bubbles. Applying post-stress ensures the removal of air bubbles and optimizes the fiber to resin ratio [8]. At the same time, better interaction between resin and fiber is also occurred. A highly compact product can be obtained by applying molding pressure. Because of this, mechanical properties are reported to improve [9].

Mechanical properties were observed to increase for coir fiber-polyester composite under various post-stress or molding pressure [10]. On the other hand secondary effect of molding pressure was observed for piassava–polyester [11] and chopped bagasse–polyester composites [12]. The aim of this project is to find the effect of post-stress on tensile properties of jute fiber reinforced polyester composites.

## 2 Experimental

### *2.1 Materials and methods*

At first, washed and sun-dried jute fiber was collected from the local market. Jute fiber used in this research was untreated.

As matrix material unsaturated polyester resin and Methyl Ethyl Ketone Peroxide hardener were used. Various amount (10 wt% and 15 wt%) of continuous jute fiber aligned in the lower mold. Then the upper mold was placed on it. Polyester resin was mixed with 2 wt% hardener. This mixture was placed in the vacuum chamber for removing primary air bubbles. This vacuumed mixture was poured into the mold. After 10 minutes, when the mixture started to coagulated, a post-stress (2.5, 5 & 10 kPa for individual sample) was applied. Each post-stress (2.5, 5 & 10 kPa) was applied for both 10 wt% and 15 wt% jute fiber reinforced composites. Composites without any post-stress (0 kg load) were also prepared for 10 wt% and 15 wt% jute fiber. For 10 wt% fiber content, 3 test specimens were prepared for each post-stress. The same approach was carried out for 15 wt% jute fiber composites. Pure polyester sample was also made.

**Figure 1.** Post-stress applied during fabrication of composite.

*2.2 Tensile test*

Universal testing machine was calibrated. Tensile properties in the longitudinal direction of jute fiber reinforced polymer matrix composites were determined according to ASTM D638 at a cross speed of 3 mm/min. Tensile properties of pure polyester sample were also measured with same standard.

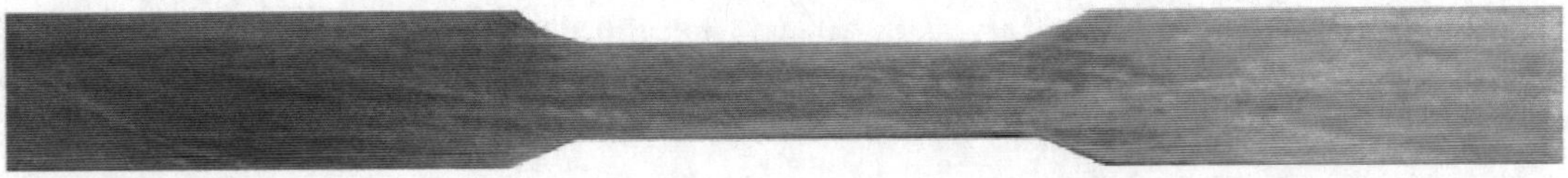

**Figure 2.** Tensile test specimen.

## 3. Results and Discussion

*3.1 Tensile strength*

The average tensile strength of pure polyester was found to be 30.45 MPa. According to the composition of their main chain, polyesters can be of different types and their tensile strengths also vary according to their compositions [13].

Normally, tensile strength of jute fiber reinforced polymer matrix composite increases with increasing pressure during fabrication. Pressure plays a significant role to obtain high tensile strength because of the good interface bonding between fiber and resin. This is due to the fact that the increase in pressure makes the fiber and resin very well compressed and resin was forced into various passages of the jute fibers [14].

Here is the tensile strength of 10 wt% jute fiber reinforced composite for various post-stresses. Standard deviation of 3 test sample of each post-stress are also given in table 1.

**Table 1.** Tensile strength of 10 wt% jute polyester composites for different post-stresses.

| Jute fiber content (wt%) | Post-stress (kPa) | Tensile strength (MPa) | Standard Deviation |
|---|---|---|---|
| | | 61.23 | |
| | 0 | 63.3 | 1.48 |
| | | 64.85 | |
| | | 80.4 | |
| | 2.5 | 82.69 | 1.73 |
| | | 84.63 | |
| 10 | | | |
| | | 88.56 | |
| | 5 | 90.94 | 1.52 |
| | | 92.25 | |
| | | 99.25 | |
| | 10 | 102.42 | 1.85 |
| | | 103.65 | |

IOP Conf. Series: Materials Science and Engineering **438** (2018) 012004 doi:10.1088/1757-899X/438/1/012004

To draw the stress-strain diagrams, we took only median value from the tensile strength of 3 specimens of each post-stress. Tensile strength of 10 wt% jute fiber reinforced polymer matrix composites increase with increasing post-stress. Here tensile strength increased from 60.53 MPa (at no post-stress) to 102.42 MPa (at 10 kPa post-stress). For other post stress conditions like 2.5 kPa and 5 kPa, average tensile strengths were 82.69 MPa and 90.94 MPa respectively.

Tensile stress vs. tensile strain curves for 10 wt% jute fiber reinforced polyester prepared under various post-stresses are plotted along with the tensile stress vs. tensile strain curve of pure polyester in figure 3.

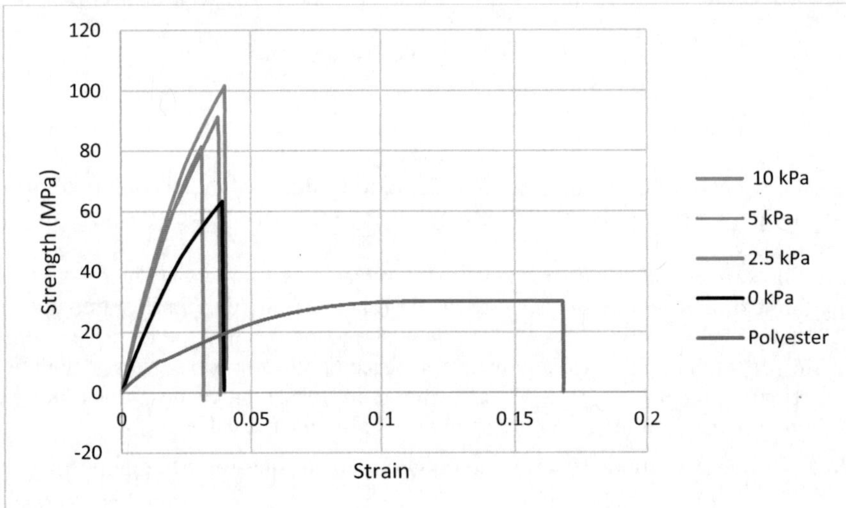


**Figure 3.** Tensile stress vs. strain curves for 10 wt% jute fiber reinforced PMC under different post stresses.

Again tensile strength of 15 wt% jute fiber reinforced composites are given below with standard deviation.

**Table 2.** Tensile strength of 15 wt% jute polyester composite for different post-stress.

| Jute Fiber Content (wt%) | Post-stress (kPa) | Tensile Strength (MPa) | Standard Deviation |
|---|---|---|---|
| | | 71.8 | |
| | 0 | 73.77 | 1.1 |
| | | 74.39 | |
| | | 78.96 | |
| | 2.5 | 79.20 | 1.13 |
| | | 81.47 | |
| 15 | | | |
| | | 90.59 | |
| | 5 | 91.13 | 1.62 |

| | | |
|---|---|---|
| | 94.27 | |
| | 108.1 | |
| 10 | 110.47 | 2.28 |
| | 113.66 | |

To draw sress-strain curves, only median value was taken for each set of specimens. Result of tensile test of 15 wt% jute fiber reinforced polyester composites under 0 kPa, 2.5 kPa, 5 kPa & 10 kPa post-stress are pronounced. Under 0, 2.5, 5 & 10 kPa post-stress tensile strength of the samples were 73.77, 79.20, 91.13 and 110.47 MPa respectively. For 15 wt%, jute fiber reinforcement effect of post-stress on tensile strength can easily be visualized.

Tensile stress vs. tensile strain curves was also plotted from above data and it's shown in figure 4.

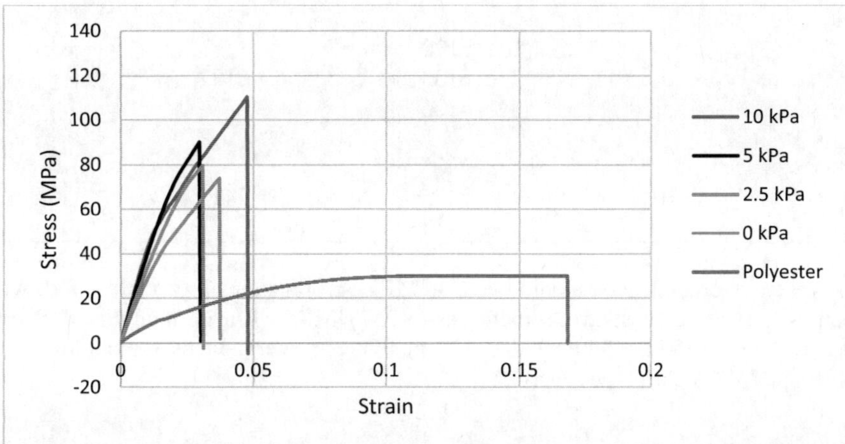


**Figure 4.** Tensile stress vs. strain curves for 15 wt% jute fiber reinforced PMC under different post-stresses.

Similar effect was found by E. Shakouri et al. carried a research for wood-plastic composite with changing die pressure [15].

*3.2 Tensile modulus*

The effect of post-stress is shown in figure 6 (for 10 wt% jute fiber) and figure 7 (for 15 wt% jute fiber). Tensile modulus increased with increasing post-stress. Normally, post-stress ensures sufficient wetting of fiber. Voids can also be reduced by using high external pressure[16,17]. At the same time, post-stress gives better fiber to resin ratio which enhances the tensile modulus of composite. E. Shakouri at el. also found that tensile modulus also increases with die pressure [15].

For 10 wt% jute fiber reinforced composite, tensile modulus was calculated by averaging the value from 3 test specimen of each post-sress. A table of average tensile modulus of 3 test specimen of each post-stress and their standard deviation given below.

**Table 3.** Average tensile modulus of 10 wt% jute polyester composite for different post-stress.

| Jute Fiber Content (wt%) | Post-stress (kPa) | Tensile Modulus (MPa) | Average Tensile Modulus (MPa) | Standard Deviation |
|---|---|---|---|---|
| 10 | 0 | 2468<br>2465<br>2480 | 2471 | 3.56 |
| | 2.5 | 5385<br>5396<br>5387 | 5392 | 3.74 |
| | 5 | 6122<br>6133<br>6126 | 6127 | 4.54 |
| | 10 | 7405<br>7415<br>7413 | 7411 | 4.32 |

For pure polyester, average tensile modulus is 746 MPa. Average tensile modulus of 10 wt% jute fiber reinforced composite (without post-stress) increases to 2471 MPa. Tensile modulus also improves with increasing post-stress. For 2.5 kPa, 5 kPa and 10 kPa post-stress, average tensile modulus values are 5392 MPa, 6127 MPa and 7411 MPa, respectively.

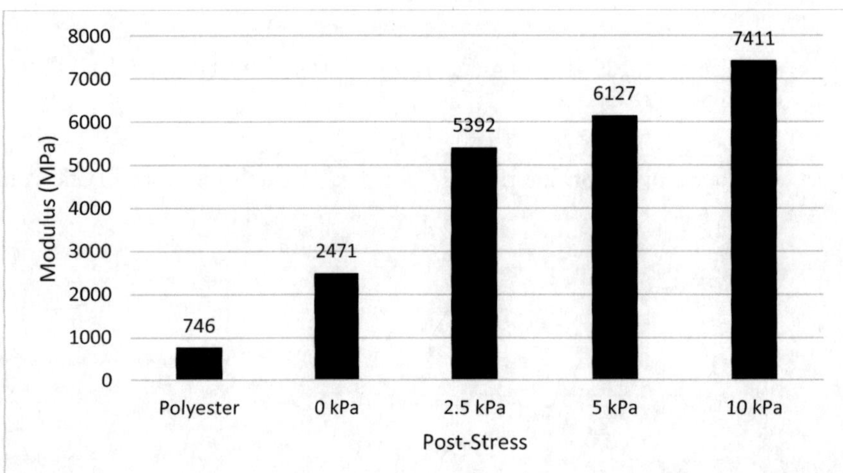


**Figure 5.** Tensile modulus vs. post stress bar chart for 10 wt% jute fiber reinforced composite.

Average tensile modulus and standard deviation were calculated for each post-stress of 15 wt% jute fiber composite. The results are given in table 4.

**Table 4.** Average tensile modulus of 15 wt% jute polyester composite for different post-stress.

| Jute Fiber Content (wt%) | Post-stress (kPa) | Tensile Modulus (MPa) | Average Tensile Modulus (MPa) | Standard Deviation |
|---|---|---|---|---|
| 15 | 0 | 4710<br>4716<br>4713 | 4713 | 2.45 |
| | 2.5 | 5055<br>5060<br>5062 | 5059 | 2.87 |
| | 5 | 6480<br>6476<br>6469 | 6475 | 4.55 |
| | 10 | 7902<br>7898<br>7909 | 7903 | 4.56 |

With increasing fiber content, tensile modulus also increases. For 15 wt% jute fiber reinforced composite, average tensile modulus is 4713 MPa. Here tensile modulus also increases with increasing post-stress. For 10 kPa post-stress, maximum average tensile modulus (7903 MPa) is observed.

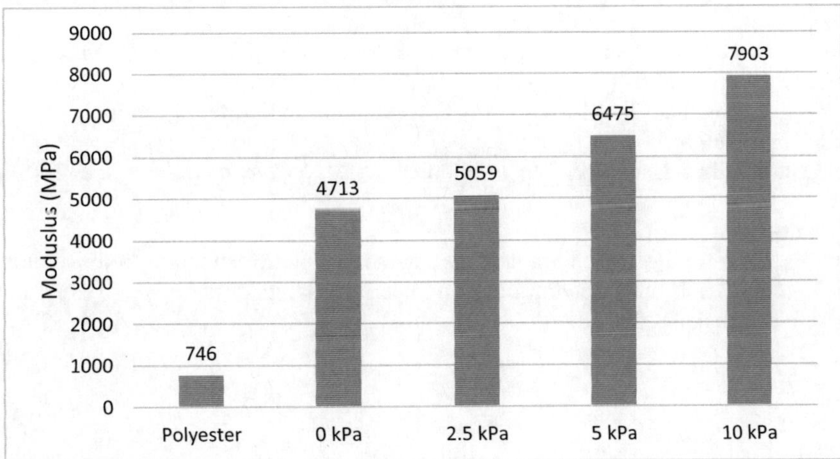


**Figure 6.** Tensile modulus vs. post stress bar chart for 15 wt% jute fiber reinforced composite.

## 4. Conclusion

In this research work, different levels of post-stresses (0, 2.5, 5 & 10 kPa) were applied to various percentages (10%, 15%) of fiber reinforcement during the fabrication of composites. The effect of post-stress on tensile strength and modulus of raw jute fiber reinforced polyester has been investigated.

IOP Conf. Series: Materials Science and Engineering **438** (2018) 012004 doi:10.1088/1757-899X/438/1/012004

- Tensile properties of composites with 10 wt% and 15 wt% of jute fiber are higher than the pure polyester sample. Tensile properties of composite increases with increasing fiber content. So it may be concluded that with increasing jute fiber content up to a certain amount causes improved tensile properties.
- Tensile properties (tensile strength and tensile modulus) of composite sample without any post-stress are lower than composite samples fabricated by applying post-stress. Tensile strength and modulus increased with increasing post-stress. Higher tensile properties for increasing post-stress were due to better impregnation of the fibers by the resin, which certainly occurred as a higher pressure is applied during the fabrication of composite.

**References**

[1] Yang HS, Kim HJ, Park HJ, Lee BJ and Hwang TS 2006 *Compos. Struct.* **72** pp 429–37
[2] Yang HS, Kim HJ, Park HJ, Lee BJ and Hwang TS 2007 *Compos. Struct.* **77** pp 45–55
[3] Choi NW, Mori I and Ohama Y 2006 *Waste Manage* **26(2)** pp 189–94
[4] S Carpenter 1997 *Reinforced Plastics* **41(11)** pp 22
[5] Vilaseca, F *et al.* 2007 *Process Biochemistry* **42** pp 329–34
[6] Planckett D, Andersen T L, Pedersen W B and Nielsen L 2003 *Compos. Sci. Tech.* **63** pp 1287–96
[7] Rana A K and Jayachandran K 2000 *Mol. Cryst. and Liq. Cryst.* **353** pp 35-45
[8] Gaston Francucci, Stuart Palmer and Wayne Hall 2017 *J. of Composite Materials* **0(0)** pp 1–13
[9] Hitoshi Takagi and Akira Asano 2008 *Composites: Part A* **39** pp 685–89
[10] S N Monteiro, L A H Terrones and J R M D'Almeida 2008 *Polymer Testing* **27** pp 591– 95
[11] J F de Deus, S N Monteiro, J R M d'Almeida 2005 *Polym. Test.* **24** pp 750
[12] M V de Sousa, S N Monteiro, J R M d'Almeida 2004 *Polym. Test.* **23** pp 253
[13] A V Ratna Prashad and K Mohana Rao 2011 *Materials and Design* **32** pp 4658–63
[14] Saravanan Kannappan and Bhaarathi Dhurai 2012 *Journal of Engineered Fibers and Fabrics* **7** pp 28-34
[15] E Shakouri, A H Behravesh, A Zolfaghart and M Golzar 2009 *J. of Thermoplastic Composite Materials* **22** pp 605-16
[16] K Tanaka, T Yamada, K Moriito and T. Katayama 2016 *WIT Transactions on The Built Environment* **166** pp 307-315
[17] Tengfei Chang, Lihua Zhan, Wei Tan and Shujian Li 2016 *Composite Interfaces* **24(5)** pp 1-12

**Acknowledgments**

The authors are grateful for a research grant in the course of this investigation to the authority of BUET and Department of Materials and Metallurgical Engineering, Bangladesh University of Engineering and Technology (BUET), Dhaka-1000.

IOP Conf. Series: Materials Science and Engineering **438** (2018) 012005 doi:10.1088/1757-899X/438/1/012005

# Flexural Property Enhancement of Jute Polyester Composite

**S.Z. Ali, M.K. Nahian and M.A. Islam**

Department of Materials and Metallurgical Engineering, Bangladesh University of Engineering and Technology, Dhaka 1000, Bangladesh
**Abstract.** Jute fiber is used as a reinforcement material for its availability, lightweight, non-toxicity, biodegradability, low cost, renewable source and moderate mechanical properties. In addition, jute fiber has longer continuous length. In this study, 10 and 15 wt% jute fiber reinforced polymer matrix composites were fabricated by hand lay-up method using polyester resin. The jute composites were developed with and without compression stress. To understand the effects of compression stress levels 2.5, 5 and 10 kPa were used. The jute fibers used here were in untreated condition (raw jute fiber). Developed composites were then characterized by three-point flexural test. Experimental results revealed that compression stress significantly enhanced the flexural properties (flexural strength and flexural modulus) of the composites. In this regard, 15 wt% jute fiber reinforced composite showed greater flexural properties.
*Keywords:* Jute fiber composite, Polyester, Compression stress, Flexural strength.

## 1. Introduction:

The burgeoning demand for a clean environment has led the innovation of green materials and utilization of natural materials. Thus, the urge for the production of high-performance engineering products from natural renewable resource is growing day by day. Natural fiber reinforced composites are among those versatile, high-performance materials which combine the unique mechanical and thermal properties that cannot be achieved in a single material. Jute fiber is one of the most common natural fiber since it is produced on a large scale in the Indian sub-continent, especially in Bangladesh, and has a minimal effect on the environment because of its biodegradable properties [1,2]. Jute is the cheapest lignocellulosic long vegetable bast fiber and abundantly available in South Asia. But Jute has an affinity towards moisture. Due to its hygroscopic nature, dimensions of the filament, as well as its mechanical and electrical properties, are changed [3]. So proper wetting of the two phases is naturally poor which does not conduct the desired properties in composite materials.

Normally thermoplastic composites are fabricated under pressure in the hot pressing machine. To carry a higher amount of applied load modulus transfer is important. Modulus transfer mainly depends on the fiber matrix bonding. Due to applied pressure, pores are also eliminated from the resin. A highly compact product can be obtained by applying molding pressure. Because of this, mechanical properties are reported to improve [4].

In hand lay-up method for thermoset composites fabrication, usually, no load or stress is applied. However, in this research work thermoset composites were made under applied stress. This was done by applying a certain amount of weight when the resin started to coagulate during the composite fabrication process. The aim of this project was to compare the enhancement in flexural properties due to compression stress for various amounts of jute fiber reinforced polymer matrix composites.

IOP Conf. Series: Materials Science and Engineering **438** (2018) 012005 doi:10.1088/1757-899X/438/1/012005

## 2. Experimental

### *2.1 Materials and methods*

In this research raw jute fiber was collected from local market of Bangladesh and then it was washed and dried in sun. The middle portion of every jute fiber was selected so that they have same diameter and possess similar properties. The polyester resin was used as matrix material and Methyl Ethyl Ketone Peroxide was used as a hardener.

The required amount of continuous and aligned jute fibers (10 &15 percent by weight) were placed on the lower part of the mold and then the upper part was closed. Resin with 2% hardener was poured into the mold and after 7-10 minutes, when the resin starts to coagulate, various compression stresses (2.5, 5 and 10 kPa) were applied. To apply compression stress, various amounts of load were imposed on the composite casting so that the whole composite solidifies under this stress. The pure polyester sample was also fabricated.

**Figure 1.** Jute-polyester composite.

### *2.2 Three-point flexural test*

Three-point flexural test was carried out for each type of sample using a Universal Testing Machine of model number INSTRON 3369. Flexural test specimens were prepared according to ASTM D 790-00 standard.

## 3. Results and Discussion

### *3.1 Flexural strength*

#### *3.1.1 Pure polyester sample*

Three-point flexural test was conducted for the pure polyester sample. According to the composition of their main chain polyester can be of different types and their flexural strength also varies with their composition. In our research, the average flexural strength of pure polyester is 51.86MPa. Ratna and Mohana also reported the same flexural strength for pure polyester [5].

#### *3.1.2 Jute polyester composite (10 wt% jute fiber)*

Flexural strength of sample without any compression stress is 57.98 MPa which is greater than pure polyester. Normally flexural strength of the composite is higher than pure polymer [6]. When this same composite is fabricated under 2.5, 5 and 10 kPa compression stresses, it is observed that flexural strength increases with increasing compression stress (table 1). Without any compression stress, jute polyester composite with 10 wt% jute fiber has a flexural strength of 57.98 MPa and flexural strength increases to 83.99 MPa for the same composite fabricated under 10 kPa compression stress. It has been seen that when pressure is applied during fabrication of coir fiber/polyester, flexural strength increases with increasing molding pressure [7].

**Table 1.** Flexural strength of 10 wt% jute polyester composite for different compression stresses.

| Compression stress (kPa) | Average Flexural Strength (MPa) |
|---|---|
| 0 | 57.98 |
| 2.5 | 65.01 |
| 5.0 | 76.57 |
| 10.0 | 83.99 |

Stress vs. strain curve for 10 wt% jute fiber reinforced polyester fabricated under different compression stresses is plotted along with the stress vs. strain curve of pure polyester in figure 2. Changes in flexural property can be easily identified from this graph.

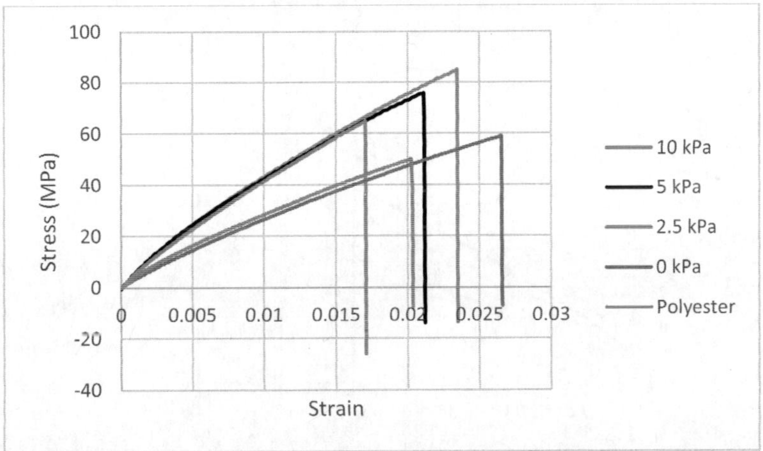


**Figure 2.** Stress vs. Strain curve for 10 wt% jute polyester composite fabricated under different compression stresses.

*3.1.3 Jute polyester composite (15 wt% jute fiber)*

Flexural strength of 15 wt% jute fiber reinforced polyester sample without any compression stress is higher than the previous sample fabricated under no compression stress, which is quite expected. It was reported that flexural strength of composite increases with increasing jute fiber loading up to 40 wt% [8,9]. At normal condition that means for no compression stress, 15 wt% jute fiber reinforced polyester composite has a flexural strength of 63.29MPa. When compression stress is applied, flexural strength also increases in this case and for 10kPa compression stress it becomes 114.38MPa which is almost double than the sample fabricated without any compression stress. For both fiber content used, there is a tendency for composites cured under higher compression stress to present corresponding higher flexural strength values. This is due to the more effective impregnation of the fibers by the resin, which certainly occurred as a higher pressure is applied during the matrix setting. Table 2 and figure 3 show the strength variation with different compression stress. From the figure, it is clear that the increase in flexural strength with increasing compression stresses is more pronounced for 15 wt% jute fiber sample than 10 wt% jute fiber reinforced polyester sample. Figure 4 shows the increase in flexural strength with compression stress for two different fiber contents. It is clear from the plot that the increase in flexural strength with compression stress is not actually linear. So compression stress applied during the fabrication has just a tendency to increase the flexural strength of composite. Still one cannot be sure that compression stress have any direct influence. Because, for both piassava–polyester composites [10] and chopped bagasse–polyester composites [11] the molding pressure was observed to have only a secondary effect on the flexural behavior. Their mechanical performance doesn't improve until the threshold molding pressure in overcome.

**Table 2.** Flexural strength of 15 wt% jute polyester composite for different compression stresses.

| Compression stress (kPa) | Average Flexural Strength (MPa) |
|---|---|
| 0.0 | 63.29 |
| 2.5 | 80.49 |
| 5.0 | 86.59 |
| 10.0 | 114.38 |

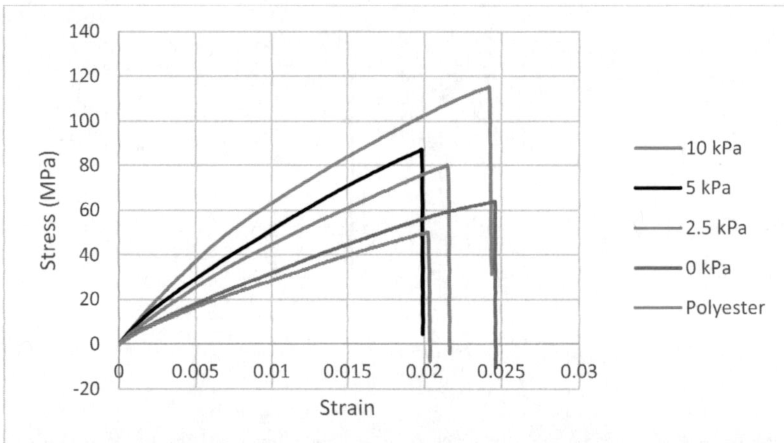


**Figure 3.** Stress vs. Strain curve for 15 wt% jute polyester composite fabricated under different compression stresses.

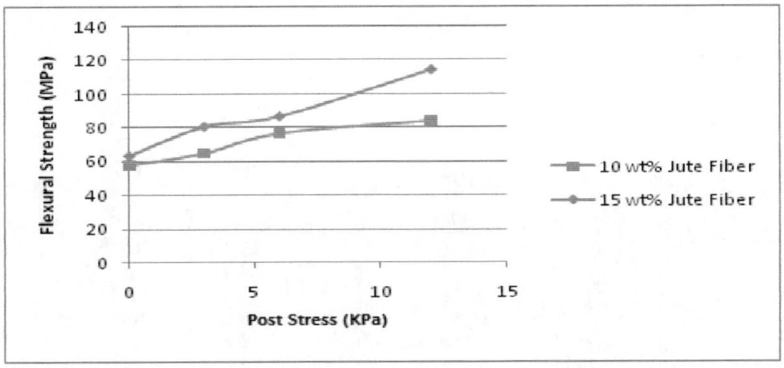


**Figure 4.** Relationship between flexural strength and compression stress for two different fiber contents.

*3.2 Flexural modulus*

Figures 5 & figure 6 show the change in flexural modulus with compression stress for both 10 wt% and 15 wt% jute fiber content. For jute-polyester composite flexural modulus was reported to increase with increasing fiber content [12]. Pure polyester has a lower flexural modulus of 1711MPa but for 10 wt% fiber content it increases to 3279MPa and for 15 wt% fiber content it increases to 5743MPa. For both amounts of fiber content, a further increase in flexural modulus is observed with increasing compression stresses. G Francucci and his co-workers observed the similar affect for flax fiber/epoxy laminates. [13]

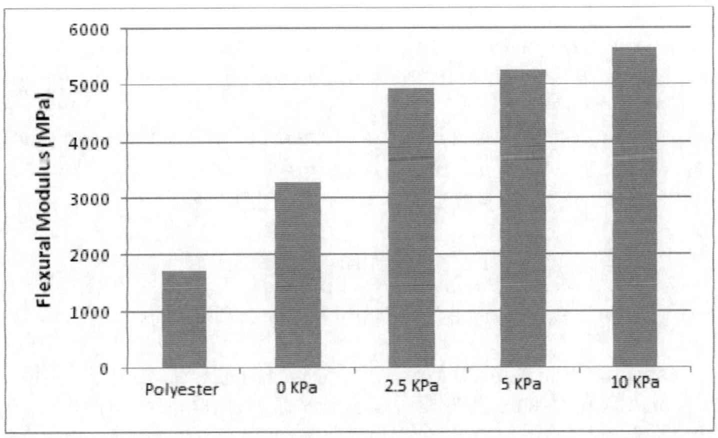


**Figure 5.** Relation between flexural modulus and compression stress for 10 wt% jute fiber reinforced polyester.

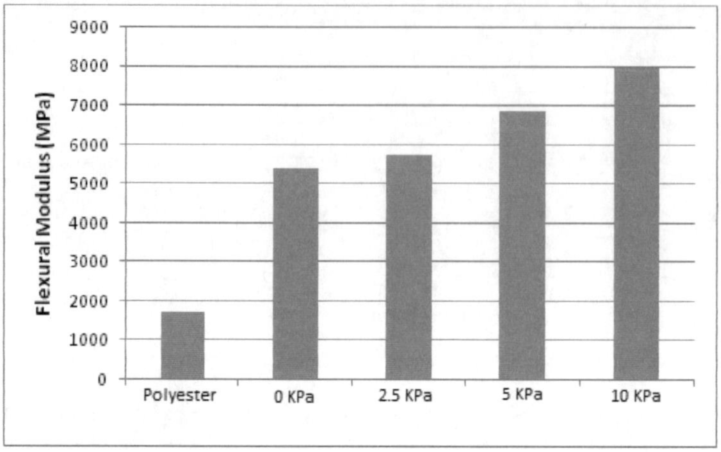


**Figure 6.** Relation between flexural modulus and compression stress for 15 wt% jute fiber reinforced polyester.

## 4. Conclusion

In this research work, different amount of compression stresses (2.5, 5 and 10 kPa) were applied to various percentages (10%, 15%) of fiber loading to fabricate composites. Following conclusions are made from this research work.

- Flexural properties (flexural strength and flexural modulus) of the composite sample without any compression stress were significantly lower than composite samples fabricated by applying compression stress. Flexural strength and flexural modulus increased with increasing compression stress.
- Higher flexural properties for increasing compression stress were due to better impregnation of the fibers by the resin, which certainly occurred as a higher pressure is applied during the matrix setting.
- Flexural properties of composites with 10 wt% and 15 wt% are higher than the pure polyester sample. Flexural properties of 15 wt% jute fiber composite are also higher than 10 wt% composite for the same amount of applied compression stress.

## 5. References

[1] Vilaseca F *et al.* 2007 *Process Biochemistry* **42** pp 329–34
[2] Planckett D, Andersen T L, Pedersen W B and Nielsen L 2003 *Composites Sci. Technol.* **63** pp 1287–96
[3] Rana A K and Jayachandran K 2000 *Mol. Cryst. and Liq. Cryst.* **353** pp 35-45
[4] Hitoshi Takagi and Akira Asano 2008 *Composites: Part A* **39** pp 685–89
[5] A V Ratna Prashad and K Mohana Rao 2011 *Materials and Design* **32** pp 4658–63
[6] M R Rahman, M M Huque, M N Islam and M Hasan 2008 *Composites: Part A* **39** pp 1739–47
[7] S N Monteiro, L A H Terrones and J R M D'Almeida 2008 *Polymer Testing.* **27** pp 591– 95
[8] E M Araújo et al. 2006 *Polímeros: Ciência e Tecnologia* **16** pp 332-35
[9] A Bindal, S Singh, N K Batra and R Khanna 2013 *Indian Journal of Materials Science* **V2013**
[10] J F de Deus, S N Monteiro and J R Md'Almeida 2005 *Polym. Test.* **24** p 750
[11] M V de Sousa, S N Monteiro and J R Md'Almeida 2004 *Polym. Test.* **23** p 253
[12] A C de Albuquerque, K Joseph, L H de Carvalho and J R M d'Almeida 2000 *Composites Sci. Technol.* **60** pp 833-44

[13] G Francucci, S Palmer and W Hall 2017 *J. Composite Materials* **0(0)** pp 1-13

**Acknowledgments**
The authors are grateful for a research grant in the course of this investigation to the authority of BUET as well as Department of Materials and Metallurgical Engineering, Bangladesh University of Engineering and Technology (BUET), Dhaka-1000, Bangladesh.

# Effect of Fiber Ratio and Chemical Treatment on Properties of Banana and Betel Nut Fiber Reinforced Hybrid Polypropylene Composites

**B. Sutradhar, M.B. Mesbah and M. Hasan**

Department of Materials and Metallurgical Engineering, Bangladesh University of Engineering and Technology, Dhaka 1000, Bangladesh

**Abstract.** The term hybrid composite refers to the composite containing more than one type of fiber material as reinforcing fillers. It has become attractive structural material due to the ability of providing better combination of properties as compared to single fiber composite. In present research, banana and betel nut fiber reinforced polypropylene composites were prepared by hot pressing technique. Raw banana and betel nut fiber were chemically treated with sodium hydroxide to increase adhesion of those fibers with polypropylene. Both raw and alkali treated fibers at 15 wt% were utilized during composite preparation. Banana and betel nut fiber ratios were varied at 1:1, 3:1 and 1:3. Scanning electron microscopic and thermogravimetric analysis, mechanical (tensile, flexural, impact and hardness) and water absorption tests of prepared composites were subsequently conducted. According to microscopic analysis, composites containing treated banana and betel nut fiber at 3:1 ratio had the best adhesion between the fibers and matrix. Polypropylene composites reinforced with banana and betel nut fiber 3:1 ratio had highest tensile and flexural properties, while composites reinforced with banana and betel nut fiber 1:3 ratio showed highest impact properties. According to water absorption test, higher banana fiber containing composite had higher water absorption as compared to higher betel nut fiber containing composite. Polypropylene composites reinforced with equal amount of banana and betel nut fiber had highest thermal stability among all prepared composites. Sodium hydroxide treatment of fibers increased flexural and hardness properties of composites as compared to raw fiber composites. On the other hand, opposite trend was observed in case of tensile, impact and water absorption properties.

## 1. Introduction

For many years composites have shown many advantages for its light weight, high corrosion resistance and many other qualities. Among the composites fiber reinforced polymer composites have grown popular now a days because of the good compatibility between the fiber and the matrix and light weight of fiber as well as the polymer. Polymer composites have been used in many important fields like automobile, aircraft, bridge, construction etc. and many other daily life uses. Natural fiber reinforced composites give some extra advantages like natural fibers are really abundant in nature, they are biodegradable so they do not cause any environmental harm, they are extremely light in weight, they have higher strengths than the matrix polymer. The experiment was done by using two kinds of natural fibers as reinforcement and polypropylene as matrix. Banana fiber and betel nut fiber were used as the reinforcing agent which are abundant in nature and have higher strength. Growing attention is being paid to banana fiber as a promising reinforcing material because it is relatively inexpensive, abundantly available and the energy requirement in their extraction is low, which has a considerable engineering advantage [1]. Betel nut husk (BNH) fiber is among a newly explored type of agro-waste fiber. The BNH fiber is extracted from betel nut fruit after a 5-day water retting process. The observed tensile properties of BNH fiber are comparable to kenaf fiber, but with significantly higher elongation at break values [2]. Polypropylene is a thermoplastic polymer and can be made by polymerizing molecules. Scientists prefer thermoplastic matrices than thermosets due to low production cycle, lower cost of processing and high reparability of thermoplastics [3]. In present research, betel nut and banana fiber were added to the polypropylene matrix as reinforcing materials to improve its properties. Mechanical and water absorption tests and microscopic and thermogravimetric analysis of the composites were performed to examine the differences among the polypropylene composites containing different ratio of banana and betel nut fiber. Effect of sodium hydroxide treatment of banana and betel nut fiber on properties of prepared composites was also evaluated.

## 2. Experimental Procedure

### *2.1 Materials*

Polypropylene was used as matrix and banana fiber and betel nut fiber were used as reinforcement in present research. Polypropylene was white in color and granular in size and its melting point was 160ºC. It was collected from market. Banana fibers were collected from bark and betel nut fibers were collected from fruit.

### *2.2 Alkali treatment*

Both banana fiber and betel nut fiber were treated with 5% NaOH to remove impurities and other greasy contents. Fibers were immersed in two separate beakers and kept at room temperature for about two and half hours. Then fibers were washed with distilled water until all NaOH were removed. Fibers then placed in an oven at 100ºC for about two hours.

### *2.3 Composite preparation*

Composites were prepared using hot press technique. Initially fiber loading was fixed at 15 wt%, while banana fiber and betel nut fiber ratio was varied at 1:1. 3:1 and 1:3 during composite preparation. Treated hybrid fiber reinforced composite was prepared using banana fiber and betel nut fiber ratio of 3:1. To produce composite fibers were cut into 3 mm size. Fibers and PP were weighted and dried in oven at 80ºC for 20 minutes. Fibers were mixed properly. Then releasing agent (silicone spray) was sprayed over the clean mold surfaces. Then PP, fibers, PP layers were made and covered by other silicone sprayed mold and placed in hot press machine. This machine was hydraulic type having capacity of maximum load of 50 KN and maximum temperature of 300ºC. The fiber matrix mixture was allowed to press at 30 KN pressure and temperature was initially raised to 160ºC and hold there for 20-25 minutes. After that temperature was raised to (190-195)ºC depending on the thickness required. Since the compression temperature was higher than PP melting point (160ºC) but lower than the fibers melting point (>220ºC) , so only matrix melted. Then water was passed through the dies and cooled to room temperature. The pressure was released and product was withdrawn from the die.

### *2.4 Mechanical tests*

Tensile, flexural, impact and hardness tests were conducted. For each test, three sample were prepared, tested and average value was taken. Tensile test were carried out according to ASTM D 638-01using an instron machine (system Id3369J8567, maximum capacity 50 KN). The three point flexural test was carried out according to ASTM D 790-0 0 using the same instron machine. The Charpy impact test of the composite was conducted using an impact tester MT 3016 and Specimen was prepared according to ASTM D 6110-97. Hardness was measured using shore durometer in its Shore D scale.

*2.5 Water absorption test*
Weight of the specimens were taken then immersed in a beaker of water for 24 hours at room temperature. Wet specimens weight were taken after removing excess water. Percentage of water absorption was then calculated.

*2.6 Scanning electron microscopy*
Surface morphology of the prepared composites was observed by using a scanning electron microscope. The surface of the fiber was made conductive by giving platinum coating using a sputtering machine. The fiber was then observed in vacuum condition into the SEM machine.

*2.7 Thermo gravimetric analysis*
TGA was carried out in a thermo gravimetric analyzer of model TGA Q50 W/FMC. Its temperature range is ambient +5°C to 800°C and heating rate is 0.1 to 100°C/min. Temperature range of 25°C to 500°C and a constant heating rate of 10°C/min was used in present research.

## 3. Results and Discussion

*3.1 Tensile properties*
Figure 1 presents tensile properties of prepared composites against fiber ratio and chemical treatment. It can be observed that for 3:1 banana:betel nut fiber reinforced polypropylene composite, tensile strength was found higher (Figure 1(a)). This is because banana has higher cellulose content as compared betel nut fiber and higher cellulose content fiber has higher tensile strength [4]. Thus higher amount of banana fiber promoted higher tensile strength to the composite. Young's modulus of banana fiber is also higher than that of the betel nut fiber [2, 4]. So, higher concentration of banana fiber incorporation demands higher stress for the same deformation and results into the increase of Young's modulus with increased banana fiber ratio [5, 6] (Figure 1(b)).

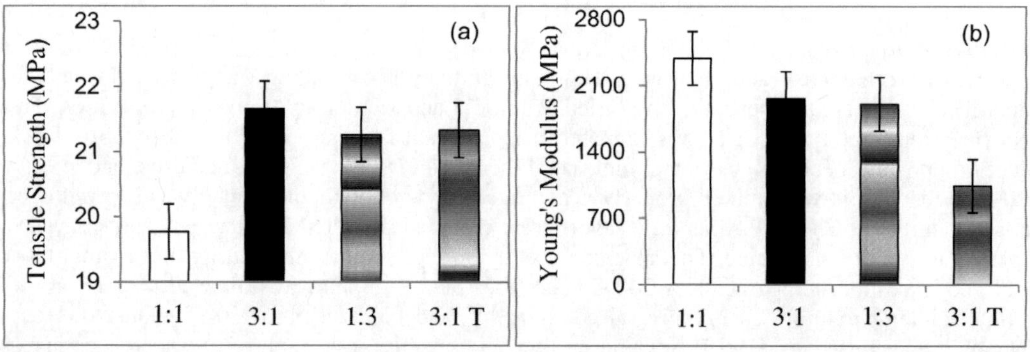


**Figure 1.** Variation of (a) tensile strength and (b) Young's modulus against banana and betel nut fiber ratio and chemical treatment.

Figure 1 also shows that due to the alkali treatment tensile strength and Young's modulus decreased. This may be because of the change in cellulose structure [7]. The alkali treatment transforms the cellulose I structure partially to cellulose II structure which is amorphous and this results in poor tensile properties of the fiber. Besides damages in the fiber structure such as increase in deep pores and thinning of the fiber cell walls might occur due to the alkali treatment, which in turn reduced the tensile properties [7, 8]. Again, the alkali treatment partially removes hemicelluloses and lignin which act as the binder substance to cement the microfibrils together. So, the removal of these substances might have created less rigid and less dense fiber and decreased the resistance of the microfibrils of the natural fiber relative to stretching [9-11]. The range of tensile strength of prepared composites was 19.5-21.5 MPa, while the same of hybrid composites containing either banana or betel nut fiber found elsewhere was 18.0-23.0 MPa [12, 13]. The range of Young's modulus of prepared composites was 900-2300 MPa, while the same of hybrid composites containing either banana or betel nut fiber found elsewhere was 680-1600 MPa [12, 13].

### *3.2 Flexural properties*

Variation of flexural properties against fiber ratio and chemical treatment is shown in Figure 2. In Figure 2 (a) it can be observed that for 3:1 banana-betel nut fiber reinforced polypropylene composite, flexural strength increased by 14.56%, which is higher than 1:3 banana-betel nut fiber reinforced polypropylene composite for which flexural strength increased by 14.07%. It is due to higher amounts of cellulose in banana than betel nut [14]. Though alkali treatment increased interfacial bonding, excess treatment increased brittleness, thus flexural strength decreased. In Figure 2 (b) it can be observed that for 3:1 banana-betel nut fiber reinforced polypropylene composite, flexural modulus increased by 20.05%, which is higher than 1:3 banana-betel nut fiber reinforced polypropylene composite for which flexural modulus increased by 16.86%. After treatment modulus increased due to increase in brittleness. The range of flexural strength of prepared composites was 37-42 MPa, while the same of hybrid composites containing either banana or betel nut fiber found elsewhere was 31-42 MPa [12, 13]. The range of flexural modulus of prepared composites was 1300-1700 MPa, while the same of hybrid composites containing either banana or betel nut fiber found elsewhere was 1200-1800 MPa [12, 13].

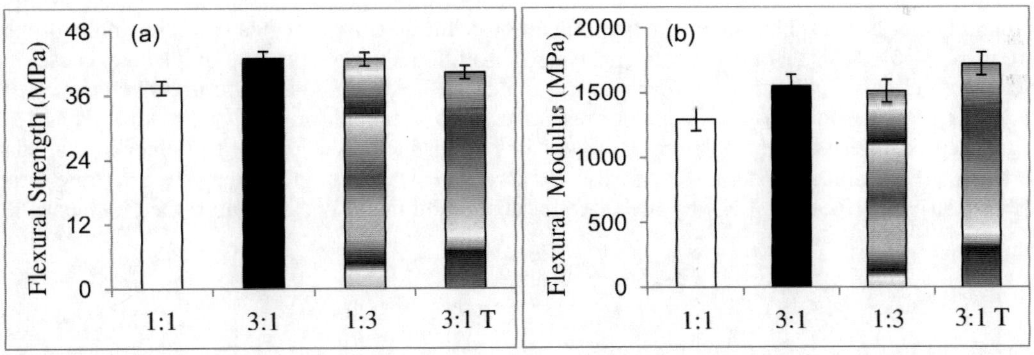


**Figure 2.** Variation of (a) flexural strength and (b) flexural modulus against banana and betel nut fiber ratio and chemical treatment.

### *3.3 Impact strength and hardness*

Due to incorporation fiber in matrix, void reduces and more force is needed to pull out the fiber. So, Charpy impact strength increases [13, 17]. Betel nut fiber has higher pull out strength than banana fiber [14]. For this reason, by incorporating banana fiber to 75% (fiber ratio 3:1) impact strength increased by 13.33%. Whereas by incorporating betel nut fiber to 75% (fiber ratio 1:3), impact strength increased by 23.33% as shown in Figure 3 (a). After alkali treatment mechanical interlocking develops between fiber and matrix, which minimizes fiber pull out. This decreased impact strength [14]. When fiber was incorporated into PP, void decreases, flexibility decreases and composite becomes more rigid. Thus, hardness increases. It can be observed in Figure 3 (b) that with increasing banana fiber to 75% (fiber ratio 3:1), hardness decreased by 0.59%, which is negligible. When betel nut fiber was increased to 75% (fiber ratio 1:3), no change in hardness was observed. After treatment, matrix-fiber adhesion increased, so hardness increased slightly (0.42%). The range of hardness of prepared composites was 72.8-73.2 Shore D, while the same of hybrid composites containing either banana or betel nut fiber found elsewhere was 64-75.2 Shore D [12, 13].

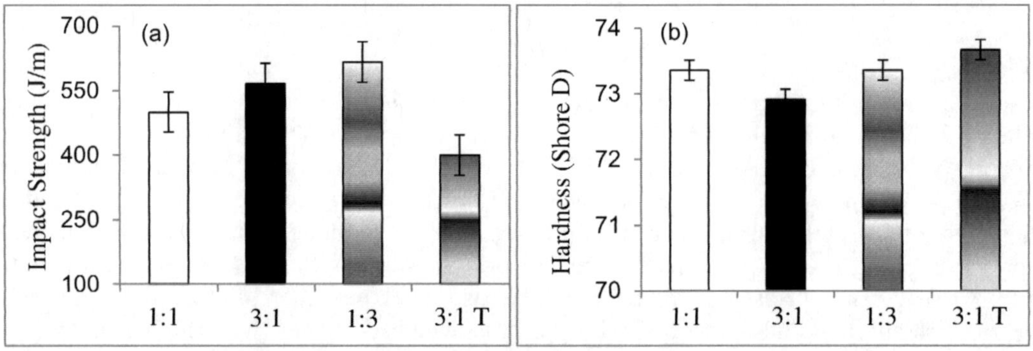


**Figure 3.** Variation of (a) impact strength and (b) hardness against banana and betel nut fiber ratio and chemical treatment.

*3.4 Water absorption characteristics*

When fiber reinforced PP is exposed to a process of water absorption, the fiber swells and then micro-cracks occur due to the increased stress at the fiber-matrix interface region. So, water can diffuse more readily through the interface of fiber and matrix [5]. Variation of water absorption of prepared composites is shown in Figure 4. For 3:1 banana-betel nut fiber reinforced polypropylene composite water absorption was found higher because banana has higher amount of cellulose content which increases the hydrophilic nature of fiber [5]. After the alkali treatment, water absorption should have been reduced because of the reduction of cellulose and hemi-cellulose content which are hydrophilic and the improvement of fiber-matrix interface adhesion [8]. However, in present research, treatment might not be done properly and over treating occurred for which micro-gaps was increased due to the removal of interfibrillar matrix material, such as lignin and pectin. For increased micro gaps water could more easily penetrate through the fiber-matrix interface. That is why higher water absorption might occurred after the treatment [6, 18, 19].

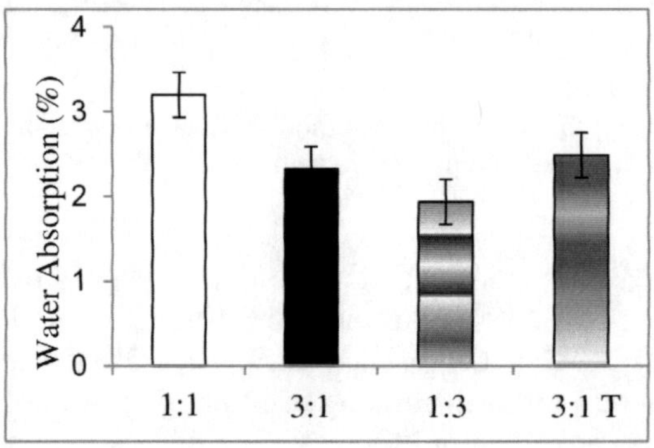


**Figure 4.** Variation of water absorption against banana and betel nut fiber ratio and chemical treatment.

*3.5 SEM analysis*

The surface morphology of composites with banana and betel nut fiber at 3:1, 1:1 and 1:3 ratios and treated banana and betel nut fiber at 3:1 ratio was observed under scanning electron microscope and is shown in Figure 5. From the SEM images, better bonding is observed for composites containing banana and betel nut fiber at 3:1 ratio as compared to other two composites. This was due to lower fiber pull out and better adhesion between fibers and matrix in case of higher banana fiber reinforced composite (Figure 5 (c)). Alkaline treatment increases fiber surface roughness resulting in better mechanical interlocking. It also increases the amount of cellulose exposed on the fiber surface, thus increasing the number of possible reaction sites. Also the main purpose of alkali treatment is to disrupt hydrogen bonding in the network structure and remove some hemicellulose, lignin, wax and oils, thereby increasing surface roughness and reducing its hydrophilic nature [19]. SEM micrographs in Figure 5 (d) show better bonding of fiber with matrix in case of treated fiber composite.

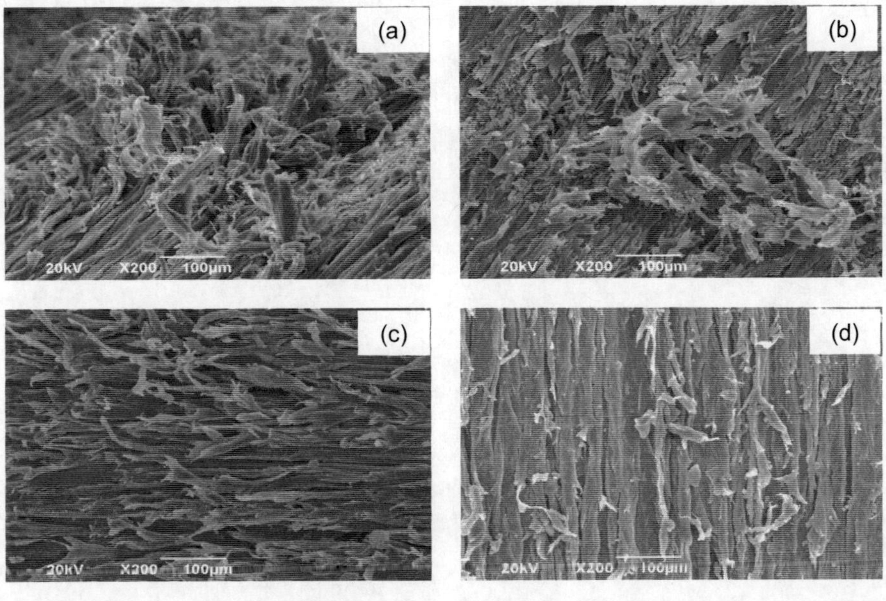


**Figure 5.** SEM micrographs of raw banana and betel nut fiber ((a) 1:3, (b) 1:1, (c) 3:1) and (d) treated banana and betel nut fiber (3:1) reinforced PP composites.

*3.6 TGA results*

From the TGA curves (Figure 6), it can be stated that an initial weight loss of 2.15% was observed in untreated 1:1 banana-betel nut fiber reinforced composite at around 231.4°C. For 3:1 and 1:3 ratio of banana-betel nut fiber reinforced composite, the initial weight loss was found to be 2.6% and 2.7% respectively at temperature of 229.5°C. So, among the three ratios, the 1:1 ratio of banana- betel nut fiber reinforced composite has the highest thermal stability. The initial weight loss is also higher for the 3:1 and 1:3 ratios; that may happen because of the presence of slightly increased amount of water. Among the 3:1 and 1:3 ratio, the 1:3 ratio of banana-betel nut fiber reinforced composite has higher initial weight loss because betel nut has higher amount of hemicelluloses which makes it more hydrophilic. NaOH catalyses pyrolysis process, Thus thermal stability decreased in case of NaOH treated fiber reinforced PP. So, the thermal degradation started to occur at lower temperature in comparison with the untreated fiber reinforced PP. The amount of cellulose decomposition or the derivative weight (%) change of cellulose got decreased in the NaOH treated fiber reinforced PP than the untreated 3:1 and the 1:3 banana-betel nut fiber reinforced PP. This is because the treatment decreased the amount of cellulose and changed the structure of the remained cellulose, which was more reactive than the previous structure. In chemical treatment, the hydrogen of hydroxyl group of cellulose was substituted by acetyl group , which is more reactive [5]. Exposure of more reactive cellulose components on the fiber surface may have promoted rough surface and more effective surface area for better bonding between fiber surface and polymer matrix via mechanical interlocking mechanism [9]. Thus decomposition of the cellulose got decreased.

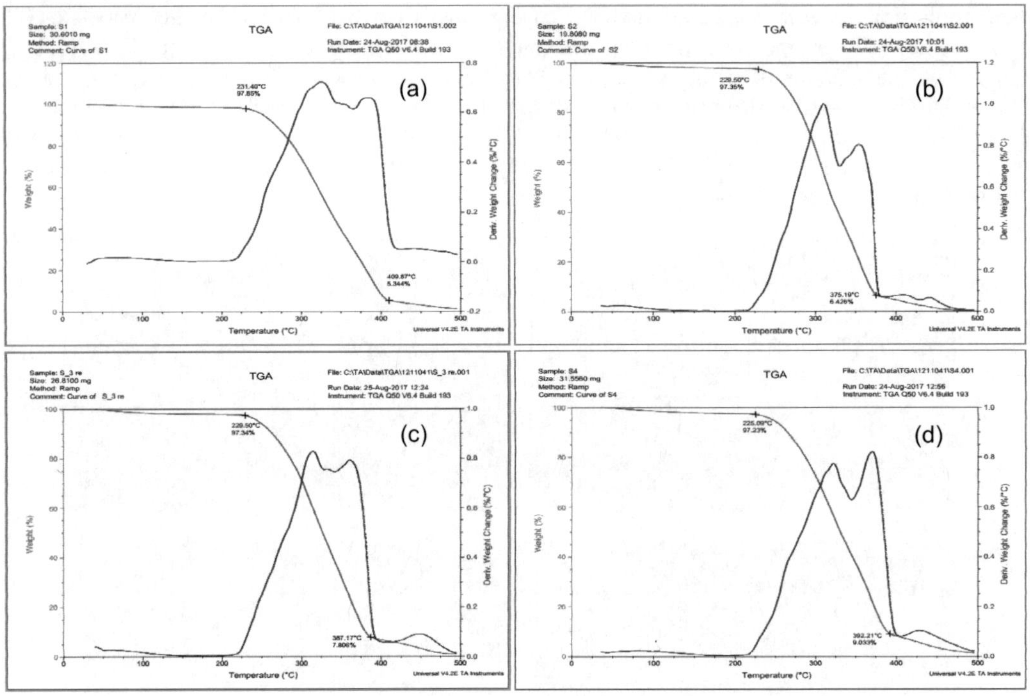


**Figure 6.** TGA curves of raw banana and betel nut fiber ((a) 1:1, (b) 3:1, (c) 1:3) and (d) treated banana and betel nut fiber (3:1) reinforced PP composites.

## 4. Conclusion

In present research, raw and chemically treated banana and betel nut fiber reinforced polypropylene composites were prepared by hot pressing technique. Both raw and alkali treated fibers at 15 wt% were utilized during composite preparation. Banana and betel nut fiber ratios were varied at 1:1, 3:1 and 1:3. Higher percentage of banana fiber containing composite showed better tensile strength, flexural strength, flexural modulus and Young's modulus than higher betel nut containing composite. on the other hand, 75% of betel nut fiber reinforcing composite showed better impact strength than 75% of banana fiber reinforcing composite. 5% NaOH treated composite showed better flexural properties and hardness than the untreated ones. Higher percentage of banana fiber containing composite showed higher water absorption than higher betel nut fiber containing composite. 50% banana fiber reinforced composite showed the highest thermal stability.

## 5. Reference

[1] Babatunde O O and Temitope G O 2016 *Utilization of banana stalk fiber as reinforcement in low density polyethylene composite,* Matéria (Rio J.), Vol. 21(4), pp. 953-965.

[2] Yusriah L, Sapuan S M, Zainudin E S, Mariatti M and Jawaid M 2016 *Thermo-physical, thermal degradation, and flexural properties of betel nut husk fiber-reinforced vinyl ester composites,* Polymer Composites, Vol. 37(7), pp. 2008-2017.

[3] Shubhra Q T H, Alam A K M M and Quaiyyum M A 2011 *Mechanical properties of polypropylene composites,* Polymer Composites, Vol. 26(3), pp. 362-391.
[4] Ramesh M, Atreya T S A, Aswin U S, Eashwar H and Deepa C 2014 *Processing and mechanical property evaluation of banana fiber reinforced polymer composites,* Procedia Engineering, Vol. 97, pp. 563-572.
[5] Nirupama P, Vijay K A and Shishir S 2016 *Banana fiber reinforced low-density polyethylene composites: effect of chemical treatment and compatibilizer addition,* Iranian Polymer Journal, Vol. 25(3), pp. 229–241.
[6] Yusriah L, Sapuan S M, Zainudin E S and Mariatti M 2012 *Exploring the potential of betel nut husk fiber as reinforcement in polymer composites: effect of fiber maturity,* Procedia Chemistry, Vol. 4, pp. 87-94.
[7] Naveen K G, P Prakash, P Kalaichelvi and K N Sheeba 2016 *The effect of temperature and hemicelluloselignin, cellulose-lignin, and cellulose-hemicellulose on char yield from the slow pyrolysis of rice husk,* Energy Sources A, Vol. 38(10). pp. 1428-1434.
[8] Rout J , Misra M, Tripathy S S , Nayak S K and Mohanty A K 2001 *The influence of fibre treatment on the performance of coir-polyester composites,* Composites Science and Technology, Vol. 61(9), pp. 1303-1310.
[9] William T Y T, Douglas J G, Carl P T and Shane C 2006 *Cellulose fiber/polymer adhesion: effects of fiber/matrix interfacial chemistry on the micromechanics of the interphase,* Journal of Adhesion Science and Technology, Vol. 20(15), pp. 1649–1668.
[10] Elammaran J, Sinin H, Md Rezaur R and Muhammad K B B 2014 *Investigation of fiber surface treatment on mechanical, acoustical and thermal properties of betel nut fiber polyester composites,* Procedia Engineering, Vol. 97, pp. 545-554.
[11] Julio C M O, Rodolfo R B and Jhon J O F 2012 *The influence of alkali treatment on banana fibre's mechanical properties,* Ingeniería e Investigación, Vol. 32(1), pp. 83-87.
[12] Akter M, Jahan E and Hasan M 2018 *Proc. Int. Conf. on Structure, Processing and Properties of Materials (Bangladesh)*, vol. 4, pp. 291-298.
[13] Das S, Rahman M and Hasan M 2018 *Proc. Int. Conf. on Structure, Processing and Properties of Materials (Bangladesh)*, vol. 4, pp. 299-306.
[14] Siddika S, Monsura F, Hasan M and Hassan A *2014 Effect of reinforcement and chemical treatment of fiber on the properties of jute-coir fiber reinforced hybrid polypropylene composites,* Fibers and Polymers, Vol. 15(5), pp. 1023-1028.
[15] Rana A K, Mandal A and Bandyopadhyay S 2003 *Short jute fiber reinforced polypropylene composites: effect of campitabilizer, impact modifier and fiber loading,* Composites Science and Technology, Vol. 63, pp. 801-806.
[16] Joseph S, Sreekala M S, Oommen Z, Koshy P and Thomas S 2002 *A comparison of mechanical properties of phenol formaldehyde composites reinforced with banana fibers and glass fibers,* Composites Science and Technology, Vol. 62, pp. 1857-1868.
[17] Mohd S J, Ishak A and Ibrahim A 2006 *Effect of rice husk filler on the mechanical and thermal properties of liquid natural rubber campatibilized high density polyethylene natural rubber blends,* Journal of Polymer Research, Vol. 13, pp. 315-321.
[18] Ramadevi P, Dhanalakshmi S, Chikkol V S and Basavaraju B 2012 *Effect of alkali treatment on water absorption of single cellulosic abaca fiber*, BioResources, Vol. 7(3).
[19] Yang H S, Kim H J, Son J, Park H J, Lee B J and Hwang T S 2006 *Water absorption behavior and mechanical properties of lignocellulosic filler polyolefin biocomposites,* Composite Structures, Vol. 72, pp. 429-437.

**Acknowledgment**
The authors would like to thank department of Materials and Metallurgical Engineering, Bangladesh University of Engineering and Technology for their support during the work.

# Mechanical Properties of Silk and Glass Fiber Reinforced Hybrid Polypropylene Composites

**M.R.H. Mazumder, F. Numera, A. Al-Asif and M. Hasan**

Department of Materials and Metallurgical Engineering, Bangladesh University of Engineering and Technology, Dhaka 1000, Bangladesh

**Abstract.** In present research, mechanical properties of silk-glass fiber reinforced hybrid polypropylene composite were evaluated. Composites were prepared using hot press machine at three levels of fiber loading (5, 10 and 15 wt.%). Tensile, flexural, impact and hardness tests were conducted for the characterization of the composites. Tensile strength decreased, whereas flexural strength increased with fiber loading. On the other hand, Young's modulus, flexural modulus and hardness increased up to 10 wt.% fiber loading. Impact strength remained unchanged with fiber loading. Based on fiber loading, 10 wt.% fiber reinforced composites had the optimum set of mechanical properties. For 10 wt.% fiber loaded composite, silk and glass fiber ratio was also varied at 1:1, 3:1 and 1:3. On the basis of fiber loading and ratio, the composite containing 10 wt.% fiber at 1:1ratio of silk and glass showed the best set of mechanical properties.

**Keywords:** Silk; Glass; Polypropylene; Hybrid composite; Properties

## 1. Introduction

Natural fiber reinforced composites are of high interest in this present day. Due to its environment friendly nature and bio-degradability, it is being incorporated in making composites [1]. Amongst the natural fiber, cellulosic or animal fiber has attracted considerable interest as a reinforcement for polymer. This is due to its properties such as high strength and stiffness, low hardness high impact strength which reduces abrasion. Silk is a natural fiber obtained from the cocoon of silkworm. It has high toughness and better mechanical properties which gives it an advantage over the other animal fibers. Glass fiber, on the other hand, is a synthetic fiber. It has properties such as high strength, stiffness. Fiber glass is lightweight. Because of their excellent property the application of glass fiber is found in different industries [1]. Silk is reactive to light, less chemically resistant and absorbs water. However it is abrasion resistant, has good elasticity and is biodegradable. On the other hand, glass is unaffected by light, does not absorb water, does not naturally degrade and is chemically inert.

IOP Conf. Series: Materials Science and Engineering **438** (2018) 012007 doi:10.1088/1757-899X/438/1/012007

However abrasion resistance of glass fiber is less than that of kevlar fiber, thus glass fiber is prone to abrasion [2]. To balance the property of the fibers and overcome their weaker properties, the hybrid composite of glass and silk fiber reinforced polypropylene matrix will have its bulk property enhanced [3]. Hybrid composites are a good combination of natural and synthetic fiber because it gives the composite structural strength which the natural fiber lacks and is environment friendly which is not true for synthetic fibers. Glass and silk fiber are both available in Bangladesh. Polypropylene is a widely used polymer matrix because of its low cost, recyclability and thermal stability [4]. Present research discusses the property enhancement of polypropylene due to addition of glass and silk fiber reinforcements and is emphasized on the mechanical strength of the composite.

## 2. Experimental

### *2.1 Materials*

In this study, thermoplastic polymer polypropylene (PP), used as matrix material, was collected from the local market in the form of granules. It has a specific gravity of 0.90-0.91, melting temperatures of around 160 °C and crystallinity of 82% [5]. Glass fiber was collected from local market and the silk fiber was collected from a silk industry in Rajshahi, Bangladesh. The die used to prepare the composite was made of aluminium. The die was made by machining aluminium to a desired shape.

### *2.2 Preparation of composites*

Hybrid composites of polypropylene matrix with varying amount of silk and glass fiber were prepared by hot press technique in a 150 x 150 x 10 mm aluminium die. A hydraulic type machine having maximum load of 35 kN and maximum temperature of 300°C was utilized. The fiber loading was varied at 5, 10 and 15 wt% with the ratio of silk to glass of 1:1 and also of silk to glass ratio of 3:1 and 1:3 for 10 wt% fiber loading. The fibers were chopped into approximate length of 3 mm. The required amount of fiber and PP were weighed in a balance. The weighed fiber was place into the oven at 100°C temperature for at least 1 hour. It was done to make the fiber dry and remove all moisture in order to prevent air bubble in composite. A clean aluminium mold was used and mold releasing agent was sprayed on it uniformly for easy removal of the finished composite. PP granules and fibers were placed into the mold die with the fibers sandwiched between the PP granules layer. This technique improved the uniformity of the fiber within the polymer and increased wettability. Then the mold die along with PP granules and fiber mixtures were placed in hot press machine at 30 kN. The temperature was initially raised to 160°C and held there for 15-20 minutes. The temperature was then raised to 195°C and held there for 5 minutes. The die was cooled to room temperature by the water cooling system. The pressure was released and the composite was withdrawn from the die.

### *2.3 Mechanical testing*

Tensile, flexural, Charpy impact and hardness tests were conducted on the prepared composites. For each test and type of composite, 3 specimens were tested and the average values are reported. Tensile tests were conducted according to ASTM D 638-01 [6] using a Universal Testing Machine situated at the department of Materials and Metallurgical Engineering, Bangladesh University of Engineering and Technology, Bangladesh. The tests were performed at a crosshead speed of 6mm/min. The dimension of the specimen used was 100 mm x 19 mm x 3 mm. Static flexural tests were carried according to ASTM D 790-00 [7] using the same testing machine mentioned above with the same crosshead speed. The dimension of the specimens used was 78 mm x 12.5 mm x 3 mm. The flexural strength and modulus were calculated using the following equations:

$$\text{Flexural strength, } \sigma_f = \frac{3PL}{2bd^2} \tag{1}$$

$$\text{Flexural modulus, } E_f = \frac{mL^3}{4bd^3} \tag{2}$$

Where P is the maximum applied load, L is the length of support span, m is the slope of the tangent, b and d are the width and thickness of the specimen respectively.

Dynamic Charpy impact tests were conducted according to ASTM D 6110-97 [8] using a Universal Impact Testing Machine. The dimension of the specimens used was 72 mm x 18 mm x 3mm. The hardness of the composite was measured using a Durometer hardness tester in Shore (D) scale.

## 3. Results and Discussion

*3.1 Effect of fiber loading*

*3.1.1 Tensile properties.* Tensile properties of composite samples were measured for each fiber content (5,10 and 15 wt.%) with the help of stress/strain curves. The tensile strength values of raw silk and glass (silk:glass=1:1) fiber reinforced hybrid polypropylene composites at different fiber loading are shown in Figure 1 (a). With the increase in fiber loading, the ultimate tensile strength of the composite decreased. As the amount of fiber added increases, contact between fiber and matrix increases to make the interfacial bonding weaker. As a result, the weak interfacial area between the matrix and fibers increase. Due to this, the ease of failure increased and thus the tensile strength decreased [9]. The same trend was also observed by other researchers [10-13]. Young's modulus values of raw silk and glass (silk:glass=1:1) fiber reinforced hybrid polypropylene composites at different fiber loading are shown in Figure 1 (b). Young's modulus increased up to 10% fiber loading and then decreased when the fiber content was further increased to 15%. Young's modulus increased because incorporation of fibers in matrix creates partially separated micro spaces, which obstruct stress propagation between the fiber and the matrix [12]. As the fiber loading increased, the degree of obstruction increased, which in turn also increased the stiffness [14].

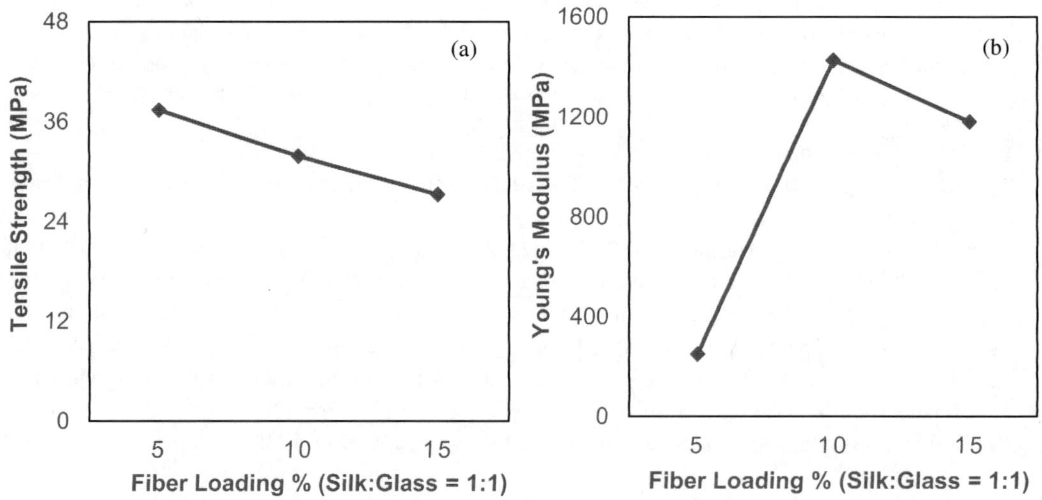


**Figure 1.** Variation of (a) tensile strength and (b) Young's modulus against fiber loading.

*3.1.2 Flexural properties.* Flexural properties were measured for samples of each fiber content (5, 10, 15 wt.%) with the help of stress/strain curves and respective equations. The flexural strength of raw silk and glass fiber reinforced hybrid polypropylene composites at different fiber loading is shown in Figure 2 (a). The flexural strength increased with an increase in fiber loading, which is in agreement with the findings by other researchers [15-17]. Owing to the suitable alignment of the polypropylene chain with the fiber, weak fiber-matrix adhesion might be easily overcome. Therefore, as fiber loading increases [12], the possibility of strong fiber-matrix adhesion increases, which results in an enhanced flexural strength [11]. Flexural modulus values of raw silk and glass (silk:glass= 1:1) fiber hybrid polypropylene composites at different fiber loading are shown in Figure 2 (b). Flexural modulus increased with an increase in fiber loading up to 10%, but decreased slightly at 15% [16, 17]. Since both silk and glass are materials having high modulus, higher fiber concentration demands higher stress for the same deformation [13] and increased fiber-matrix adhesion provides increased stress transfer from the matrix to the fiber. This result can be attributed to the incorporation of rigid glass fiber into the soft PP matrix. As a result, high modulus fiber incorporation into the soft polypropylene matrix increased flexural modulus.

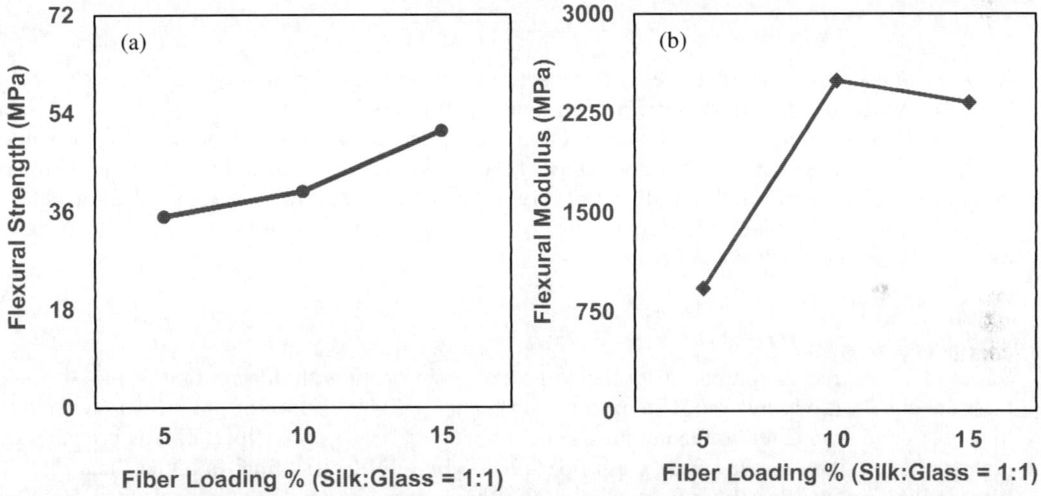


**Figure 2.** Variation of (a) flexural strength and (b) flexural modulus against fiber loading.

*3.1.3 Impact strength.* Impact strength of a material provides information regarding the energy required to break a specimen of given dimensions; the magnitude of which reflects the ability of the material to resist a sudden impact. Variation of Charpy impact strength with fiber loading for raw silk and glass (silk:glass= 1:1) fiber reinforced hybrid composite is shown in Figure 3 (a). As shown in the figure, impact strength of the composites remained unchanged with an increase in fiber loading. The impact strength of the fiber reinforced polymeric composites depend on the nature of the fiber, polymer and fiber-matrix interfacial bonding [18, 19]. Poor interfacial bonding induces micro-spaces at the fiber-matrix interface. The micro-spaces cause micro-cracks when impact occurs, resulting in crack propagation and giving no change in impact strength with increase in fiber loading.

IOP Conf. Series: Materials Science and Engineering **438** (2018) 012007 doi:10.1088/1757-899X/438/1/012007

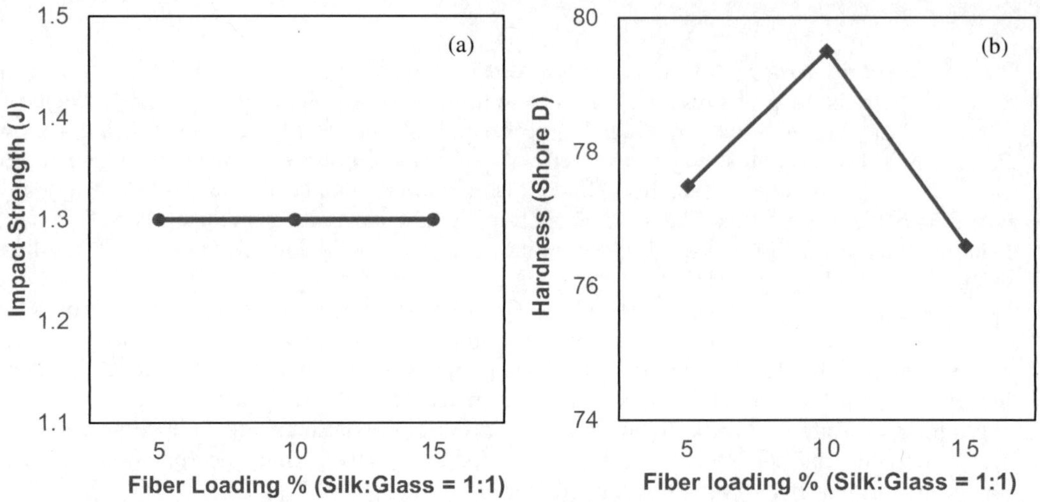


**Figure 3.** Variation of (a) impact strength and (b) hardness against fiber loading.

*3.1.4 Hardness values.* Hardness of a composite depends on the distribution of fiber into the matrix [13, 20]. Variation of hardness with fiber loading for raw silk and glass (silk:glass= 1:1) fiber reinforced hybrid composites is shown in Figure 3 (b). Hardness increased up to 10% fiber loading, however it decreased for 15% fiber loading. Hardness increased because incorporation of fiber into the polypropylene matrix reduced the flexibility of the matrix resulting in more rigid composite. With 15% fiber loading, hardness decreased because of the poor dispersion of the fiber into the matrix with more voids between matrix and fiber.

*3.2 Effect of fiber ratio*

Variation of mechanical properties of the composites reinforced with different ratios of silk and glass fiber are shown in Figures 4-6. Composite with higher weight fraction of glass fiber showed higher impact strength compared to the composite with higher fraction of silk fiber. This is probably due to high strain to failure of the single glass fiber [20]. Strength of glass fiber is higher than that of silk fiber and as a result, glass fiber is capable of absorbing more energy due to impact force. Both tensile strength and Young's modulus increased with increase in silk fiber loading. The chemical composition of the fiber and its internal structure affected tensile strength [21]. With the same fiber loading ratio of 1:1, the flexural strength, flexural modulus and hardness showed the highest value. This is due to the proper adhesion between the fibers and the matrix. Fibers were also uniformly distributed throughout the matrix with the same fiber loading ratio.

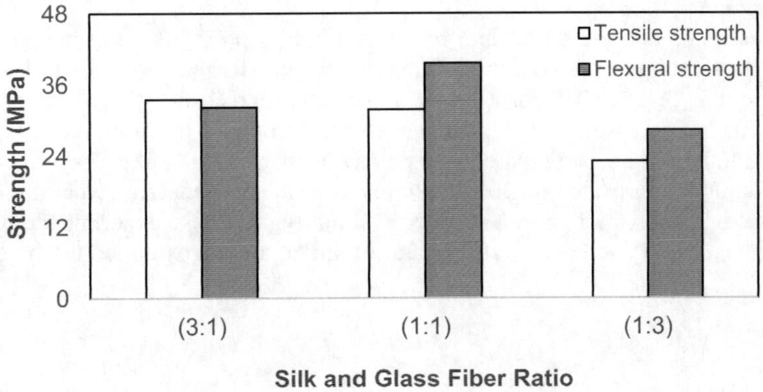


**Figure 4.** Variation of tensile and flexural strength at different silk and glass fiber ratio.

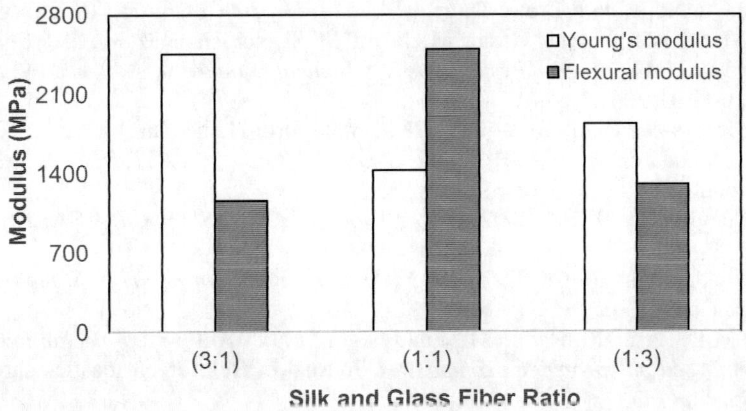


**Figure 5**. Variation of tensile and flexural modulus at different silk and glass fiber ratio.

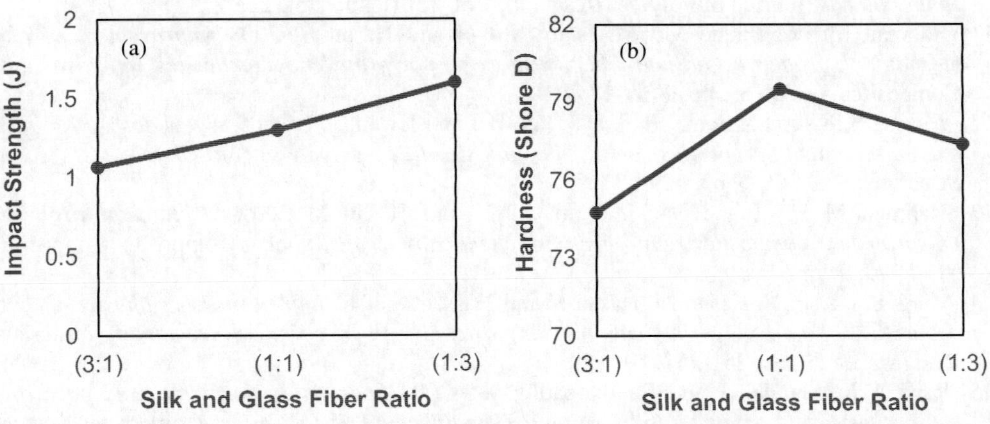


**Figure 6.** Variation of (a) impact strength and (b) hardness at different silk and glass fiber ratio.

## 4. Conclusion

In present research, silk and glass fiber reinforced hybrid polypropylene composites were prepared using hot press technique at three levels of fiber loading (5, 10 and 15 wt.%). Silk and glass fiber ratio was varied at 1:1, 3:1 and 1:3 for 10 wt% fiber loading. Tensile strength of prepared composites decreased with an increase in fiber loading, whereas Young's modulus increased up to 10% fiber loading. Flexural strength of prepared composites increased with fiber loading, on the other hand flexural modulus increased up to 10% fiber loading. Impact strength remained unchanged but the hardness (Shore D) increased up to 10% fiber loading. Based on the experimental results and analysis, composite containing 10 wt% fiber at 1:1 ratio of silk and glass exhibited the best set of mechanical properties.

## References

[1] Ashik K P and Sharma R S 2015 *A review on mechanical properties of natural fiber reinforced hybrid polymer composites*, Journal of Minerals and Materials Characterization and Engineering, Vol. 3, pp. 420-426.

[2] Fibre Glass Development corporation, https://www.fibreglast.com/product/the-fundamentals-of-fiberglass , Date of Access: 16 May 2018.

[3] John M J, Anandjiwala R D and Thomas S (2014) *Hybrid Composites* (South Africa).

[4] Haque M M, Islam M N, Hasan M, Islam M S and Islam M S 2010 *Coir fiber reinforced polypropylene composites: physical and mechanical properties*, Advanced Composite Materials, Vol. 19, pp. 191-206.

[5] Brydson J A 1999 *Plastics Materials* (UK: Butterworth-Heinemann).

[6] ASTM Standard D 638-01 2002 *Annual Book of ASTM Standard* (USA: ASTM International),Vol. 08.

[7] ASTM Standard D 790-00 2002 *Annual Book of ASTM Standard* (USA: ASTM International), Vol. 08.

[8] ASTM Standard D 6110-97 2002 *Annual Book of ASTM Standard* (USA: ASTM International),Vol. 08.

[9] Rahman M R, Hasan M, Haque M M and Islam M N 2010 *Physico-mechanical properties of jute fiber reinforced polypropylene composites*, Journal of Reinforced Plastics and Composites, Vol. 29(3), pp. 445-453.

[10] Jamil Md S, Ahmed I and Abdullah I 2006 *Effects of rice husk filler on the mechanical and thermal properties of liquid natural rubber compatibilized high-density polyethylene/natural rubber blends,* Journal of Polymer Research, Vol. 13(4), pp. 315-321.

[11] Rahman Md R, Haque Md M, Islam Md N and Hasan M 2008 *Improvement of physico-mechanical properties of jute fiber reinforced polypropylene composites by post-treatment,* Composites A, Vol. 39, pp. 1739-1747.

[12] Yang H S, Kim H J, Son J, Park H J, Lee B J and Hwang T S 2006 *Water absorption behavior and mechanical properties of lignocellulosic filler–polyolefin bio-composites,* Composite Structures, Vol. 72(4), pp. 429-437.

[13] Rahman M R, Huque M M, Islam M N and Hasan M 2009 *Mechanical properties of polypropylene composites reinforced with chemically treated abaca*, Composites A, Vol. 40, pp. 511-517.

[14] Mir S S, Nafsin N, Hasan M, Hasan N and Hassan A 2013 *Improvement of physico-mechanical properties of coir-polypropylene biocomposites by fiber chemical treatment*, Materials and Design. Vol. 52, pp. 251-257.

[15] Rana A K, Mandal A and Bandyopadhyay S 2003 *Short jute fiber reinforced polypropylene composites: effect of compatibiliser, impact modifier and fiber loading,* Composites Science and Technology, Vol. 63(6), pp. 801-806.

[16] Joseph S, Sreekala M S, Oommen Z, Koshy P and Thomas S 2006 *A comparison of the mechanical properties of phenol formaldehyde composites reinforced with banana fibers and glass fibers,* Composites Science and Technology, Vol. 62(14), pp. 1857-1868.

[17] Reis J M L and Carneiro E P 2012 *Mechanical characterization of sisal fiber reinforced polymer mortars: Compressive and flexural properties*, Journal of Reinforced Plastics and Composites, Vol. 31(23), pp. 1662-1669.

[18] Lin J C, Chang L C, Nien M N and Ho H L 2006*Mechanical behavior of various nanoparticle filled composites at low-velocity impact,* Composite Structures, Vol. 74(1), pp. 30-36.

[19] Bajpai P K, Singh I and Madaan J 2012*Comparative studies of mechanical and morphological properties of polylactic acid and polypropylene based natural fiber composites,* Journal of Reinforced Plastics and Composites, Vol. 31(24), pp. 1712-1724.

[20] Premalal H G B, Ismail H and Baharin A 2002 *Comparison of the mechanical properties of rice husk powder filled polypropylene composites with talc filled polypropylene composites,* Polymer Testing, Vol. 21(7), pp. 833-839.

[21] Hatta N and Akmar N 2008*Proc. CUTSE Int. Conf. (Miri, Sarawak, Malaysia).*

**Acknowledgement**
The authors would like to thank Department of Materials and Metallurgical Engineering, Bangladesh University of Engineering and Technology (BUET), Dhaka, Bangladesh for their support during present research.

# Synthesis and Characterization of Reduced Graphene Oxide Reinforced Polymer Matrix Composite

**V C Saha[1], M M A Sabuj[2], P Shams[3], S Rahman[4], M R Qadir[5], M R Islam[6] and F Gulshan[7]**

1, 2, 3, 4, 7 Department of Materials and Metallurgical Engineering, Bangladesh University of Engineering and Technology, Dhaka-1000, Bangladesh.

5 Pilot Plant and Process Development Centre, Bangladesh Council of Scientific and Industrial Research, Dhaka-1000, Bangladesh.

6 Department of Physics, Bangladesh University of Engineering and Technology, Dhaka-1000, Bangladesh.

**Abstract**: Graphene and its derivatives, owing to their superior properties have been being used as reinforcement to improve the properties of polymer matrix. In this work, the influence of Reduced Graphene Oxide (RGO) on the properties of Poly Vinyl Alcohol (PVA) has been studied. RGO was synthesized from graphite powder starting with synthesis of Graphene Oxide (GO) by Modified Hummer's Method followed by the reduction of GO with Hydrazine Hydrate. Synthesized RGO was characterized by X-ray diffraction (XRD), Scanning Electron Microscopy (SEM) and Energy-Dispersive X-ray Spectroscopy (EDS). PVA composites with different RGO loading were fabricated by simple solution mixing method. The fabricated composites were characterized by Optical Microscopy, XRD, UV-Visible Spectroscopy, Thermo Gravimetric Analysis (TGA), Differential Thermal Analysis (DTA) and Shore hardness test. The reports from those tests have ensured that after incorporating RGO in PVA matrix, band gap energy was decreased, crystallinity and shore hardness were increased and thermal stability was improved up to a certain level of RGO loading.

## 1. Introduction

Graphene has become a topic of both intensive and extensive research owing to its extraordinary mechanical, electrical, optical and thermal properties. Much credit goes to its unique molecular structure. It is a 2-D sheet of hexagonally arranged C atoms where each atom is $sp^2$ hybridized [1]. Due to $sp^2$ hybridization of C atoms, graphene molecule has free pi electrons located above and below the 2D sheet [1]. These electrons act like massless particle and are responsible for high electronic mobility and conductivity [2]. Besides, these pi electrons overlap and help to enhance the strength of C-C valence bond [1] which is responsible for high mechanical strength. Other important factors that make graphene so special are small C-C bond length [1], zero band gap, ballistic conductivity [2], thermal conductivity, impermeability and so on. All these factors working together make graphene a special material and create its demand in various fields like composite materials, bioengineering, nanotechnology, energy technology and so on [3]. Among its versatile applications, one of the alluring applications is to use it to improve the properties of current materials and substances [3]. Graphene has been using as reinforcement to improve various properties of many polymer. Well dispersed graphene in polymer matrix forms very interactive network structure which enhances different properties of polymer [4].

IOP Conf. Series: Materials Science and Engineering **438** (2018) 012008 doi:10.1088/1757-899X/438/1/012008

Polyvinyl alcohol (PVA) is a polymer which offers many properties to be improved. The only challenging aspect is well dispersion of graphene in PVA matrix on which efficiency of reinforcement depends [4]. One judicious solution is to introduce oxygenated functional groups to the surface of graphene which enhance dispersion of graphene in polymer matrix [4]. This special molecular structure of graphene having oxygenated functional on its surface is called Reduced Graphene Oxide (RGO) [5]. The properties of this graphene derivative is very similar to that of graphene except the dispersion behavior. This study is aimed at fabrication of RGO/PVA composites and observing the influence of RGO on various properties of PVA.

## 2. Experimental

*2.1 Synthesis of Graphene oxide (GO)*

Modified Hummer's Method [6] was conducted for the synthesis of GO from graphite powder. 5g Graphite powder was mixed with 2.5gm $NaNO_3$ and the mixture was taken in a volumetric flask containing 115 ml 98% $H_2SO_4$. After that the flask was kept under magnetic stirring for around 30 minutes. Then it was taken to the ice bath to maintain the temperature <20$^0$C, where 15 g $KMnO_4$ was added to the solution very slowly. The solution was then stirred at 35$^0$ for 1 hour. Then the solution was taken to oil bath where 250 ml de-ionized $H_2O$ was slowly added to the solution for 15 minutes at 95$^0$C. Then $H_2O$ & $H_2O_2$ (30%) were added to this solution and stirred at 65$^0$C for 2 hrs. The solution was changed to olive green color. The solution was then washed and filtrated and centrifuged at 8000 rpm for 15 minutes. The obtained gel was dried and again mixed with DI water to ultra-sonicate for 1 hour. The obtained product was dried at 70$^0$C for 24 hours to get the GO powder.

*2.2 Synthesis of Reduced Graphene oxide (RGO)*

At first, GO solution was prepared with 260 mg of obtained GO and 260 ml de-ionized water by ultra-sonication for 1 hour. Subsequently, 5 ml of 99% Hydrazine Hydrate was added to solution and heated for 8 hours at 70$^0$C. Finally the product was dried at 60$^0$C for 24 hours and RGO was obtained shown in 'figure 1(a)'.

*2.3 Synthesis of RGO reinforced PVA*

1 gm PVA for each sample was dissolved in 100 mL DI water and heated at 90$^0$C for 30 minutes to prepare PVA solution. For the preparation of RGO solution, calculated amount of RGO was dissolved in 100 mL DI water and heated at 70$^0$C. Then the solution was ultra-sonicated for 1 hour. At last the PVA solution and the RGO solution were mixed and heated at 60$^0$C followed by ultra-sonication for 1 hour. The obtained solution was poured into cultured dish and dried at 70$^0$C for around 24 hours. Obtained product was RGO/PVA composites shown in 'figure 1(b)'.

**Figure 1.** (a) RGO powder (b) RGO/PVA composites in cultured dish

## 3. Results and Discussion

*3.1 SEM with EDS*

Scanning Electron Microscopy (SEM) was conducted to observe the surface morphology of RGO. SEM topography shows presence of wrinkles and crumples on RGO surface. Wrinkle [7] shown in 'figure 2(a)' is a common phenomenon of a 2D material [7] and forms mainly due to negative thermal expansion of RGO [7]. Crumples [7] are 'paper-ball like' shape shown in 'figure 2(b)' which occur due to an isotropic compression and thermal reduction [7].

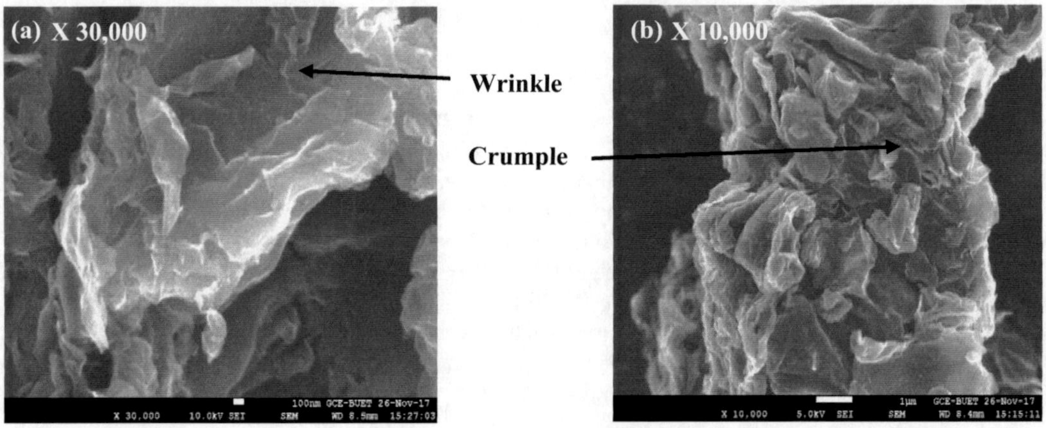


**Figure 2.** SEM topography images of RGO (a) wrinkles (b) crumples

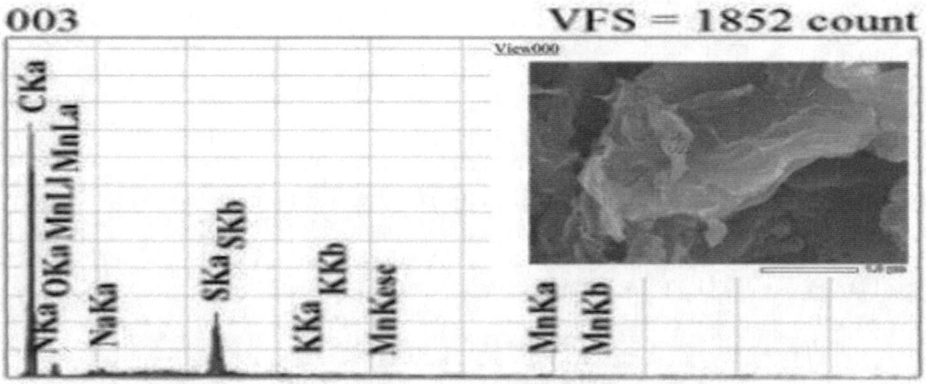


**Figure 3.** EDS report of RGO

Energy-dispersive X-ray spectroscopy (EDS) was performed to find out the composition of RGO. 'Figure 3' indicates the presence of considerable amount of C, S, O and Na. From the EDS report, the amount (mass %) of C, S, O and Na is 89.08, 5.09, 2.64 and 1.94 respectively. The amount of other elements is negligible. Molecular structure of RGO confirms that presence of C and O is very expected. The intruder elements S and Na were incorporated into RGO molecule from $H_2SO_4$ and $NaNO_3$ respectively during the synthesis. This indicates that washing during the synthesis of RGO was not conducted properly.

### *3.2. Optical microscopy*

Optical Microscopy was conducted to reveal the dispersion behaviour of RGO in PVA matrix. The dark spot appeared in pure PVA matrix is inclusion 'figure 4(a)', though it is very negligible. In case of 3% RGO loading, dispersion seems to be homogeneous, although a very negligible amount of agglomerated RGO is observed. In case of 5% RGO loading, considerable amount of agglomeration is observed 'figure 4(c)'. In case of 10% RGO loading, highest agglomeration is observed 'figure 4(d)'. 'Figure 4(e)' shows agglomerated RGO in 10% RGO/PVA composite.

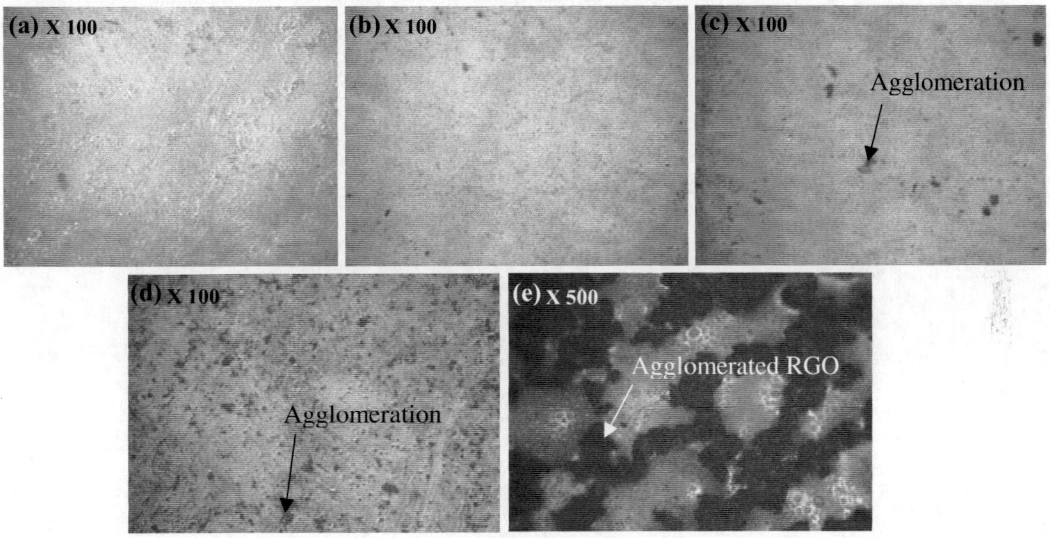


**Figure 4.** Optical microstructure of (a) Pure PVA (b) 3% RGO/PVA (c) 5% RGO/PVA (d) 10% RGO/PVA (e) Agglomeration in 10% RGO/PVA

### *3.3. XRD*

Cu Kα having wavelength of 1.5406 $A^0$ was used at 40 kV in this test. For graphite, GO and RGO XRD was conducted to get a good idea about d-spacing of those substances. From 'figure 5' it is evident that graphite shows its characteristic peak at around $26.5^0$ whose corresponding d-spacing is 0.33 nm. GO shows a broad peak at around $11^0$ and its corresponding d-spacing is 0.8 nm. The reason behind this increase in d-spacing is intercalation of oxygen containing functional groups in the interlayer of graphite during oxidation of graphite to get GO. GO also shows a weak peak at around $26.5^0$ which is nothing but an indication of incomplete oxidation. RGO shows a very strong broad peak at around $25.8^0$ and corresponding d-spacing is 0.35 nm. The value is smaller than that of GO but larger than that of graphite. The reason behind this is during reduction of GO, some oxygenated functional groups were removed from the interlayer of GO but still contains some of that which is totally absent in case of graphite. RGO also shows a weak peak at around $43^0$ which is an indication of incomplete reduction of GO.

Intensity peak of pure PVA is almost flat having a very weak peak at around $20^0$ which is shown in 'figure 6'. This peak is an indication of non-crystalline nature of pure PVA. As the RGO loading increases, the peak height also increases and the peak width becomes narrower which indicates that crystallinity of PVA increases. However, all RGO/PVA composites show peak at around $20^0$.

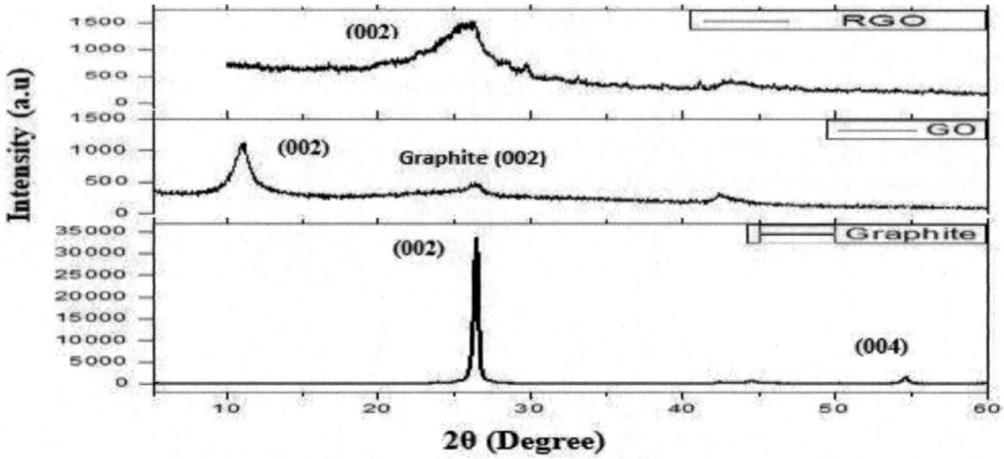


**Figure 5.** XRD report of Graphite, GO and RGO

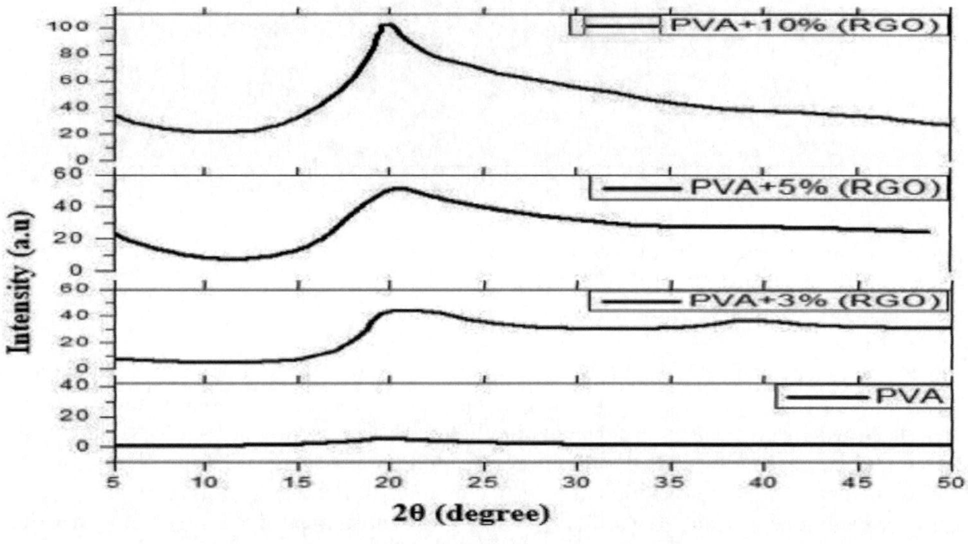


**Figure 6.** XRD report of pure PVA and RGO/PVA composites

### *3.4. UV-Visible Spectroscopy*

This test was conducted to observe the absorbance of pure PVA and RGO/PVA composites. Absorbance values have been presented graphically in 'figure 7(a)'. From the graph it is cleared that with increasing RGO loading absorbance of PVA increased. 'Figure 7(b)' shows Tauc's plot method [8] which was accomplished with the help of obtained absorbance value and the thickness of the materials [8]. This method was used to determine the band gap of the samples. Obtained band gaps of different samples by Tauc's plot have been presented in 'figure 8'. It is very clear that pure PVA had the highest band gap of 3.57 eV and with RGO loading band gap of PVA composites decreased. Therefore, electric conductivity of PVA composites increased with increasing RGO loading as it is inversely proportional to band gap energy [9].

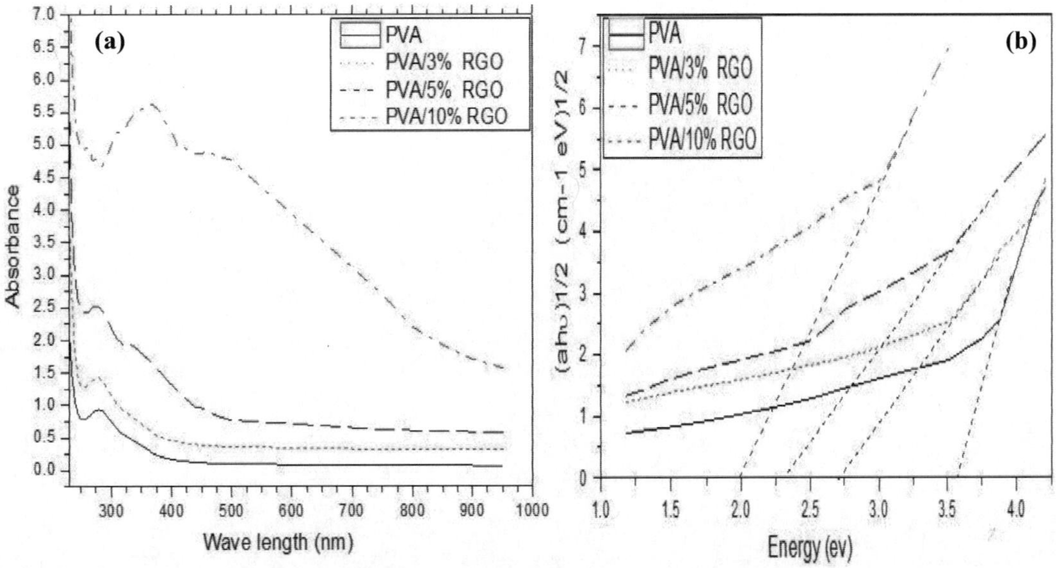


**Figure 7.** (a) Absorbance of PVA with different RGO loading (b) Tauc's Plot to determine band gap

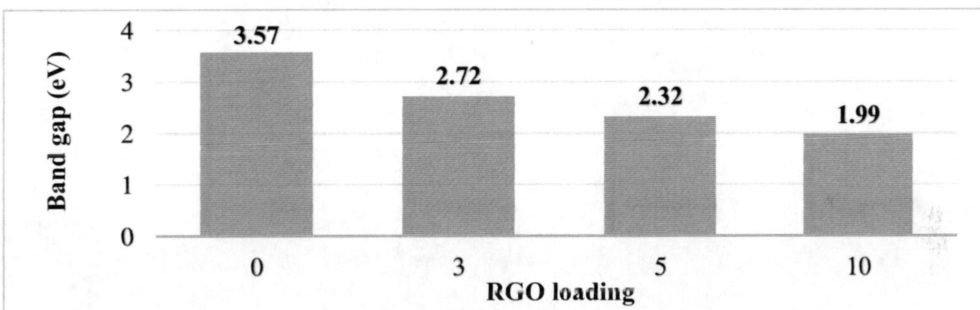


**Figure 8**. Band gap comparison among PVA samples with different RGO loading

*3.5. TGA and DTA*

TGA and DTA were carried out to observe the effect of RGO on the thermal property of PVA. From 'figure 9' it is clear that decomposition of pure PVA and PVA composites took place through several stages. The maximum decomposition rate of pure PVA, 3%RGO/PVA, 5% RGO/PVA and 10% RGO/PVA is 0.74% /$^{0}$C, 0.66% /$^{0}$C, 0.71% /$^{0}$C and 0.78% /$^{0}$C respectively and corresponding temperature is 491.71$^{0}$C, 526.14$^{0}$C, 417.54$^{0}$C and 230.36$^{0}$C. From the results it is clear that up to 3% RGO loading, maximum decomposition rate decreased and corresponding decomposition temperature increased which is a good indication of improved thermal stability. But beyond 3% RGO loading, maximum decomposition rate started to increase and corresponding decomposition temperature decreased. The reason behind this behaviour is that beyond 3% RGO loading, interfacial bonding between RGO and PVA matrix was poor [4] due to considerable amount of RGO agglomeration which has been cleared from Optical Microscopy.

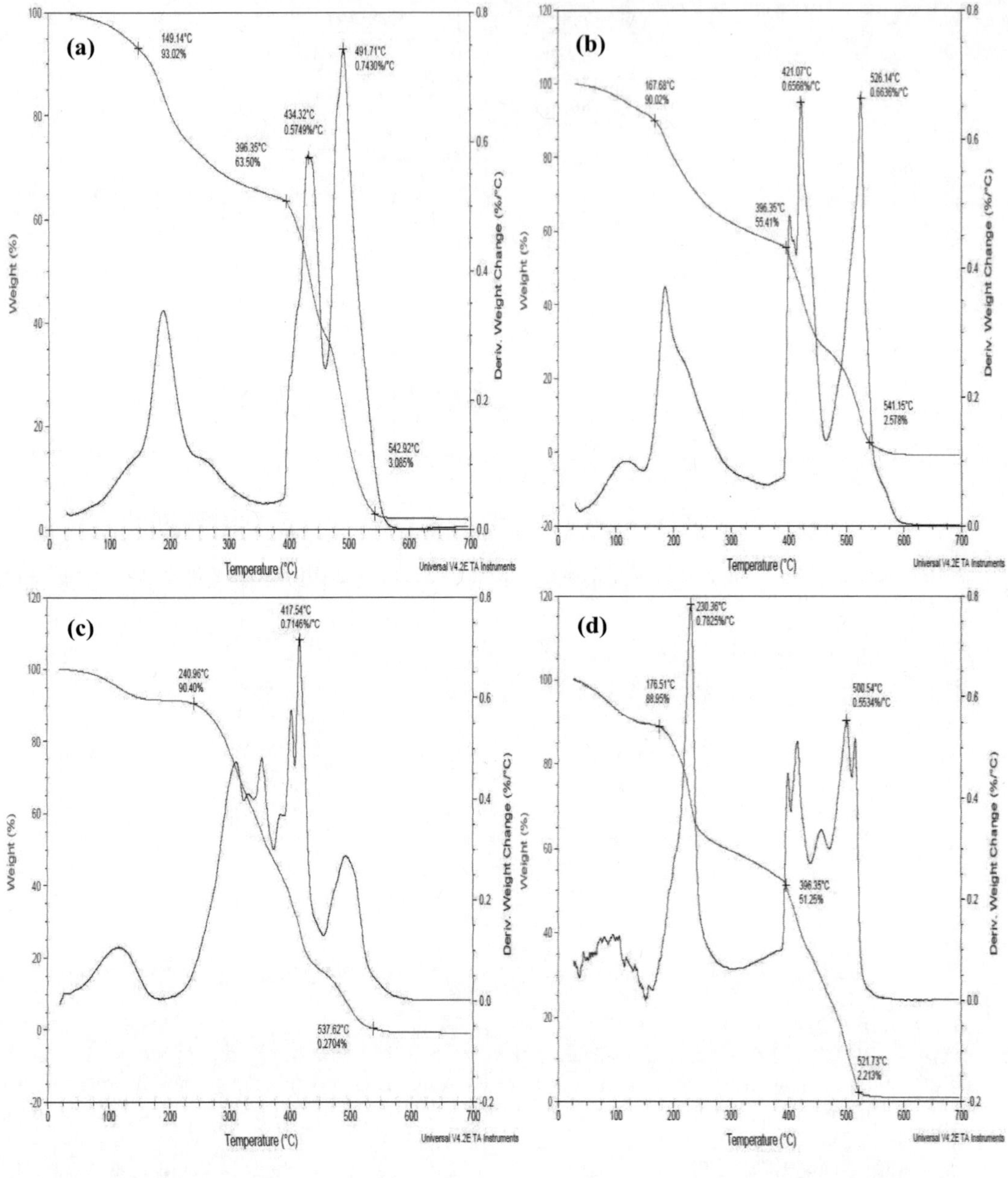


**Figure 9.** TGA and DTA report of (a) Pure PVA (b) 3% RGO/PVA (c) 5% RGO/PVA (d) 10% RGO/PVA

### *3.6. Shore hardness test*

Hardness values of pure PVA, 3%RGO/PVA, 5% RGO/PVA and 10% RGO/PVA obtained from this test are 35.6, 44.8, 50.2 and 60.4 (in D scale) respectively shown in 'figure 10'. The reason behind this increase in hardness value with increasing RGO loading is, when RGO was incorporated to PVA matrix, it increased the crystallinity of PVA [10] which has been evident from XRD report.

IOP Conf. Series: Materials Science and Engineering **438** (2018) 012008 doi:10.1088/1757-899X/438/1/012008

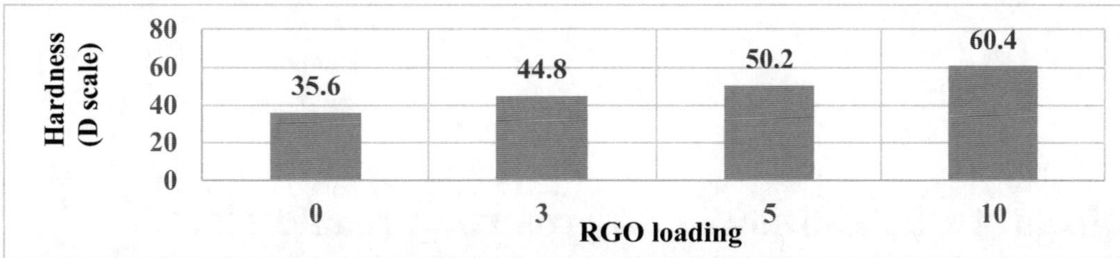


**Figure 10.** Hardness of PVA with different RGO loading

## 4. Conclusion

RGO synthesis from graphite powder was successfully carried out. Fabrication of RGO/PVA composites by solution casting was possible but good dispersion of RGO was achieved up to 3% RGO loading and beyond that agglomeration of RGO occurred which has been evident from optical microscopy. This agglomeration affected the interfacial bonding between RGO and PVA matrix which in turn affected the thermal property of PVA confirmed by TGA and DTA. However, with increasing RGO loading, crystallinity of PVA increased which has been cleared from XRD report and due to this, shore hardness of PVA increased with RGO loading. In addition, UV-Vis spectroscopy has confirmed that absorbance of PVA increased and band gap decreased with RGO loading.

## 5. References

[1] E F Sheka 2016 *The uniqueness of physical and chemical natures of graphene: their coherence and conflicts* ISTCP IX 2016 Conference (North Dakota)

[2] Phaedon A, Christos D 2012 *Graphene: synthesis and applications* Materials Today vol 15 p 86-97

[3] Edward P, Randviir, Dale A.C. Brownson, Craig E B 2014 *A decade of graphene research: production, applications and outlook* Materials Today vol 17 p 426-432.

[4] WANG X, SONG M 2013 *Toughening of polymers by graphene* Nanomaterials and Energy 2(5) p 265-278.

[5] Cristina Gómez-Navarro, Jannik C M, Ravi S S, Andrey C,Simon K, Marko B, KlausK, Ute K 2010 *Atomic Structure of Reduced Graphene Oxide* Nano Letters 10(4) p 1144-1148.

[6] N.I. Zaaba, K.L.Foo, S J Tanb, W W Liu, C.H.Voon 2017 *Synthesis of Graphene Oxide using Modified Hummers Method: Solvent Influence* Procedia Engineering vol 184 p 469-477.

[7] Wenjuan Zhu, Tony L, Vasili P, A A Bol, Yu Zhu, H Yan, Jerry T, P Avouris 2012 *Structure and electronic transport in graphene wrinkles* Nano Letters vol 12 p 3431-3436.

[8] Lei Jiang, X P Shen, J L Wu, K C Shen 2010 *Preparation and Characterization of graphene/poly (vinyl alcohol) nanocomposites* Applied Polymer Science vol 118 p 275-279.

[9] M W BARSOUM 2003 *FUNDAMENTALS OF CERAMICS* (New York) chapter 7 p 202.

[10] Thomas Roy Crompton 2006 Polymer Reference Book (Shropshire) chapter 14 p 433.

**Acknowledgements**

The authors are thankful to the Department of Materials and Metallurgical Engineering, BUET; the Department of Glass and Ceramics Engineering, BUET and the Pilot Plant and Process Development Centre, BCSIR for providing the necessary support.

# Effects of Ca Addition on Microstructure and Hot Deformation of AZ61 Magnesium Alloy

**I A F Rabbi, M M A Shawon and H M M A Rashed**

Department of Materials and Metallurgical Engineering, Bangladesh University of Engineering and Technology, Dhaka 1000, Bangladesh

Corresponding Author: hrashed@mme.buet.ac.bd

**Abstract.** Magnesium and its alloys are potential candidate in the transportation sector being the lightest structural material. In this work, effects of calcium addition in a as-cast and homogenised AZ61 (aluminium 6 wt.% and zinc 1 wt.%) magnesium alloy was investigated. Phase prediction based on thermodynamic data indicated that formation of $Al_2Ca$ phase could effectively suppress the formation of $Mg_{17}Al_{12}$ phase. This $Al_2Ca$ phase was present in the microstructure in almost the actual stoichiometry, as found in SEM EDS. The formation of this phase was also confirmed by DSC. During hot compression, recrystallization took place irrespective of temperature. However, at elevated temperature, slight grain growth of the refined grains was evident. This was due to the thermal energy from the test temperature. Nevertheless, such grain growth was not very effective due to the presence of nano-size $Al_2Ca$ phase, which provided pinning effect by positioning at the grain boundaries.

## 1. Introduction

Magnesium (Mg) is one of the lightest structural materials with low density (1.74 g/cm$^3$) as well as high specific strength. Magnesium alloys generally indicate weight reduction which results in reduction in fuel and energy consumption [1]. As they inherit low density, high specific strength, good thermal and electrical conductivities, they have been used for a wide variety of applications [2-4]. Magnesium alloys improve fuel efficiency of vehicles and reduce $CO_2$ emission due to their high specific strength and stiffness which have been a great fall back for steel and aluminium alloys in the transportation industries [3, 5-7]. They can be a considerable substitution of plastics in the field of electronic and computer industries as well [8]. In spite of these promising properties, most magnesium alloys have low formability which limits their uses in the automotive and aerospace sectors.

The addition of calcium (Ca) Mg-Al-Zn alloys generally guides to the formation $Al_2Ca$ (Melting Temperature-1352 k) [9]. In case of AZ31 alloy, the strength and creep properties can be improved through particle strengthening due to the presence of this hard and thermally stable $Al_2Ca$ phase [10]. When the microstructure was refined by thermomechanical processing (extrusion, rolling etc.), a surpassing superplasticity is revealed by the calcium containing AZ31 alloy in comparison with the bare AZ31 alloy. This unique characteristic is unveiled due to the effective pinning effect of $Al_2Ca$ particles that impedes grain coarsening during the high temperature deformation [11-13].
In this current work, effect of Ca addition in AZ61 magnesium alloy was investigated during hot deformation to determine temperature and strain rate dependency on deformation. Thermodynamic modelling was carried out to understand the dominating phases at different temperatures.

## 2. Experimental

The experiments were conducted by using AZ61 alloy which contains 6% Al, 1% Zn, 2% Ca. It was prepared by melting in an induction furnace in a mild steel mould in presence of Ar gas flow after being preheated at 300°C. After casting, a rectangular bar was produced having a dimension of 30 cm in length, 14 cm in width 15 mm in thickness. The bar was carefully cooled in air after casting. Then after machining, four rectangular bars having thickness of 12 mm from the initially produced cast bar.

X-ray Fluorescence (XRF) was conducted as well to detect the abundances of elements that are present in the cast samples.

Homogenization treatment was carried out at 420°C for 18 hours followed by water quenching. The hot compression specimens had cylindrical height of 15 mm and diameter of 10 mm. After sample preparation, hot compression tests were conducted in a Testometric UTM machine (M500-50CT) at four distinct temperatures—300 °C, 350 °C, 400 °C and 450 °C. Specimens were heated up to the test temperature and maintained at that temperature for 30 min prior to hot compression. The specimens were lubricated with graphite to reduce friction at the punch-specimen interface. The initial height of the test specimens were reduced to 75% at two separate true strains: i) $\varepsilon_1$=0.125, ii) $\varepsilon_2$=0.063. Then the specimens were air cooled and the deformed lengths were measured.

For microstructure analysis, standard sample preparation techniques were applied to the deformed specimens followed by etching in acetic picral. Images were acquired in an optical microscope OPTIKA B-600MET equipped with an image analyser for microstructure examination.

Scanning electron microscopy (SEM) equipped with Energy-Dispersive Spectrometer (EDS) was carried out by JSM-7640F field emission scanning electron microscope. SEM characterisation with optimum resolution at optimum depth of field was applied for observing second phase particles. The quantitative analysis of the chemical composition of the phases in the homogenised alloy was performed using EDS in back-scattered electron mode.

Differential scanning calorimetry (DSC) was performed on the as-cast sample (27 gm) to measure temperature difference in various phase transitions. For this test, starting and finishing temperatures were room temperature and 550 °C respectively with a heating rate of 5 °C/min.

## 3. Results and Discussion

### *3.1 Microstructure of the Alloy*

The chemical composition of the alloy is given in Table 1. In the initial calculations, the amounts of alloying elements were fixed at Al 6 wt.%, Zn 1 wt.% and Ca 2 wt.%.

**Table 1.** Chemical compositions of the alloy

| Element | Wt% |
|---|---|
| Mg | 90.75 |
| Al | 5.99 |
| Ca | 2.19 |
| Zn | 1.05 |
| Fe | 0.03 |

Optical micrographs of the alloy are shown in Figure 1 in as-cast and homogenised conditions. The extent of coring is evident in the as-cast microstructure. Dendrites of the matrix α phase is also clearly visible. After homogenisation, the amount of coring is reduced and the microstructure is more uniform.

IOP Conf. Series: Materials Science and Engineering **438** (2018) 012009 doi:10.1088/1757-899X/438/1/012009

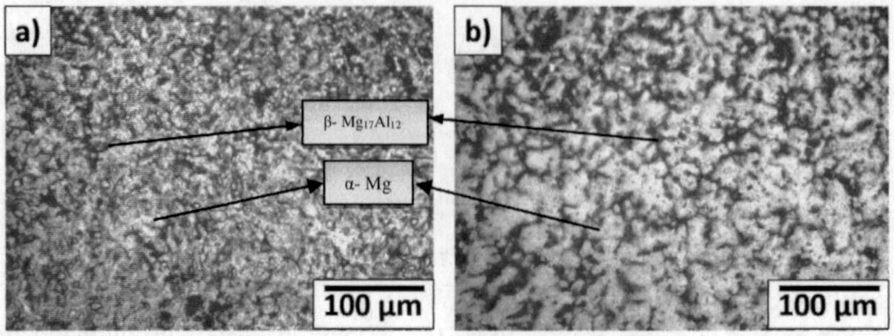


**Figure 1.** Optical micrograph of a) as-cast and b) homogenised AZ61-Ca alloy

*3.2 DSC Analysis*

Figure 2 shows DSC analysis of AZ61-Ca alloy. The sample shows an endothermic peak at 532.5°C. The temperature indicates the characteristic property of the sample. At this temperature phase changing reaction occurs. Since DSC is carried out up to 600 °C, complete melting of the alloy is not represented in the curves. The peak obtained at approximately 532 °C is explained in the following section.

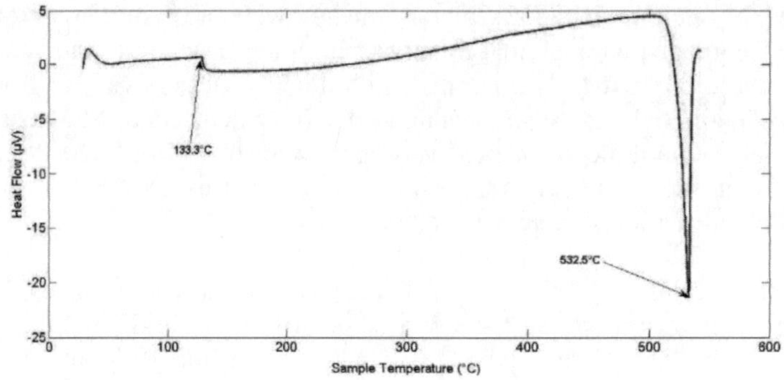


**Figure 2.** DSC curves for as cast Ca-AZ61 alloy

*3.3 Effects of Ca on Phase Formation*

In AZ61 alloy, the amount of Zn is approximately 1%. This little amount of Zn is completely dissolved in α-Mg matrix. In this context, Zn is totally present in the α-Mg matrix and does not form any considerable individual phase.

Thermodynamic modelling calculations were carried out for 700 to 200 °C. Generally, it is considered that Scheil-Gulliver solidification model provides a better prediction of the solidification condition. However, 18 hours of homogenisation at 420 °C is considered to be adequate to bring the system towards equilibrium and hence, phase predictions were made under the equilibrium condition. In Figure 3, calculations were made for the same alloy system with and without Ca addition and room temperature was considered as the phases present at 200 °C. This is due to inaccuracy of the predictions at lower temperatures.
Without Ca addition, solidification started at 604.36 °C and completed at 509.42 °C. When Ca was added to the system, these temperatures were altered slightly. Solidification started at 615.7 °C and completed at 501.48 °C.

Predicted phases are $Al_2Fe$, $Al_5Fe_2$, $Al_3Fe$, $Mg_{17}Al_{12}$, $Al_2Ca$ and $Al_2Ca$. Amongst these phases, a few Fe containing phases are less than 0.10%. These are omitted from discussion as their stability may not be accurate.

According to the prediction, the characteristic $Al_2Ca$ phase started to form at approximately 530 °C. This is in close-match with the experimental results of DSC.

The calculated area fraction of the β $Mg_{17}Al_{12}$ was 4.6 wt.% and $Al_2Ca$ (denoted by $Al_2Re$) was 5.14 %. Surprisingly 2% Ca has a profound impact on the stability of β phase. Without Ca, the phase content was approximately 8.44%. It implies that Al has more affinity towards Ca to form $Al_2Ca$ phase. Generally, addition of Ca to the AZ61 alloy leads to the formation of $Al_2Ca$ or $Al_4Ca$ phase with a decrease in the amount β $Mg_{17}Al_{12}$.

When Ca is added, it leads to the formation of $Al_2Ca$ or $Mg_2Ca$ intermetallics. As the electronegativity difference between Ca and Al (ΔEN=0.61) is higher than that of Ca and Mg (ΔEN=0.31), $Al_2Ca$ is likely to form in this circumstances [14]. The formation of $Mg_2Ca$ phase can be obtained due to the presence higher amount Mg (background pickup during EDS). Figures 4(a), 4(b), 4(f) contains large amounts of Mg and Ca. As the amount of Al is much lower, $Mg_2Ca$ phase is formed rather than $Al_2Ca$ phase (Table 2). However, from Figures 4(c), 4(d) and 4(e), it is evident that there is considerable amount of Ca and Al. So, $Al_2Ca$ phase is formed in this situation due to the electronegativity effect (Table 2).

Owing to these circumstances, Al tends to react with Ca rather than Mg. With the increase of Ca content, the amount of $Al_2Ca$ precipitates increases which concludes the reduction of the available Al content in the alloy during solidification. This incident sums up with the reduction in the β $Mg_{17}Al_{12}$.

From further observation in SEM, some small particles were observed which were embedded on the matrix (Figures 5(a), 5(b) and 5(c)). In Figure 5(a), $Al_2Ca$ particles are clearly seen to be present both at grain boundaries and in grain interior. The particle sizes also vary. Particles of micron sizes are present. Moreover, nano-size $Al_2Ca$ of sizes less than 100 nm were also vividly embedded in the matrix. These embedded particles consist of $Al_2Ca$ phase. They impede grain growth due to the Zener pinning effect.

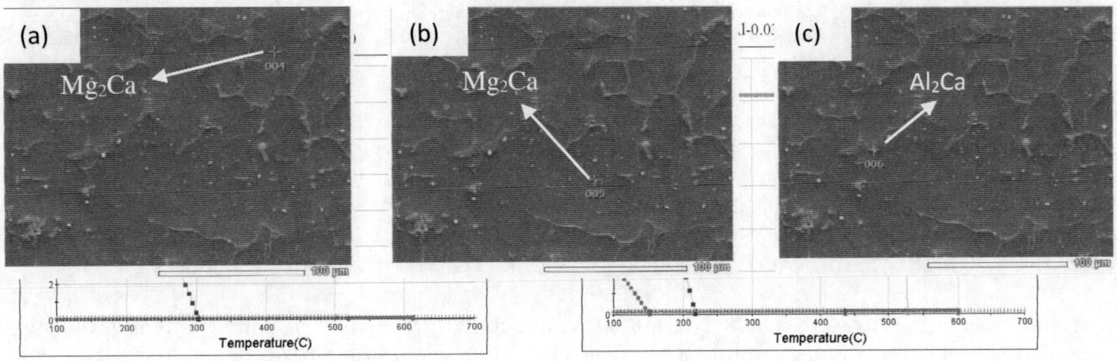


**Figure 3.** Equilibrium Model of AZ61 alloy with and without 2% Ca addition

IOP Conf. Series: Materials Science and Engineering **438** (2018) 012009 doi:10.1088/1757-899X/438/1/012009

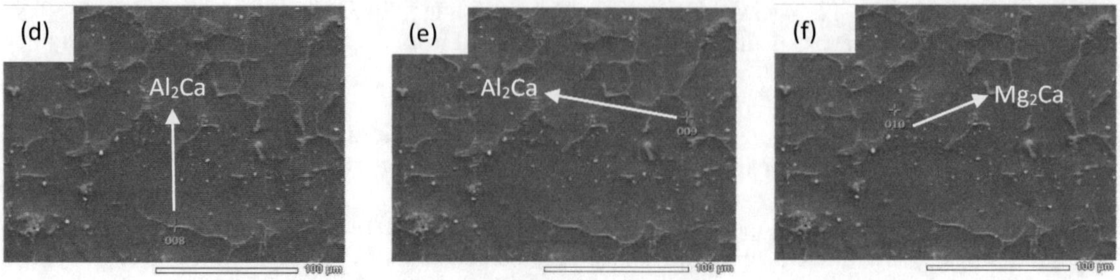


**Figure 4.** Particle present in the microstructure of the AZ61-Ca alloy

**Table 2** The EDS analysis result on the SEM images in Figure 4

| Figure | Mg (at.%) | Al (at.%) | Ca (at.%) | Zn (at.%) | Phase |
|---|---|---|---|---|---|
| a | 94.16 | 1.77 | 3.67 | 0.40 | $Mg_2Ca$ |
| b | 91.28 | 1.72 | 6.28 | 0.72 | $Mg_2Ca$ |
| c | 50.02 | 8.38 | 40.67 | 0.92 | $Al_2Ca$ |
| d | 64.13 | 15.11 | 19.36 | 1.41 | $Al_2Ca$ |
| e | 68.23 | 22.46 | 8.87 | 0.44 | $Al_2Ca$ |
| f | 92.68 | 2.05 | 4.61 | 0.66 | $Mg_2Ca$ |

*3.4 Stress-Strain Curves during Hot Compression*

The flow curves are shown in Figure 6. The curves represent typical dynamic recrystallization behaviour. At the initial stage, the flow stress increases up to a maximum stress due to strain hardening by continuous accumulation of dislocations and an increase in dislocation density. Afterwards the stress decreases with increasing compressive strain. Such softening behaviour is observed due to dynamic recrystallization and nucleation and coarsening of β particles after a critical strain [15-17]. Such behaviour is attributed to both temperature and strain rate.

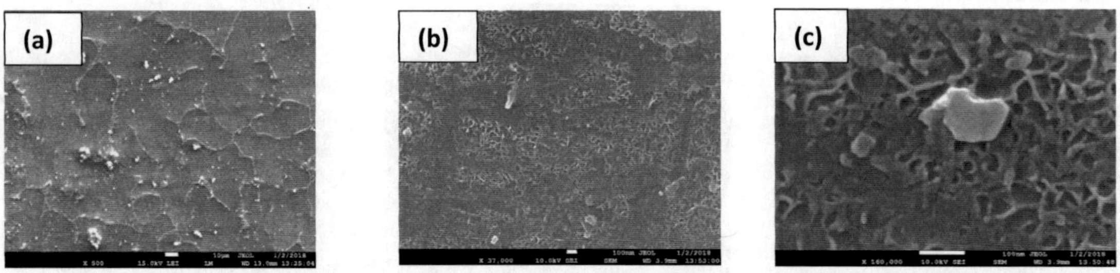


**Figure 5.** Embedded particle on the matrix (a) 500x (b) 37000x (c) 160000x

At low temperatures as well as at high strain rates, hardening followed by softening is more promising, whereas at high temperatures and low strain rates, a dynamic equilibrium between hardening and softening occurs at the very beginning of deformation. At high strain rates, dislocation multiplication is more rapid than at low strain rates and contributes to higher strain hardening effect. Figure 6 shows such steady stress continues more in case of 0.125 $s^{-1}$ strain rate than that for 0.063 $s^{-1}$ strain rate for a certain strain range. This is as well noticeable that after the steady stress, flow stress increases with increasing strain due to hindrance in deformation by increase in the size of recrystallized grains and their area fraction decreases by increasing test temperature. Such behaviour is more prominent at low strain rate and begins at lower testing temperature.

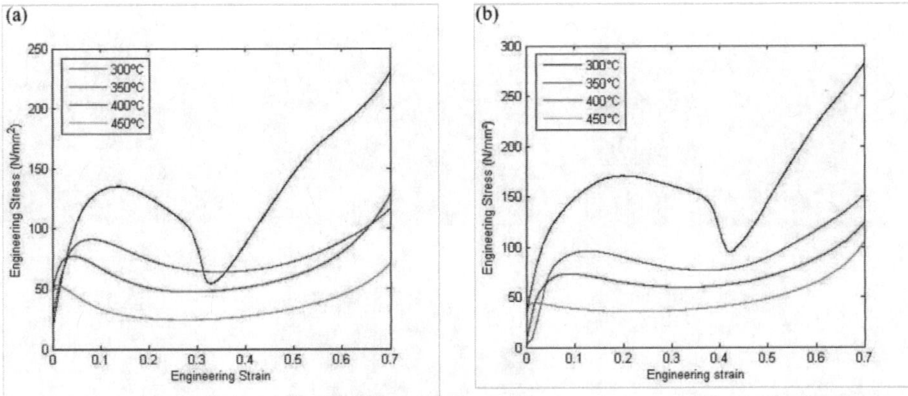


**Figure 6.** Compression curves for the homogenised specimens. (a) At a strain rate of 0.063 $s^{-1}$ and various temperatures; (b) At a strain rate of 0.125 $s^{-1}$ and various temperatures

### *3.5 Microstructural Changes during Hot Compression*

Figure 7 shows the evolution of microstructure at strain rates of 0.063 and 0.125 $s^{-1}$ for various temperatures. Comparing with the as-cast and homogenised microstructures (Figure 1), the grains are much smaller, indicating refinement of grains. With increasing temperature, grain growth had occurred to some extent. However, due to the presence of $Al_2Ca$ phase, such growth is not substantial. Nano-sized particles of this phase is believed to effectively block grain growth even at elevated temperature due to pinning effect.

At 300 °C, the grains are broken, elongated and grain distribution is heterogeneous. The microstructure is more dendritic type at lower temperatures and higher strain rate. Dendritic type microstructure changes to planar type as temperature and strain is increased. As gradients near the grain boundary provides suitable site for nucleation, so recrystallization occurs at higher temperatures. At a constant strain, the average grain size increases with increasing the deformation temperature due to the increase in the restoration rate [15]. Grain refinement occurs at high temperatures (e.g. 450 °C). Refinement of grains and growth is more prominent at the higher strain rate. After refinement grains are homogeneous. Ca addition to the AZ61 alloy causes effective refinement to the microstructure of the alloy. From the SEM analysis, Figure 5 indicates that $Al_2Ca$ is extensively present in the alloy. As testing temperature is increased, the size of recrystallized grains increases and their volume fraction decreases. Therefore, finer grains can be gained by increasing strain without increasing too much temperatures.

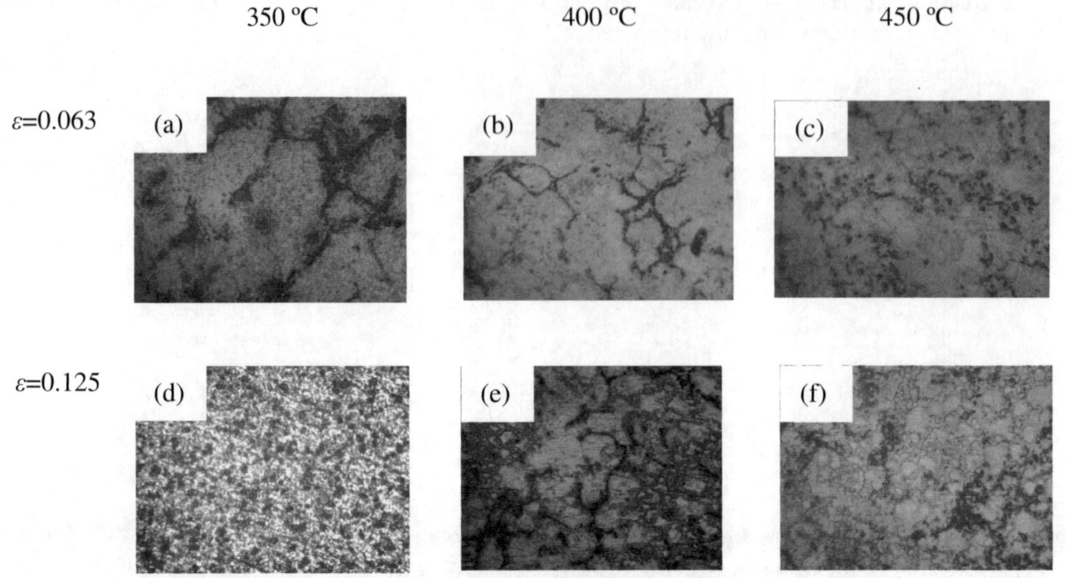


**Figure 7.** Optical microstructure evolution of AZ61 with 2% Ca addition during hot compression tests at $\varepsilon$=0.063 and temperatures of (a) 350 °C, (b) 400 °C and, (c) 450 °C and tests at $\varepsilon$=0.125 and temperatures of (d) 350 °C, (e) 400 °C and, (f) 450 °C

## 4. Conclusions

In this work, effects of Ca addition in AZ61 alloy was investigated during hot compression.

- Addition of calcium led to the formation of $Al_2Ca$ phase. This phase has effectively suppressed the stability of $Mg_{17}Al_{12}$ phase.
- DSC curves correlates well with the predicted results of phase formation from thermodynamic simulation.
- $Al_2Ca$ phase is present in the microstructure in stoichiometric composition of slight variation. SEM results confirmed that the phase can be present in different sizes—particles of size less than 100 nm was also detected.
- During hot compression, recrystallization has occurred intensely, irrespective of temperature. However, grain growth was evident at elevated temperatures. As expected, higher temperature was responsible for grain growth. Nonetheless, grain growth did not occur profoundly due to the presence of nano-size particles.
- Higher deformation rate is beneficial for formation of refined grains.

## Acknowledgments

This work was funded by HEQEP Sub-Project CPSF 3117. The authors are grateful to Bangladesh University of Engineering and Technology for providing laboratory facilities.

## Reference

1. Friedrich HE, Mordike BL. Magnesium technology: Springer; 2006.

2. Avedesian MM, Baker H. ASM specialty handbook: magnesium and magnesium alloys: ASM international; 1999.

3. Čížek L, Greger M, Dobrzański L, Juřička I, Kocich R, Pawlica L, et al. Mechanical properties of magnesium alloy AZ91 at elevated temperatures. Journal of Achievements in Materials and Manufacturing Engineering. 2006;18(1-2):203-6.

4. Kainer KU. Magnesium alloys and technology: John Wiley & Sons; 2003.

5. Xu S, Kamado S, Matsumoto N, Honma T, Kojima Y. Recrystallization mechanism of as-cast AZ91 magnesium alloy during hot compressive deformation. Materials Science and Engineering: A. 2009;527(1):52-60.

6. Greger M, Kocich R, Čížek L. Superplastic properties of magnesium alloys. Journal of Achievements in Materials and Manufacturing Engineering. 2007;22(2):83-6.

7. Pérez-Prado M, Del Valle J, Contreras J, Ruano O. Microstructural evolution during large strain hot rolling of an AM60 Mg alloy. Scripta Materialia. 2004;50(5):661-5.

8. Faraji G, Dastani O, Mousavi SA. Microstructures and mechanical properties of Al2O3/AZ91 surface nanocomposite layer produced by friction stir processing. Proceedings of the Institution of Mechanical Engineers, Part B: Journal of Engineering Manufacture. 2011;225(8):1331-45.

9. Ozturk K, Zhong Y, Liu Z-K, Luo AA. Creep resistant Mg-Al-Ca alloys: Computational thermodynamics and experimental investigation. JOM Journal of the Minerals, Metals and Materials Society. 2003;55(11):40-4.

10. Kim JH, Kang NE, Yim CD, Kim BK. Effect of calcium content on the microstructural evolution and mechanical properties of wrought Mg–3Al–1Zn alloy. Materials Science and Engineering: A. 2009;525(1):18-29.

11. Shang L, Yue S, Verma R, Krajewski P, Galvani C, Essadiqi E. Effect of microalloying (Ca, Sr, and Ce) on elevated temperature tensile behavior of AZ31 magnesium sheet alloy. Materials Science and Engineering: A. 2011;528(10):3761-70.

12. Kim Y, Kim W. Flame-resistant Ca-containing AZ31 magnesium alloy sheets with good mechanical properties fabricated by a combination of strip casting and high-ratio differential speed rolling methods. Metals and Materials International. 2015;21(2):374-81.

13. Kim W, Lee K, Park J, Kim M, Wang J, Yoon U. Microstructure and superplasticity of Mg–Al–Ca electromagnetic casting alloys after hot extrusion. Materials Science and Engineering: A. 2008;494(1):391-6.

14. Jiang B, Liu W, Qiu D, Zhang M-X, Pan F. Grain refinement of Ca addition in a twin-roll-cast Mg–3Al–1Zn alloy. Materials Chemistry and Physics. 2012;133(2):611-6.

15. Ion S, Humphreys F, White S. Dynamic recrystallisation and the development of microstructure during the high temperature deformation of magnesium. Acta Metallurgica. 1982;30(10):1909-19.

16. Ma L, Liu Z, Jiao S, Yuan X, Wu D. Dynamic recrystallization behaviour of Nb-Ti microalloyed steels. Journal of Wuhan University of Technology-Mater Sci Ed. 2008;23(4):551-7.

17. Maksoud IA, Ahmed H, Rödel J. Investigation of the effect of strain rate and temperature on the deformability and microstructure evolution of AZ31 magnesium alloy. Materials Science and Engineering: A. 2009;504(1):40-8.

IOP Conf. Series: Materials Science and Engineering **438** (2018) 012010 doi:10.1088/1757-899X/438/1/012010

# Effect of $Mg_2Sn$ Particles on Microstructure and Mechanical Properties of aged AZ Series Magnesium alloy.

**M.E. Hossain[1], H.M.M.A. Rashed[1], M.A. Zubair[2]**

[1]Department of Materials and Metallurgical Engineering, Bangladesh University of Engineering and Technology, Dhaka1000, Bangladesh
[2]Department of Glass and Ceramic Engineering, Bangladesh University of Engineering and Technology, Dhaka1000, Bangladesh

Corresponding Author: hrashed@mme.buet.ac.bd

**Abstract.** Magnesium alloys have received a considerable research focus in transportation sector due to its lightness and environmental legislation to reduce green-house effects. In this work, a magnesium alloy containing aluminium, zinc and tin as major alloying addition was studied. After casting, the alloy was homogenised and hot-rolled at 400 °C. This temperature was sufficient to promote extensive dynamic recrystallization as evident from microstructure. Texture was obtained as pole figures. The pole figures clearly revealed basal texture which also ensures the rotating of the crystals along the rolling directions. Aging of the alloy was conducted up to 30 hr at 170 °C. Surprisingly, this T5 treatment showed increased mechanical properties with time. This behaviour was studied in scanning electron microscopy. It was found that spherical-shaped $Mg_2Sn$ particles were precipitated as both nano-size particles and micron-size particles. These particles were responsible to age-strengthening of the alloy. Since particle stimulated nucleation is size dependent phenomenon, only $Mg_{17}Al_{12}$ phase effectively participated in grain refining. Aging response was entirely due to the presence and coarsening of $Mg_2Sn$ particles.

## 1. Introduction

Magnesium (Mg) alloys have drawn a great attention in automobiles and transportations due to their lightweight and low densities. Lightweight automobiles and transports are environment-friendly with higher fuel economy [1,2]. Also lightweight constructions, automotive industries, aerospace industries and other fields widely used Magnesium and their alloys that attract researchers to work with various Mg alloys [3, 4]. The low yield strength, poor ductility at room temperature, higher cost than steels and aluminium alloys however makes their application limited [5-7]. Among various hardening methods, magnesium alloys can be strengthened by solid solution strengthening or by precipitation hardening [8].This paves a way of strengthening with addition of Sn, Zn, Ca, Ce or some other element to Magnesium or Mg-Al[9,10] or Mg-RE[11-13] binary systems. Mg-Al binary system contains BCC $\beta$-$Mg_{17}Al_{12}$ phase that influences the mechanical properties of the alloy system [14-16]. In the case of Mg-Sn binary phase diagram, solubility of Sn in $\alpha$-Mg decreased from 14.85wt% at 561°C to 0.17wt% at room temperature. Sn addition creates stable $Mg_2Sn$ phase that has FCC crystal structure with a high melting temperature(770°C).Addition of Sn to the Mg-Al system promotes grain refinement and improves mechanical properties [17,18].

Mg-Al-Zn-Sn alloy system shows a better combination of strength and ductility at room temperature or elevated temperature along with lower cost than other Mg alloys[19-24].An improved age hardening response of this alloy system attracts researchers to develop new age hardenable alloys. In this work, age hardening behaviour of Mg-5Al-1Zn-2Sn alloy rolled-sheet will be investigated after hot rolling followed by T5 heat treatment. The ability of Sn nano-particles to promote or hinder recrystallization will also be highlighted.

## 2. Experimental

A plate of dimension 250×120×20 mm is cast from commercially pure Mg, Al, Zn and Sn ingots. The entire alloying elements were melted in an induction furnace at 800°C in an inert atmosphere. A continuous flow of pure argon (99.995%) created the inert atmosphere. After complete melting, the alloying mixture was agitated for 5 minutes and then poured into a metal mould which was preheated to 200°C. The composition was measured by X-Ray Fluorescence (XRF) technique which is given below in Table 1. The cast plates were then machined off and homogenized for 18 hrs at 410°C followed by water quenching. Two thick plate of individual dimension 100×50×13 mm were cut from the homogenized cast plate and hot rolled. Hot rolling was performed at 400°C with a rolling speed of 0.25m/sec. The thickness of the plate was reduced from 13mm to 2.5mm with 80% of total reduction. The reduction rate was 0.66mm per pass. The rolling temperature was made uniform by reheating the plate after every pass. Tensile samples with gauge length of 20mm were prepared. These samples were further T5 heat treated at an aging temperature of 170°C. Aging time of 3hr, 6hr, 12hr, 18hr, 24hr and 30hr were selected for every group of samples. Samples for optical microscopy (OM) and scanning electron microscopy (SEM) were prepared by following standard metallographic techniques. Samples for OM were etched with acetic picral (Picric acid 4.2g, Acetic acid 10mL, Distilled water 10mL, Ethanol 70mL). Optical micrographs were observed with Optical microscope (OPTIKA B-600MET) and the SEM micrographs were observed after aging with Field-Emission Scanning Electron Microscope (JEOL JSM-7600F) with an Energy-Dispersive Spectrometer (EDS). The Texture analysis was carried out with XRD. Tensile tests at room temperature were carried out with a Universal Testing Machine (INSTRON 3369). A FV-800 Vickers hardness tester with pyramidal indenter and 3kg load was used to determine the Vickers hardness at several points of the rolled sheet after aging.

**Table 1**
The chemical composition of the investigated alloys (%wt).

| Al | Zn | Sn | Si | Fe | Mg |
|---|---|---|---|---|---|
| 5.033 | 0.993 | 2.617 | >0.025 | >0.025 | Bal. |

## 3. Results and discussion

### *3.1. Microstructures of the as-cast and homogenized alloy*

Fig.1a Show the as-cast microstructure of the Mg-5Al-1Zn-2Sn alloy taken from the middle of the cast plate. The optical microstructure shows coarse dendrites of primary α-Mg with discontinuous second phase at the inter-dendritic region, Fig. 1b. After 18hrs of homogenization treatment all the second phases were dissolved and the coarse dendritic structure turned into a coarse grain structure with distinct grain boundary (Fig. 1c).

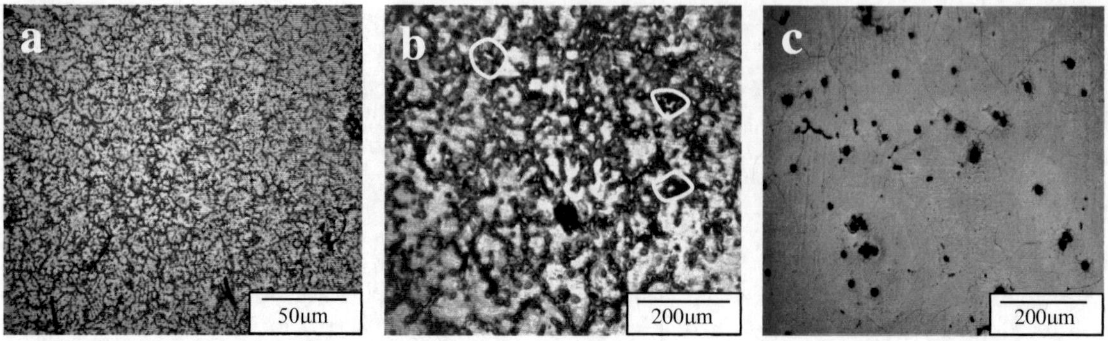


Figure 1.Optical microstructure of cast alloy a) shows the coarse dendritic structure of the alloy, b) shows the inter-dendritic second phase particles (inside yellow circle) c) homogenized at 410°C for 18hrs.

*3.2 Texture of the Alloy*

The pole figures of the rolled alloy were constructed from XRD data. For hexagonal metals, the texture is commonly represented by the orientation of the {0001} plane. The texture obtained for basal{0001}, prismatic $\{10\bar{1}0\}$ and pyramidal $\{10\bar{2}0\}$ planes are shown in terms of pole figures in Fig. 2.A strong basal texture is observed. Most of the poles are aligned parallel to the sheet thickness. It implies that the c-axis,< 0001 >, direction of the hcp magnesium crystals lies perpendicular to the rolling direction (RD). No splitting of the texture in the rolling direction, as sometimes observed in magnesium alloy sheet was detected.

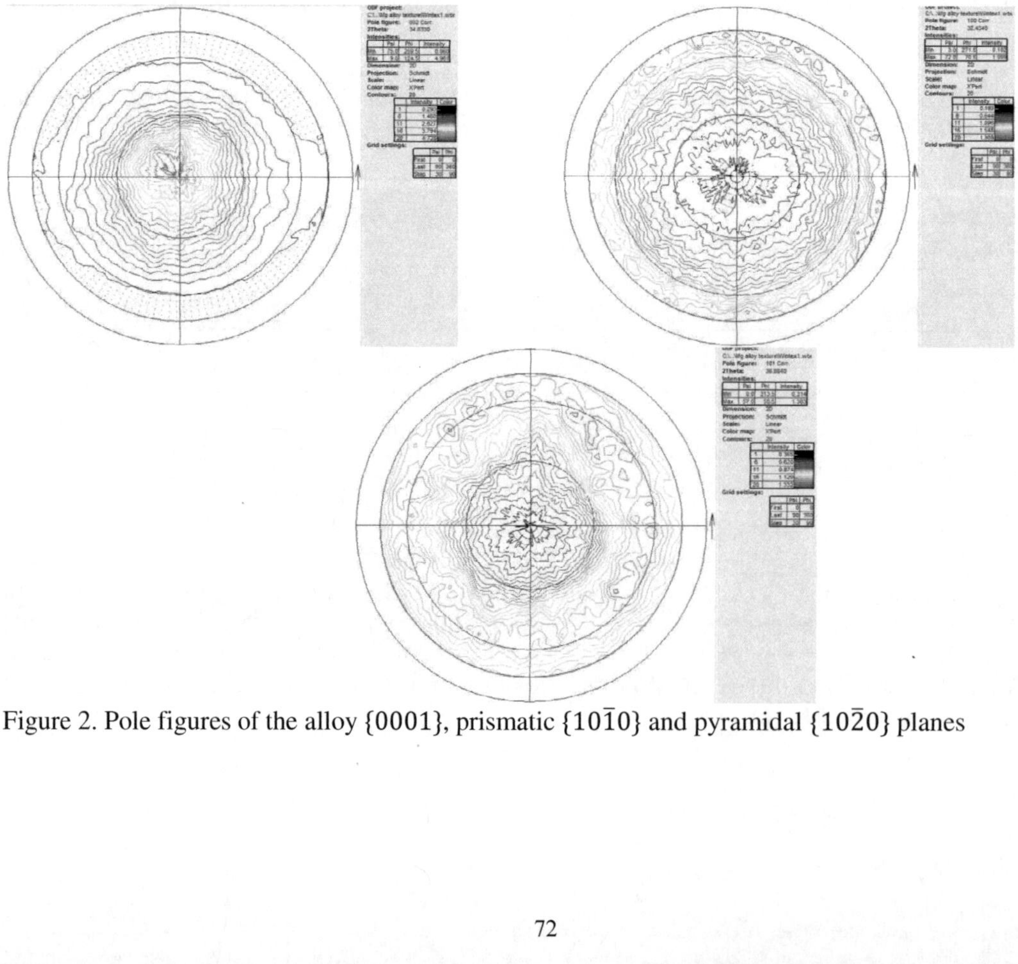

Figure 2. Pole figures of the alloy {0001}, prismatic $\{10\bar{1}0\}$ and pyramidal $\{10\bar{2}0\}$ planes

### *3.3. Microstructures of hot rolled alloy*

Dynamic recrystallization (DRX) was observed during hot rolling of the homogenized thick plate. Fig. 3 shows the microstructure of the rolled sheet with a rolling reduction of 25% showing dynamic recrystallization. Initiation of new grains along grain boundaries were observed with increasing rolling reduction. Final microstructure after 80% rolling reduction consists of fine equiaxed grains uniformly distributed over the structure (Fig. 3b). Deformation modes in the normal direction during rolling were associated mostly with grain boundaries as grain boundary sliding, bulging of grain boundary and grain boundary separation (Fig. 3a). Twinning contributes slightly in this direction though it is the main deformation modes along the surface, parallel to rolling direction (Fig. 3c).

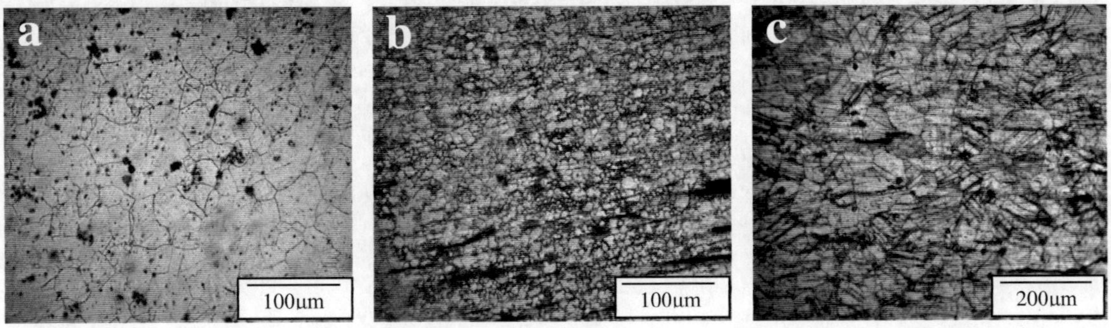


Figure 3.Optical microstructure after hot rolling at 400°C a) showing dynamic recrystallization, b) equiaxed microstructure after 80% reduction c) twinning at the surface of the sheet after 80% reduction.

### *3.4. Microstructures after aging*

Aging effect of the rolled sheet can be seen from the Fig. 4 after aging at 170°C for different amount of aging time. SEM micrographs show small amount of second phase particles embedded over the matrix after 3hr aging. The spherical bright particle reveals $Mg_2Sn$ phase after EDS analysis also the irregular particles reveals as $Mg_{17}Al_{12}$. Particle coarsens, and new particles appear with aging time. Irregular $Mg_{17}Al_{12}$particles and bright spherical $Mg_2Sn$ particles precipitate along the grain boundaries and some of them precipitate inside grains. Spheroid like $Mg_2Sn$ precipitates along grain boundaries mainly produces pinning effect.

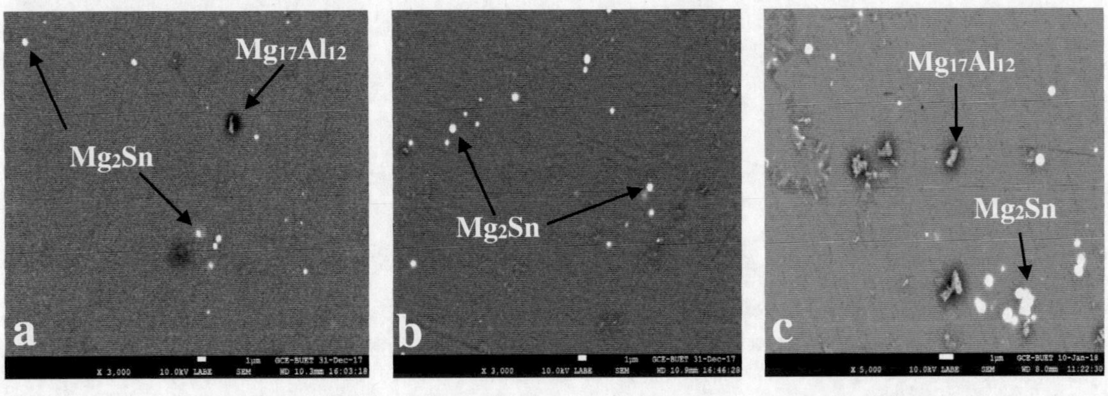

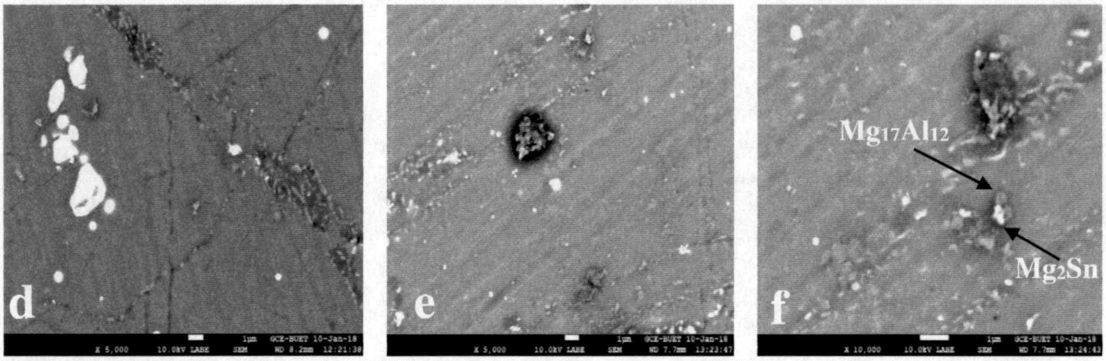


Figure 4.SEM BSE micrograph of the rolled sheet after aging at 170°C for a) 3hr, b) 6hr, c) 12hr, d) 18hr, e) 24hr, f) 30hr.

With increasing aging time, the volume fraction of second phase particles was also increased; new particles formed along grain boundaries and interior of the grains. Microstructure after 24hr aging shows new bright $Mg_2Sn$ particles formation, these particles also appeared associated with$Mg_{17}Al_{12}$ phase after 30hr of aging.

### *3.5. Mechanical behaviour after aging*

Mechanical behaviour of the alloy after T5 heat treatment depends on the degree of precipitation hardening. So the size, shape and distribution of the precipitate determine the mechanical behaviour of AZT512 alloy after aging. The variation of yield strength, ultimate tensile strength and ductility with increasing aging time is shown in Fig.5.

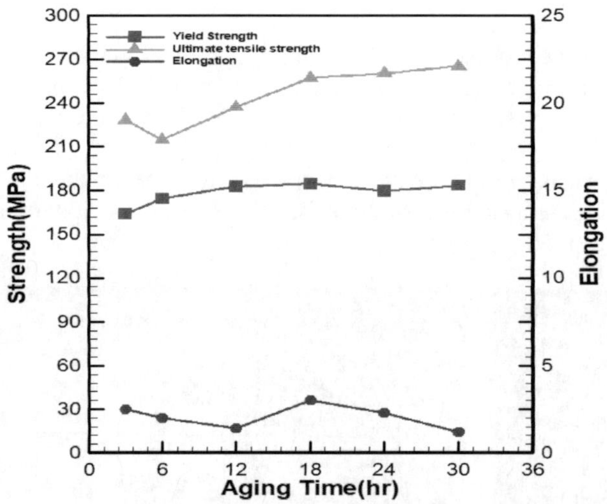


Figure 5. Tensile curve of the alloy aged at 170°C

Yield strength (YS) of the alloy increases from an initial value of 164.2MPa to a peak value of 184.5MPa corresponding with an aging time of 18hrs. YS then decreased slightly with aging time. Ultimate tensile strength (UTS) decreases from an initial value of 228.3MPa to 214.6MPa and then increases gradually with aging time, and holds a peak value of 265.2MPa for 30hrs of aging. This can be simply associated with the increasing number of precipitates with increasing aging time. From Fig. 5, it is obvious that the amount of spherical $Mg_2Sn$ particles increased with aging time. According to

Nie [25, 26], spherical precipitates are effective in blocking dislocation gliding on basal, prismatic, and pyramidal planes.

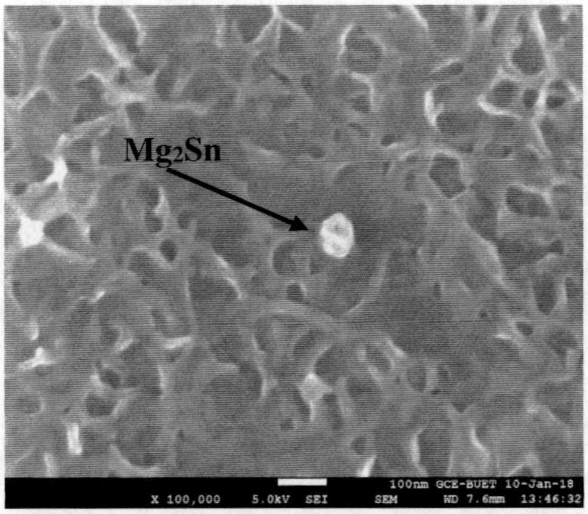


Figure 6.Embryo of $Mg_2Sn$ particle embedded in α-Mg matrix.

Fig. 6 shows an embryo of $Mg_2Sn$ particle embedded in the α-matrix. These embryos help in cluster formation and the spherical $Mg_2Sn$ particle cause pinning of the dislocation in the grain boundary during plastic deformation. As a consequence, the difference between YS and UTS increases with time. Ductility changes merely with aging time.

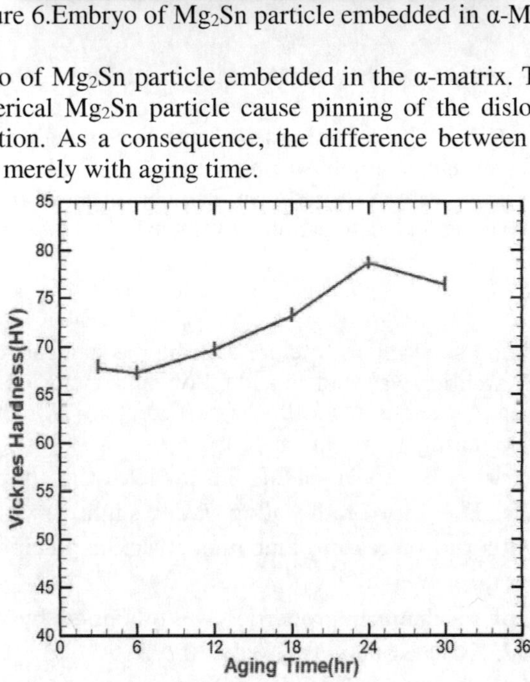


Figure 7. Age hardening curve of the alloy aged at 170°C

Fig.7 shows the hardness variation with aging time of the alloy. 68HV is the initial hardness value obtained with 3hr of aging; same value of hardness obtained with 6hr of aging and then increase gradually to 70HV for 12hr, 73HV for 18hr and 78HV for 24hr and decrease to 76HV for 30 hr of aging time. The peak hardness value obtained is 78HV correspond to 24hr of aging, 10HV larger than initial value. According to sasaki et al. [27], increase in the hardness value with aging is apparently attributed to the homogeneous distribution of $Mg_2Sn$ precipitates and increase in number of precipitates. The micrographs confirm a uniform distribution of fine precipitates of the spherical $Mg_2Sn$ after aging (Fig. 4) and also observed an increase in the amount of precipitate with increasing aging time.

### *3.6 Effects of Sn on Recrystallization*

Addition of Al and Sn has profound effect on developed microstructure after rolling. As-cast and homogenized microstructure has an average grain size of 200 µm, whereas after hot rolling, the grain size was reduced to 10 µm. In the microstructure, it is evident that a few grains have grown abnormally to a size comparable to as-cast structure. However, most grains were less than 40 µm in size. This implies that after recrystallization of the alloy during hot rolling, grains were extensively refined. Only a few grains were observed to grow to a larger size.

Second phase particles of size larger than 1 µm is typically considered to promote recrystallization by particle stimulated nucleation (PSN). Fig. 4 confirms such presence of particles—$Mg_{17}Al_{12}$ and $Mg_2Sn$. They are the obvious candidate for favourable condition of grain refining. In Fig. 6, one nano-size $Mg_2Sn$ particle is shown. It is believed that these particles will hinder growth of grain boundary and thus retard grain growth. However, they will not be suitable for grain refinement. Recrystallization was favoured by twinning and $Mg_{17}Al_{12}$ particles. $Mg_2Sn$ particles can act as nucleation sites for new grains. However, it is more likely that nano-size $Mg_2Sn$ particles will delay recrystallization due to pinning effect as they are more likely to be positioned in the prior grain boundaries.

### *3.7 Effects of Sn on Aging*

$Mg_2Sn$ particles act as pinning particles as they are nucleated on grain boundaries. $Mg_{17}Al_{12}$ phase is apparently not sufficient to restrict deformation after aging. In contrast, the size of $Mg_2Sn$ particles has increased considerably during aging. These particles still provide enough hindrance to dislocation movement. This is confirmed through strength and hardness data. With aging, the particles were coarsened. These coarsened particles were mostly $Mg_2Sn$. They have grown from the tiny nano-size to micron-size level and yet effective to block dislocation movements.

## 4. Conclusion

The cast, homogenised, rolled and aged microstructure and the mechanical properties after aging of AZ series alloy with 2 wt. % Sn addition was studied and following conclusion could be deduced.

a. Addition of 2% Sn to an AZ series Mg alloy is sufficient to promote grain refinement by dynamic recrystallization during hot rolling at 400°C.

b. Increasing aging time coarsen the spherical $Mg_2Sn$ particles that improves the mechanical properties of the alloy. Ductility of the alloy varies slightly with aging time. Vickers Hardness increases with the increasing amount of $Mg_2Sn$ precipitate and also uniform distribution of precipitate.

c. The best combination of mechanical properties was obtained by aging for 18hr. YS of 184.5MPa and UTS of 257MPa along with 3% ductility.

d. Nano-size $Mg_2Sn$ particles were confirmed in the microstructure which is effective in blocking dislocation motions. These particles were coarsened during aging and yet provide strong influence on hindrance of dislocation movement.

## 5. Acknowledgments

This work was funded by HEQEP Sub-Project CPSF 3117. The authors are grateful to Bangladesh University of Engineering and Technology for providing laboratory facilities. The Authors would also like to thank colleagues Shaon, Siddik, Istiaq, Jahir and Sajal for their support and cooperation.

## 6. Reference

[1]Pan H, Ren Y, Fu H, Zhao H, Wang L, Meng X, Qin G. Recent developments in rare-earth free wrought magnesium alloys having high strength: a review. Journal of Alloys and Compounds. 2016 Apr 5;663:321-31.
[2] Bettles C, Gibson M. Current wrought magnesium alloys: strengths and weaknesses. Jom. 2005 May 1;57(5):46-9.
[3] Sasaki TT, Elsayed FR, Nakata T, Ohkubo T, Kamado S, Hono K. Strong and ductile heat-treatable Mg–Sn–Zn–Al wrought alloys. Acta Materialia. 2015 Oct 15;99:176-86.
[4]Zhang J, Guo ZX, Pan F, Li Z, Luo X. Effect of composition on the microstructure and mechanical properties of Mg–Zn–Al alloys. Materials Science and Engineering: A. 2007 May 15;456(1-2):43-51.
[5] Jung JG, Park SH, Yu H, Kim YM, Lee YK, You BS. Improved mechanical properties of Mg–7.6 Al–0.4 Zn alloy through aging prior to extrusion. ScriptaMaterialia. 2014 Dec 15;93:8-11.
[6] Qi F, Zhang D, Zhang X, Xu X. Effect of Sn addition on the microstructure and mechanical properties of Mg–6Zn–1Mn (wt.%) alloy. Journal of Alloys and Compounds. 2014 Feb 5;585:656-66.
[7] Zhu SQ, Yan HG, Liao XZ, Moody SJ, Sha G, Wu YZ, Ringer SP. Mechanisms for enhanced plasticity in magnesium alloys. Acta Materialia. 2015 Jan 1;82:344-55.
[8] Callister WD, Rethwisch DG, Materials Science and Engineering: AnIntroduction, 8th ed., 2010, J. Wiley, USA
[9] Clark JB. Age hardening in a Mg-9 wt.% Al alloy. Acta Metallurgica. 1968 Feb 1;16(2):141-52.
[10] Celotto ST. TEM study of continuous precipitation in Mg–9 wt% Al–1 wt% Zn alloy. Acta materialia. 2000 May 11;48(8):1775-87.
[11] Nie JF, Gao X, Zhu SM. Enhanced age hardening response and creep resistance of Mg–Gd alloys containing Zn. ScriptaMaterialia. 2005 Nov 1;53(9):1049-53.
[12] Smola B, Stulıková I, Von Buch F, Mordike BL. Structural aspects of high performance Mg alloys design. Materials Science and Engineering: A. 2002 Feb 15;324(1-2):113-7.
[13] Ping DH, Hono K, Nie JF. Atom probe characterization of plate-like precipitates in a Mg–RE–Zn–Zr casting alloy. ScriptaMaterialia. 2003 Apr 14;48(8):1017-22.
[14] Leil TA, Hort N, Dietzel W, Blawert C, Huang Y, Kainer KU, Rao KP. Microstructure and corrosion behavior of Mg-Sn-Ca alloys after extrusion. Transactions of Nonferrous Metals Society of China. 2009 Feb 1;19(1):40-4.
[15] Kleiner S, Beffort O, Wahlen A, Uggowitzer PJ. Microstructure and mechanical properties of squeeze cast and semi-solid cast Mg–Al alloys. Journal of Light Metals. 2002 Nov 1;2(4):277-80.
[16] Song GL. Effect of tin modification on corrosion of AM70 magnesium alloy. Corrosion Science. 2009 Sep 1;51(9):2063-70.
[17] Hort N, Huang YD, Kainer KU. Intermetallics in magnesium alloys. Advanced Engineering Materials. 2006 Apr 1;8(4):235-40.
[18] NayebAA,Hashemi, ClarkJB, Mg-Sn (Magnesium-Tin), Binary Alloy Phase Diagrams, second ed., T.B. Massalski (Ed.), 1990, 3, 2549–2552.
[19] Kim YK, Sohn SW, Kim DH, Kim WT, Kim DH. Role of icosahedral phase in enhancing the strength of Mg–Sn–Zn–Al alloy. Journal of Alloys and Compounds. 2013 Feb 5;549:46-50.
[20] Mahmudi R, Moeendarbari S. Effects of Sn additions on the microstructure and impression creep behavior of AZ91 magnesium alloy. Materials Science and Engineering: A. 2013 Mar 20;566:30-9.
[21] Elsayed FR, Sasaki TT, Mendis CL, Ohkubo T, Hono K. Compositional optimization of Mg–Sn–Al alloys for higher age hardening response. Materials Science and Engineering: A. 2013 Mar 20;566:22-9.
[22] Wang B, Pan F, Chen X, Guo W, Mao J. Microstructure and mechanical properties of as-extruded and as-aged Mg–Zn–Al–Sn alloys. Materials Science and Engineering: A. 2016 Feb 22;656:165-73.
[23] Yoon J, Park S. Forgeability test of extruded Mg–Sn–Al–Zn alloys under warm forming conditions. Materials & Design. 2014 Mar 1;55:300-8.
[24] Kabir AS, Sanjari M, Su J, Jung IH, Yue S. Effect of strain-induced precipitation on dynamic recrystallization in Mg–Al–Sn alloys. Materials Science and Engineering: A. 2014 Oct 20;616:252-9.
[25] Nie JF. Precipitation and hardening in magnesium alloys. Metallurgical and Materials Transactions A. 2012 Nov 1;43(11):3891-939.
[26] Nie JF. Effects of precipitate shape and orientation on dispersion strengthening in magnesium alloys. ScriptaMaterialia. 2003 Apr 14;48(8):1009-15.
[27] Sasaki TT, Oh-Ishi K, Ohkubo T, Hono K. Enhanced age hardening response by the addition of Zn in Mg–Sn alloys. ScriptaMaterialia. 2006 Aug 1;55(3):251-4.

# Development of a High Strength Al–$Mg_2$Si-Mg-Zn Based Cast Alloy

**M. A. Liza and A. K. M. Bazlur Rashid**

Department of Materials and Metallurgical Engineering, Bangladesh University of Engineering and Technology, Dhaka 1000, Bangladesh

**Abstract.** This paper explores the possibility of developing an Al-$Mg_2$Si-Mg-Zn based sand cast alloy having improved mechanical properties. The effect of Zn and free Mg after combining with Si to form $Mg_2$Si on microstructure and mechanical properties was investigated under as-cast condition. Eight different alloys were prepared by varying Mg(4.15 – 12.0 wt.%) and Zn(2.58 – 6.22 wt.%) contents and 0.2 proof strength, tensile strength, elongation at failure and hardness were measured under as-cast condition. The microstructural investigation was conducted by optical microscopy and all alloys were found to contained α-Al, Al-$Mg_2$Sieutectic and $Mg_2$Siphases. The addition of Mg generally improves the proof strength and decreases the elongation, while the tensile strength shows a peak for alloys having Mg contents between the range 3 – 6 wt.%. Addition of Zn content also has the similar effect in improving strength and hardness and decreasing ductility. Comparing all alloys, alloy 3having 5.41wt.%Mg, 1.85%Si and 5.45%Zn has the best combination of properties.

## 1. Introduction

Al–Si foundry alloys are widely used in the automotive industry due attractive foundry properties, good mechanical properties with high strength-to-weight ratio, corrosion resistance and ease of recycling [1]. Enhancement of mechanical properties by proper heat treatment procedure also offers attractive advantage [2]. Among the sand cast varieties of the currently available sand cast aluminium alloys, the heat-treatable Al-7Si-Mg alloys are still a popular choice for automobile and similar application. The enhanced mechanical properties of these alloys are attributed to the precipitation of $Mg_2$Si particles in the α-Al matrix.

In the past two decades, various manufacturing industry had started to use Al- $Mg_2$Si (Al-Mg-Si) alloys in structure and such other applications. Alloying elements present in Al– $Mg_2$Si–Mg–Zn based alloys change properties significantly by alteration of the morphology, structure, size distribution and size of grains and precipitates. For this, greater vehicle fuel efficiency, more weight-savings, and decreased $CO_2$emissions in transport have been achieved quickly [2]. Al- $Mg_2$Si alloy is such an alloy which gives desirable high strength by $Mg_2$Siparticles. The past studies [2]on the Al-$Mg_2$Sialloy are mainly based on the hyper-eutectic alloys for die casting applications and little information could be found for hypo-eutectic alloys for manufacturing sand cast quality automotive components.

In this paper, the focus is to develop a cast hypo-eutectic Al-$Mg_2$Si alloy and compare its properties with those for the popular Al-7Si-Mg alloys. The effect Zn and excess Mg addition on structure and properties will also be investigated. This is because, several investigators found that aluminium alloys containing both zinc and magnesium developed substantially higher strengths than those containing any of these alloying elements added singly, and significantly higher strengths were obtained from these alloys [3-7].

## 2. Experimental

In this work, commercially pure aluminium ingot, pure magnesium ingot, pure zinc ingot, Al–50wt% Si master alloy ingot and Al–80wt% Mn master alloy purchased locally were used raw materials. Eight different alloy compositions were produced from these starting materials. A melt of 5 kg was prepared each time in gas-fired clay bonded graphite crucible. The melt was held for about 15 minutes at about 720°C for homogenisation. After degassing and drossing, the temperature of the melt was raised quickly to 750°C and then poured in Y-block sand mould.

Optical emission spectroscopy was performed to analyse each casting. From the Y-block casting of 200*100*20 mm dimension, tensile specimens following ASTM E8 standards were prepared. Tensile test was conducted at room temperature using Shimadzu universal testing machine to determine 0.2 proof strength, tensile and elongation at failure of the sample. Three specimens were tested for each condition and average values were reported. Brinell hardness of the specimens was measured by using Universal Hardness Testing machine using 5 mm diameter steel ball and 500 kg load. Finally, the optical microstructure of each sample was examined to analyse their structure. Standard metallographic techniques were used to prepare each sample.

## 3. Results and discussion

$Mg_2Si$ is a compound of Mg and Si elements with atom ratio of Mg:Si = 2:1 and the weight ratio of Mg:Si = 1.73:1. The$Mg_2Si$ crystal occurs during solidification because the maximum solids solubility of Si into Mg is less than 0.03%[5]. Al–$Mg_2Si$ can be regarded as a simple pseudo-binary alloy, Figure 1. This is particularly useful to simplify the thermodynamics of Al–Mg–Si system with a specific ratio in a binary phase diagram. However, the present experiments were carried out using individual element and it is easier to understand using Mg and Si. To balance the need in thermodynamic calculation and experimental confirmation, the corresponding compositions of each alloy is reported in the form of total Mg and Si, $Mg_2Si$, excess Mg (after forming $Mg_2Si$) are listed in Table 1.

**Table 1.**Chemical composition of the studied eight alloys.

| | %Mg | %Si | %Zn | %Mn | %Fe | Al |
|---|---|---|---|---|---|---|
| Alloy 1 | 4.15 | 1.46 | 6.22 | 0.19 | 0.16 | Balance |
| Alloy 2 | 4.29 | 1.93 | 2.58 | 0.06 | 0.31 | Balance |
| Alloy 3 | 5.41 | 1.85 | 5.45 | 0.02 | 0.25 | Balance |
| Alloy 4 | 5.83 | 2.19 | 4.91 | 0.02 | 0.24 | Balance |
| Alloy 5 | 6.94 | 2.55 | 5.26 | 0.19 | 0.18 | Balance |
| Alloy 6 | 7.40 | 4.19 | 5.01 | 0.02 | 0.16 | Balance |
| Alloy 7 | 11.6 | 6.58 | 3.83 | 0.14 | 0.18 | Balance |
| Alloy 8 | 12.0 | 8.84 | 4.17 | 0.17 | 0.14 | Balance |

### *3.1 General Properties of Al-$Mg_2Si$-Mg-Zn Alloys*

A summary of mechanical properties of the eight tested alloys is given in Table 2. Compared with the mechanical properties of commonly used sand-cast Al-7Si-0.3Mg (LM25 or A356) alloys in the as-cast conditions (0.2 Proof strength = 90 MPa, Tensile strength = 180 MPa, Elongation = 5 %), all Al-$Mg_2Si$-Mg-Zn alloys showed improved strength and decreased ductility with Alloy 3 (5wt% $Mg_2Si$ and 2.2% excess Mg) showed the highest value.

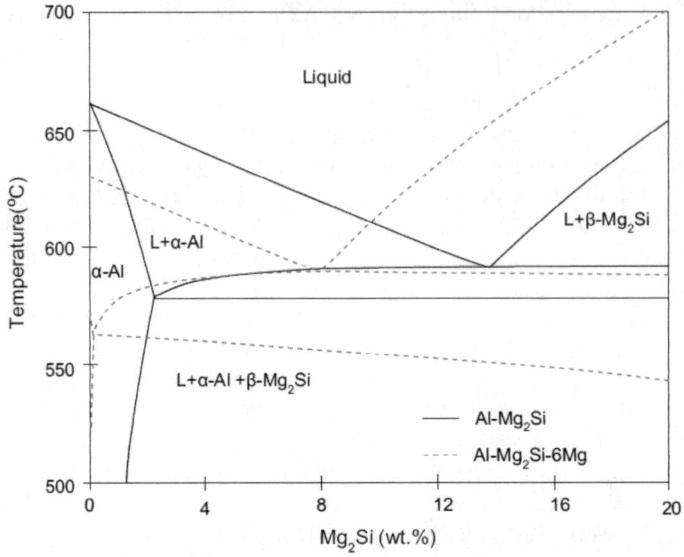


**Figure 1:** The pseudo-binary phase diagram of Al-$Mg_2Si$ system.

**Table 2.** Mechanical properties of studied eight alloys.

| | 0.2 Proof Strength (MPa) | Tensile Strength (MPa) | Elongation (in 50 mm) (%) | BHN (500 kg, 5 mm) |
|---|---|---|---|---|
| Alloy 1 | 154 | 196 | 2.4 | 98 |
| Alloy 2 | 150 | 194 | 3.0 | 76 |
| Alloy 3 | 229 | 237 | 1.0 | 108 |
| Alloy 4 | 215 | 224 | 2.2 | 99 |
| Alloy 5 | 215 | 217 | 2.1 | 101 |
| Alloy 6 | 211 | 215 | 2.1 | 55 |
| Alloy 7 | 124 | 187 | 1.9 | 101 |
| Alloy 8 | 121 | 181 | 1.7 | 99 |

Normally, reduced porosity and improved microstructural uniformity in the castings result in the decrement of sources for brittle fracture and improve the mechanical properties. Pores present in the microstructure favour crack initiation and propagation therefore are bad for the mechanical performance of the material [2].The reduced defects can retard the initialisation of cracks and improve the mechanical properties. More Mg and Si in the alloy create more eutectic Al-$Mg_2$Siphases, which result in increased strength and decreased ductility. By analysing thermodynamics and the phase diagram, we understand that although Mg content can be varied in large range, it is preferred in a range of 5-8wt.% in Al-Mg-Si based alloy for getting relatively narrow solidification range [8], and a better mechanical property. In the present work, this is also evident as Alloy 1 to 8, containing $Mg_2Si$ content in the range of 5 – 11 wt.% produce higher mechanical strength. Alloy 2 is a special case, as it contains lower Zn content.

### *3.2 Effect of Mg Content on the Mechanical Properties*

The pseudo-binary Al–$Mg_2Si$ system (Figure1) normally behave like a normal binary eutectic system with a eutectic point at 13.9 wt% Mg2Si (Al–8.81 wt%Mg–5.09 wt%Si).With extra Mg in the Al–

$Mg_2Si$ system, the eutectic point was shifted towards a lower $Mg_2Si$ concentration and the area of co-existence for the three phases (L + α-Al + $Mg_2Si$) was enlarged. This means that, for a given $Mg_2Si$ content, an extra Mg addition resulted in a decrease of the volume fraction of the primary α-Al phase and an increase of the volume fraction of Al–$Mg_2Si$ eutectic phase.

Figures 2(a) and 2(b) showed the variation of the 0.2 proof strength, tensile strength, elongation and hardness of the as-cast Al-$Mg_2Si$-Mg-Zn alloys with Mg contents. In general, the proof strength and tensile strength were improved with increasing Mg content. Conversely, the elongation was decreased with increasing Mg in the alloys .This may be attributed to the fact that alloys having lower Mg content have a much higher solid fraction of α-Al phase, which maintained their ductility. With increasing Mg content, Al-$Mg_2Si$ eutectics were formed, which improved strength and reduced ductility. When the coarse primary $Mg_2Si$ phase was formed at a high Mg content (alloy 7), both the ductility and tensile strength were decreased markedly.

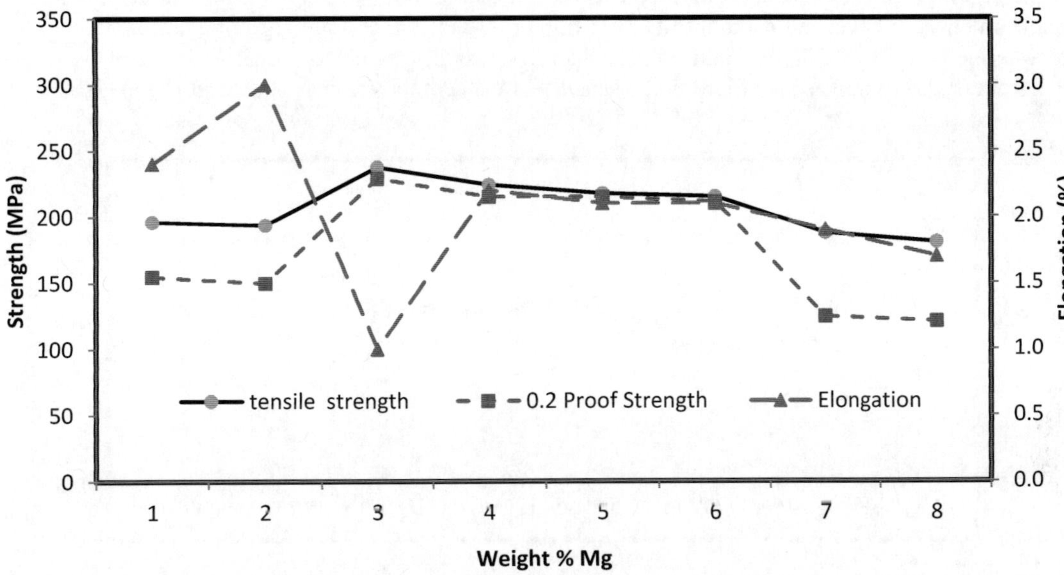


**Figure 2(a):** Variations of 0.2 Proof Strength, tensile strength and elongation with Mg content.

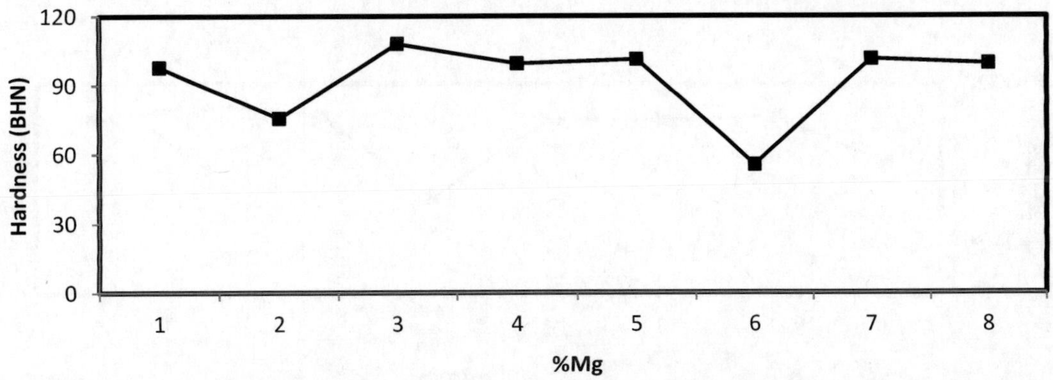


**Figure 2(b):** Variations of hardness with Mg content.

IOP Conf. Series: Materials Science and Engineering **438** (2018) 012011 doi:10.1088/1757-899X/438/1/012011

The solid fraction of the primary α-Al is a key factor which affecting the alloy ductility. The increase in yield strength was associated with decrease in elongation. When considering equilibrium condition, maximum solid solubility of Mg in Al is 17.4wt% and $Mg_2Si$in Al is 1.85wt% [10].It should be noted that when the α-Al/$Mg_2Si$ eutectic phase started to form, Mg atoms were rejected from the α-Al phase. Therefore, the combination between $Mg_2Si$ and extra Mg content helped to maximize the strengths with proper ductility. Same occurrence can also be found in Al-Si eutectic alloys, in which case the primary Si phase can be observed [11-13].

*3.3Effect of Zn Contents on the Mechanical Properties*

Figure 3 showed the effect of Zn addition on the mechanical properties of Al-$Mg_2$Si-Mg-Zn alloys. The addition of Zn resulted in an increase in proof strength and hardness and a decrease in ductility of the alloys, while the tensile strength remained mostly constant. The low ductility of high-zinc contained alloys was attributed to the coarse grain structure of these alloys [4]. From Figure 3(a) zinc was the 2nd highest (i.e., 5.45%), ductility was the lowest. Although 6.22% Zn(which was highest Zn content among the seven alloys),ductility was higher(2.4%) due to low Mg content (4.15%). When zinc was increased to 2.58, 4.91 and 5.45%, ductility was decreased to 3.0, 2.2 and 1, respectively. The hardness variation was also found to be sensitive to Zn composition as shown in Figure 3(b).

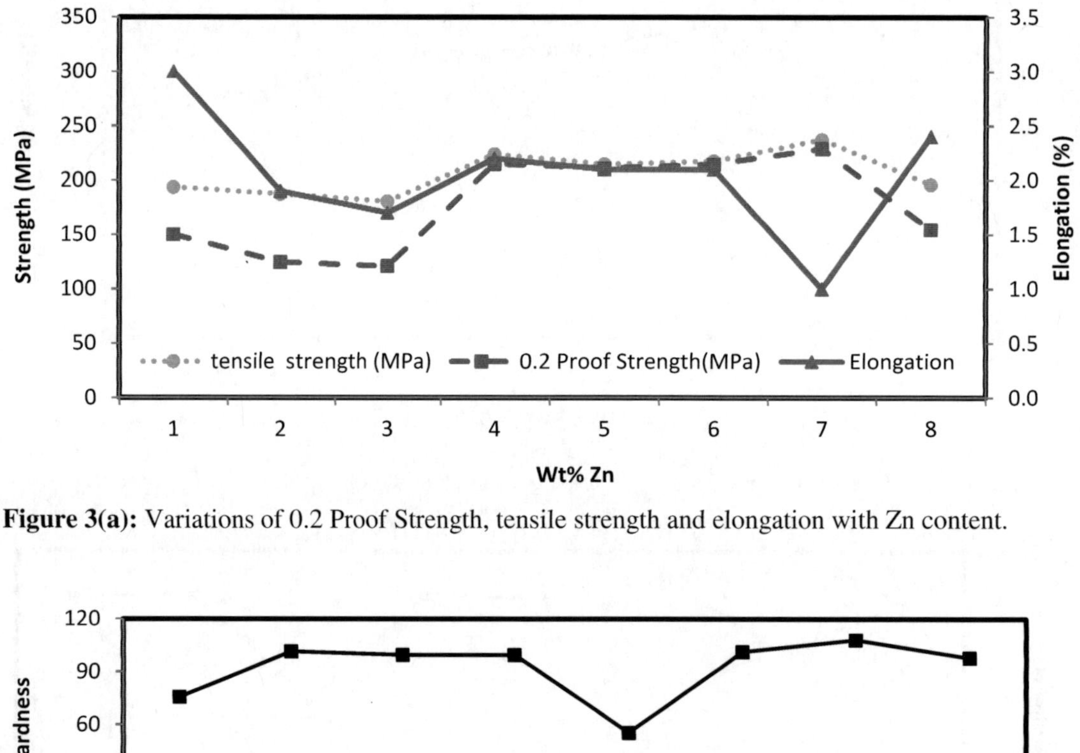


**Figure 3(a):** Variations of 0.2 Proof Strength, tensile strength and elongation with Zn content.

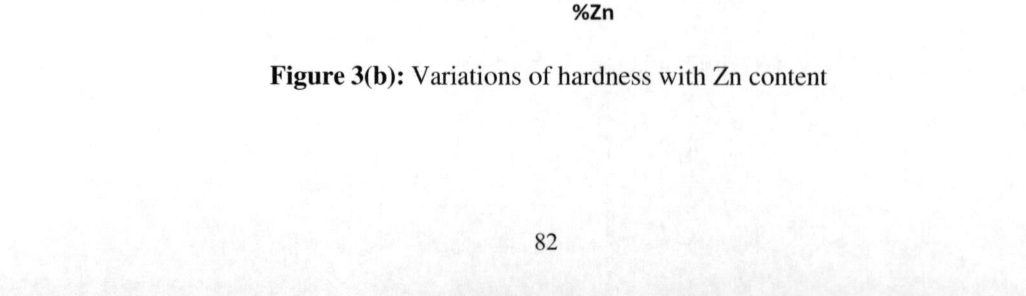


**Figure 3(b):** Variations of hardness with Zn content

When Zn is added to these alloys, there was no Zn-rich primary phase observed, even in the alloy with 4.3wt.% Zn. This was because Zn has a high solubility in aluminium in both liquid and solid states [14]. Zn addition to alloy 2 (5.45% Zn) was expected to react with the Mg and form the $Mg_2Zn$ phase, which improved the mechanical properties under as-cast conditions.

### *3.4 Effect of Mg on Microstructure*

Alloys having a lower Mg content showed predominantly coarse α-Al matrix in their structures. Few small Al-$Mg_2$Sieutectics were also seen. According to Feng Yan [5], increasing Mg can increase the Al-$Mg_2Si$ eutectics in the Al-$Mg_2Si$ alloy at the hypo-eutectic composition. With increasing Mg content, the microstructure consisted of fine primary α-Al phase surrounded by Al-$Mg_2Si$ eutectic phase. Upon further increase in Mg level, polygon shaped primary $Mg_2Si$ phase was appeared in the structure. It is clear from Figures 4(g) and 4(h) that most primary $Mg_2Si$ particles exhibited polygon shape when the Mg contents were 11.6%, such as hexagon, rectangle and trapezoid. However, with raising the Mg contentto 12%, as shown in Figure 4(h), the shape of primary $Mg_2Si$ particles changed to polygon shapes with hole [9]. The $Mg_2Si$ particles had an average size of 61.5 μm and were randomly distributed.

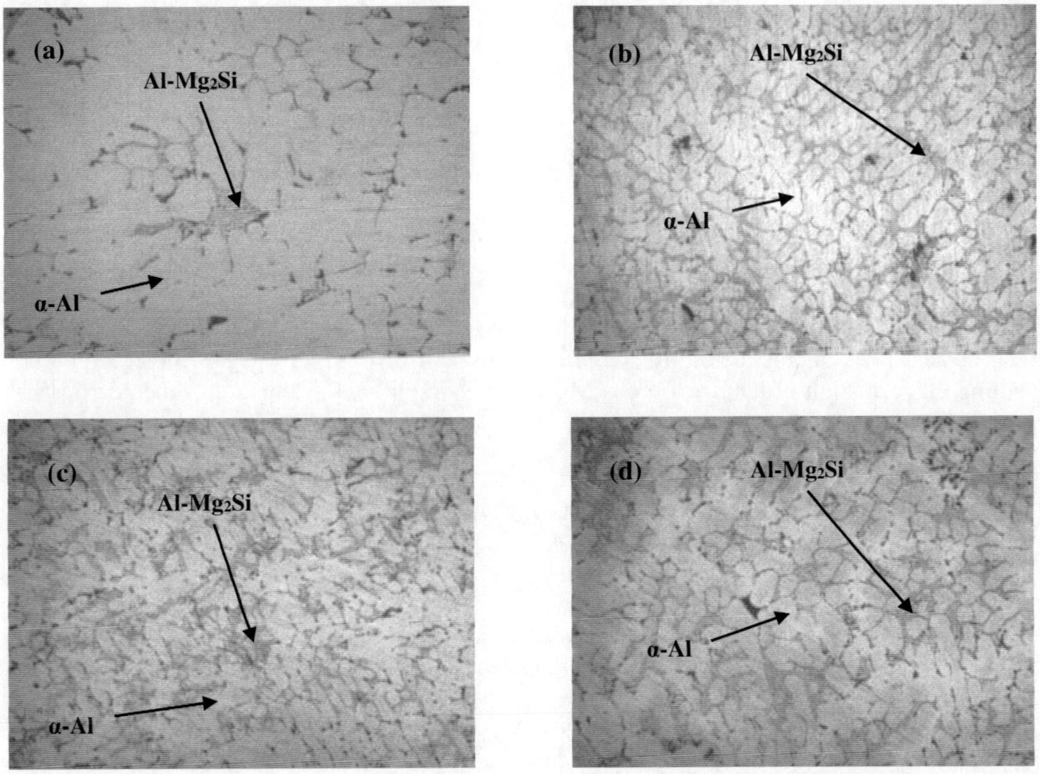


**Figure 4:** Optical micrographs of (a) alloy 1, (b) alloy 2, (c) alloy 3, (d) alloy 4.

## 4. Conclusion

Amongst the 8 alloys studied in this work, Al-5.45Zn-5.41Mg had the best combination of strength and ductility, offering 0.2 proof strength of 229MPa, tensile strength of 237MPa and elongation of 1%.Greater Mg and Zn contents in the alloys generally enhance strength and hardness and reduce

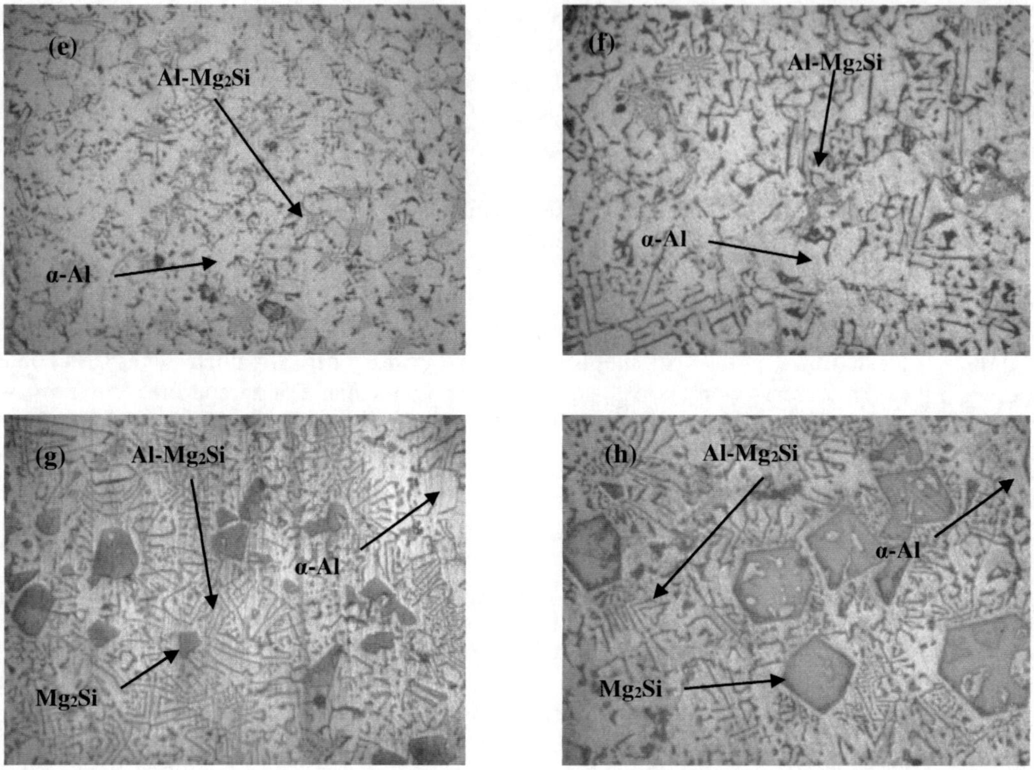


**Figure 4 (contd..):** Optical micrographs of (e) alloy 5, (f) alloy 6, (g) alloy 7, and (h) alloy 8.

ductility. A lower Si content could prevent the formation of the primary $Mg_2Si$ phase. With varied Mg and Zn contents, the microstructure of the alloys consisted of α-Al, Al- $Mg_2Si$, $Mg_2Si$ phases. By promoting the formation of laminar/fibrous $Mg_2Si$ eutectic phase, the strength of the Al-$Mg_2Si$-Mg-Zn based alloy was expected to increase.

## 5. Reference

[1] J. G. Kaufman, E. L. Rooy. *Aluminum Alloy Castings Properties, Processes, and Applications*, USA: ASM International; 2004. pp 3-48.

[2] D. Watson, *Microstructure and Mechanical Properties of Ductile Die-cast Al-Mg-Si-Mn Alloys,* 2015; Brunel University, London.

[3] N. R. Rathoda, J.V. Manghanib, *Effect of Modifier and Grain Refiner on Cast Al-7Si Aluminum Alloy: A Review.* International Journal of Emerging Trends in Engineering and Development. 2012; 5(2)

[4] W. K. Krajewski, J. Buraś, P. K. Krajewski, A. L. Greer, K. Faerber, P. Schumache, *New Developments of Al-Zn Cast Alloys.* Materials Today: Proceedings.2015; 2(10): 4978-4983

[5] S. Ji, F. Yan, Z. Fan, *Development of High Strength Al-$Mg_2Si$-Mg Based Alloy for High Pressure Die casting Process*. 2015; Brunel University, London.

[6] S. Ji, W. Yang, F. Gao, D. Watson, Z. Fan, *Effect of iron on the microstructure and mechanical property of Al–Mg–Si–Mn and Al–Mg–Si diecast alloys.* Materials Science & Engineering: A. 2013; 564: 130–139.

[7] E. E. Schmid, K. V. Oldenburg,G. Frommeyer, *Microstructure and properties of as-cast intermetallic $Mg_2Si$-Al alloys*. Zeitschrift für Metallkunde.1990; 81.

[8] S. Ji, D. Watson, Z. Fan , M. White, *Development of a super ductile die cast Al–Mg–Si alloy.* Materials Science and Engineering: A. 2012; 556: 824-833.

[9] Z-L Huang, K. Wang, Z-M. Zhang, B. LI, H-S. Xue, D-Z. Yang. *Effects of Mg content on primary $Mg_2Si$ phase in hypereutectic Al−Si alloys.* Transactions of Nonferrous Metals Society of China.2015; 25(10): 3197-3203.

[10] G.E. Totten, D.S. MacKenzie, editors. *Handbook of Aluminium: Vol. 1: Physical Metallurgy and Processes,* USA: CRC Press; 2003

[11] D. C. Tsai, L. H. Chen, T. S. Lui, *Effect of Silicon Particles on the EDM Characteristics of Al-Si Alloys.* Materials transactions, JIM.2002; 43(2): 199.

[12] X. Jian, Q. Han, *Formation of Hypereutectic Silicon Particles in Hypoeutectic Al-Si Alloys Under the Influence of High-Intensity Ultrasonic Vibration.* China Foundry, 2013; 10(2): 118-123.

[13] S. Wang, R. Ma, Y. Z. Wang, L. Y. Yang, *Growth Mechanism of Primary Silicon in Cast Hypoeutectic Al-Si Alloys.* Transactions of Nonferrous Metals Society of China, 2012; 22(6): 1264-1269.

[14] H. Löffler, D. Bergner, *Structure and Structure Development of Al-Zn Alloys*, Berlin: Akademie; 1995.

**Acknowledgments**

The financial support from the Department of Materials and Metallurgical Engineering of Bangladesh University of Engineering and Technology (BUET) is gratefully acknowledged.

IOP Conf. Series: Materials Science and Engineering **438** (2018) 012012 doi:10.1088/1757-899X/438/1/012012

# Theoretical and Experimental Evidence of Modified Structure, Magnetism and Optical Properties in Ba and Mn Co-Substituted $BiFeO_3$

**Tilottoma Saha[1], Fabia Farlin Athena[1], M A Matin[2] and Ahmed Sharif[1]**

[1]Department of Materials and Metallurgical Engineering, Bangladesh University of Engineering and Technology, Dhaka 1000, Bangladesh
[2]Department of Glass and Ceramic Engineering, Bangladesh University of Engineering and Technology, Dhaka 1000, Bangladesh

**Abstract.** To improve the vacancy related defects associated with pure $BiFeO_3$ (BFO) different Ba-Mn co-doped samples ($Bi1_{-x}Ba_xFe_yMn_yO_3$; x=0.1; y=0.0, 0.05, 0.1, 0.15) were prepared from sol-gel method. The influence of co-doping on structural, magnetic and optical properties of $BiFeO_3$ has been studied in different experimental conditions. Particle size and magnetic properties were measured from SEM and VSM analysis respectively. Again band gap was calculated from UV analysis. The co-doped samples showed reduced particles size as well as better surface morphology. M-H loops suggest that the presence of $Mn^{2+}$ increased ferromagnetic quantity at some levels. Still room temperature occurrence of ferromagnetism obtained from M-H loops of all samples showed weak ferromagnetic behaviour. In addition the incorporation of $Ba^{2+}$ at $Bi^{3+}$ site and $Mn^{2+}$ at $Fe^{3+}$ site decreased the band gap which was evidenced by increased optical property.

**Keywords**: Multiferroics, BFO, Ferromagnetism, Band gap, Surface morphology

## 1. *Introduction*

Multiferroic materials are those which can exhibit more than one ferroic order simultaneously such as ferroelectricity, ferromagnetism and ferroelasticity. It is possible to control the degree of magnetism using electric field and vice versa. It is only possible due to their magnetoelectric coupling. In recent years these materials have achieved a great deal of attention due to their multidimensional applications such as data storage media, spintronics and multistate memories. Bismuth ferrite (BFO) is one of the most promising multiferroic materials that can exist in room temperature. It shows ferroelectric ordering below curie temperature (1103K) and space-disproportionate antiferromagnetic property below the neel temperature(643K) [1]. It also offers a coupling interaction between polarization and magnetization called magnetoelectric coupling. However $BiFeO_3$ shows weak ferromagnetic property at room temperature because of the combined action between exchanges and spin orbit interactions. For this reason a vanishing magnetization is obtained in BFO by spin canting away from perfect antiferromagnetic ordering and spiral spin arrangement. From the last decade's vigorous research documents,it can be inferred that, the properties of BFO can be tailored towards an ideal multiferroic material through suitable doping. Co-doping can be an effective way to tailor the multiferroic properties of BFO [2]. In this research BFO is synthesized using modified sol-gel method technique and the influence of co-doping (Ba-Mn) on structural, ferroelectric, ferromagnetic and optical properties of BiFeO3 have been studied in different experimental condition [3]. Recently a large number of investigations have already been conducted on $BiFeO_3$ co-doped with various element combinations in which Ba-Mn co-doping shows very promising result. In this paper, some surprising findings are revealed for Ba-Mn co-doped BFO.

## 2. Experimental

### 2.1. Materials and Method

Co-doped $Bi_{1-x}Ba_xFe_{1-y}Mn_yO_3$ [x=0.1; y=0, 0.5, 0.10, 0.15] samples were prepared by modified sol-gel method. Weighed amount of $Bi(NO_3)_3.5H_2O$ (Merck[@], India), $Fe_2(NO_3)_2.9H_2O$ (Merck[@], India), $Ba(NO_3)_2.6H_2O$ (Merck[@], India), $Mn(NO_3)_2.4H_2O$ (Sigma Aldrich, USA) were dissolved in 250ml deionized water under continuous stirring at 70°-75°C. The reaction media was made acidic by using citric acid. In the next step 5ml Ethylene glycol was added as chelating agent. After turning the solution blackish-red ethylene glycol was again added as polymerization agent and stirring kept continuous until the formation of dried gel. The temperature of the dried gel was further slightly increased only to be ignited at any point of time. After ignition, the dried gel began to burn at a self-propagating combustion reaction with vigorous flames until the gel was completely burnt out to form ash like flakes. Then these flakes were dried out at 120°C for 24hr. After that these flakes were neatly collected and collapsed by using spatula or glass rod to obtain the fluffy fine loose powder of BBFO or BBFMO nanoparticles. After that samples were calcined at 600°C for 2hrs and annealed at 500°C with a heating and cooling rate 3°C/min to remove any impurity and improved crystallization.

### 2.2. Characterization

To observe the morphology of the synthesized powder, field emission scanning electron microscope, FESEM (JEOL JSM 7600F) was used. Powder samples were adhered with carbon tape that was attached to a Copper Stub. Samples were made conductive, coated with platinum by ion sputtering method. Then sample holder was mounted on a holder and inserted in FESEM. The micrographs obtained from FESEM analysis were used for microstructural study. Energy dispersive X-ray spectroscopy (EDS) and SEM were simultaneously allowed for targeted analysis of sample surfaces. X-ray diffraction analysis was conducted using the XRD (3040-X"Pert PRO, Philips) facility in order to determine the phases present in the synthesized powders. The high intensity Cu K-α radiation was focused on the sample in the scanning from 10° to 90° [4]. The magnetization response of the samples with changing magnetic field i.e. the M-H loop was determined using a vibrating sample magnetometer (VSM: Model EV-9, Microsense LLC, USA). A small piece of the sample (15-20mg) was broken off and put in the instrument after weighing. VSM was used to determine the magnetic property of the synthesized powders up to applied field of 21 kOe. The band gap refers to the energy difference between the top of the valance band to the conduction band [5]. Band gap was determined from UV-Vis absorption value where energy and wavelength were calculated.

## 3. Results and discussion

### 3.1. XRD Analysis

Figure 1 showed the XRD patterns of 10 mol% Ba and 0, 5, 10, 15 mol% Mn Co-doped BFO samples. All the prominent peaks in the XRD patterns confirmed single phase formation in all the Mn doped BBFO nanoparticles without any trace of impurity phase. From the XRD peaks it was observed that, co substitution of both A and B site atom didn't change the parent structure of the BFO samples. Some peaks near 25°-30° represented γ- $Bi_2O_3$ phase in 0 mol% Mn doped sample while the others were in good agreement with the BFO reference profile. It was very difficult to obtain pure BFO nanoparticles.

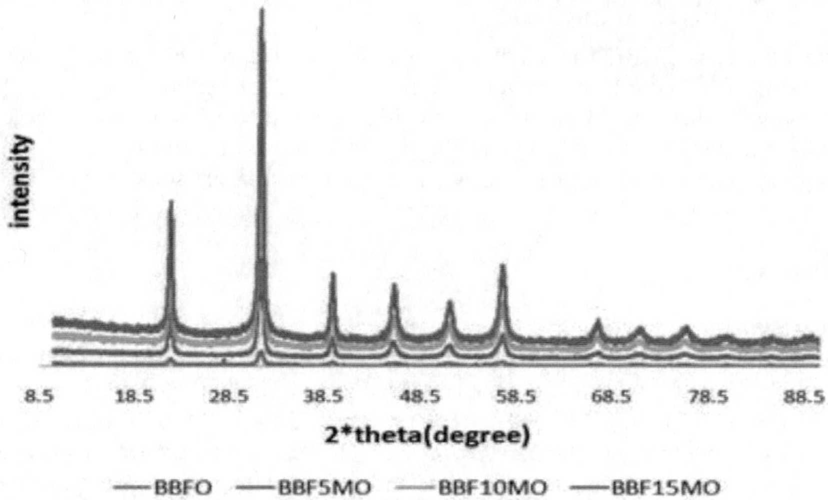


**Figure 1** XRD pattern of Ba and Mn doped BFO

due to kinetics of formation. However, it could be seen from Figure 1 that the highest peaks of [104] and [110] at around 30° appeared overlapped, which was a common phenomenon in BFO nanoparticles. The A-site substitution by barium ion ($Ba^{2+}$) has been attempted by many authors in recent years to enhance the multiferroic properties of $BiFeO_3$. $Ba^{2+}$having a large ionic radius than $Bi^{3+}$ distorted the original structure of BiFeO3 and changes it's cycloidal spin structure to a canted spin structure resulting in net magnetization at room temperature [6]. Moreover, addition of Ba also significantly reduced the oxygen vacancy related defects (leakage current) by occupying evaporated Bi sites and increased the electrical resistivity of BFO by controlling grain size. Further incorporation of Mn in the Fe site in BFO had been marked by a slight decrease in the impurity phase peaks near 27°, 32° or 35° due to reduce chances of bismuth evaporation.[7]

*3.2 Microstructural analysis:*

Figure 2 showed the FESEM image of BBM-4 samples with EDS analysis sintered at 600°C. It revealed the dense and uniform distribution of particles having grain size near 100nm had been obtained by sol-gel method. EDS analysis proved the successful doping of Ba and Mn from the composition of the product. In the doped sample the amount of Ba and Mn detected by EDS analysis was a little less (table 1) than those actually added during the synthesis of the sample but an approximate idea can be developed from these values. Increase of Mn as well as decrease of Fe content told that co-doping took place for the samples [8].

**Figure 2:** SEM images of (a) 0% (b) 5% (c) 10% and (d) 15% Mn doped BBFO.

The introduction of dopants had a little effect on the morphology of the particles [6, 7]. The particles after sintering became porous due to the decomposition of citric acid, and the diameter shrank to about

100 nm. From the FESEM micrographs of the sintered pellets, it was clear that the average grain size of pure BFO was relatively large; whereas comparatively smaller sizes of grains were observed in the co-doped samples [9]. It had been observed that grain growth was depended on the concentration of oxygen vacancies and diffusion rate of the ions. In pure BFO, evaporation of Bi (highly volatile) generated large amount of vacancies; which made it easy for the ions to diffuse and thus resulted in larger sized grains [10]. Microstructure of the pellets was overall dense; but few scattered pores were observed. The different Ba-Mn content had pronounced effect on average grain size, density and homogeneity of the grains with increasing Ba-Mn content, the grain size of the doped samples had a tendency of decreasing, which was due to the inhibition of grain growth brought by Ba-Mn codoping. From the SEM analysis it is observed that the minimum grain size was observed for Mn content 10% but with the increase of Mn content oxygen vacancies were formed for charge compensation and grain size increased at 15% Mn content [11].

**Table 1.** EDS observed components quantity

| Element | Wt% (0%Mn) | Wt% (5%Mn) | Wt % (10%Mn) | Wt % (15%Mn) |
|---|---|---|---|---|
| O | 4.28 | 3.10 | 3.76 | 4.20 |
| Mn | 0 | 1.89 | 2.23 | 3.56 |
| Fe | 28.41 | 17.53 | 16.28 | 15.54 |
| Ba | 6.55 | 5.85 | 4.03 | 5.07 |
| Bi | 60.77 | 71.63 | 73.71 | 71.63 |

*3.3 Magnetic property*

The magnetic hysteresis (M-H) loops of Ba and Mn doped BFO nanoparticles measured at room temperature were shown in Figure 3 which indicated soft ferromagnetic behavior of the samples. The magnetization increased with increasing the Mn content in BBFO samples. Remnant magnetization (2Mr) of 0.124, 0.217, 0.259, 1.203 emu/gm were observed for 0, 5, 10, 15% Mn doped BBFO samples respectively. The coercive field (2Hc) and saturation magnetization (Ms) were shown in the table 2. With the increase of Mn content, larger magnetization at room temperature was observed. The origin of spontaneous magnetization was due to crystal structure of BFO sample [12]. With the decrease of particle size, the helical order in the spin can be suppressed. It had been already been theoretically proven that suppression of helical order with a periodicity about 62nm or canted antiferromagnetic order between successive (III) ferromagnetic planes, might have resulted higher magnetization. In this way, with the increase of Mn content, distortion was created in the rhombohedral structure and magnetization was increased [13]. Nevertheless, an enhancement in ferromagnetic properties of BFO with increasing Ba concentration may be expected due to increasing structural distortion. However, the substitution of $Ba^{2+}$ for $Bi^{3+}$ may cause two parallel phenomena with respect to concentration of oxygen vacancies, one was decrease in concentration of oxygen vacancies by filling the probable vacant volatized $Bi^{3+}$ sites and the second was the creation of oxygen vacancies to neutralize the charge produced by substituting $Ba^{2+}$ for $Bi^{3+}$ [14]. Hence, leakage current of BFO decreased up to an optimum Ba doping and then again increased on further doping. It had been reported that, above 0.2 mole fraction $Ba^{2+}$ addition (x=0.2) the leakage current effect became prominent again for this reason $Ba^{2+}$ content remained fixed (x=0.1) in this research while percentages of $Mn^{2+}$content were changed (y=0, 0.05, 0.1, 0.15).Generally BFO showed weak ferromagnetic property due to distorted rhombohedral pervoskite structure. With the Mn doping, distortion was increased in the BFO lattice which was confirmed by morphology analysis. However addition of $Mn^{2+}$ion to the BFO required charge compensation which may be achieved by the formation of $Fe^{4+}$ or oxygen vacancies. If $Fe^{4+}$ exists, the statistical distribution of $Fe^{3+}$ and $Fe^{4+}$ ions in octahedral may also lead to the net magnetization and ferromagnetism [15]. The increased magnetization in BBM samples may be attributed to the structural phase transition from rhombohedral to orthorhombic phase resulting in destruction of spiral spin structure which in turn led to release of latent magnetization. The helical order in nanocrystals can be well suppressed through decreasing the particle size which actually caused higher magnetization. The remnant magnetization increased with the increase in the substitution concentration and reached a maximum at y=0.05, which indicated almost all broken spin cycloid structure. The further increase in substitution concentration resulted in the reduction of remnant magnetization due to the appearance of complete antiferromagnetic ordering in the orthorhombic structure because of the significant contribution from the crystallographic phase of Pbnm space group. The substitution and other effects can break the cycloid in BFO, which had been reported to affect the magnetic properties of the ceramics but presence of magnetic impurities ($Fe_2O_3$ or $Fe_3O_4$) may be the cause for the same. In our case, the presence of $Fe_2O_3$ magnetic impurity had not been detected from XRD patterns. Moreover, we had not observed ferromagnetic hysteresis loop for BFO which had largest amount of impurity phase among all ceramics. Also, the area of ferromagnetic hysteresis loop and coercivity increased with the increase in substitution percentage up to 5% and then decreased due to appearance of complete antiferromagnetic ordering in the orthorhombic structure.

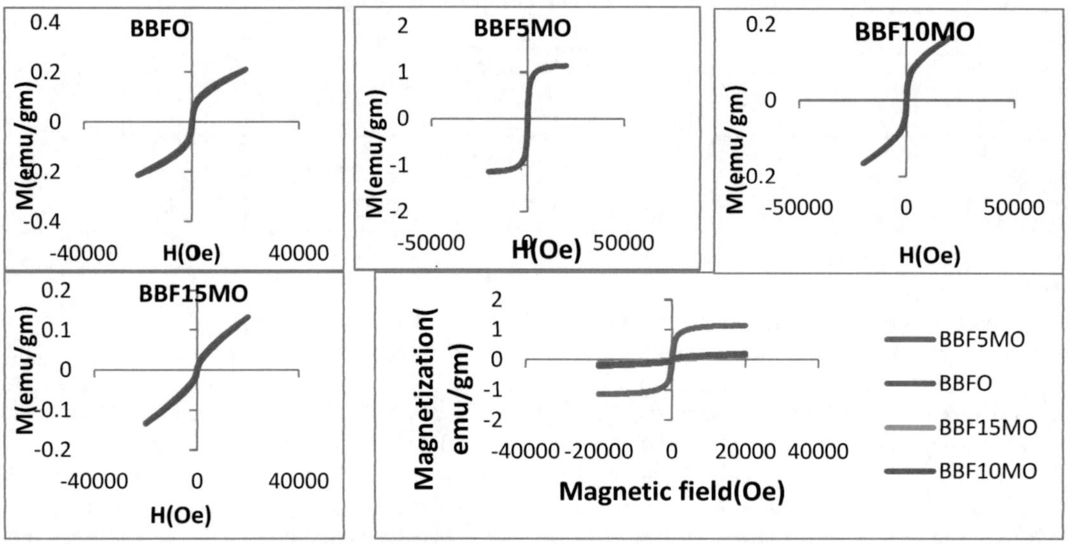


**Figure 3:** Magnetic property of different amount of Ba and Mn on magnetization of BFO samples.

**Table 2** The magnetic parameters for Ba and Mn doped BFO samples

| Sample (BBFO) | Coercive field, 2Hc (Oe) | Remnant magnetization, 2Mr (emu/gm) | Spontaneous magnetization, Ms (emu/gm) |
|---|---|---|---|
| 0 mol% Mn | 2100 | 0.08 | 0.23 |
| 5 mol% Mn | 2100 | 0.487 | 1.203 |
| 10 mol% Mn | 2100 | 0.073 | 0.17 |
| 15 mol% Mn | 2100 | 0.02 | 0.14 |

*3.4 Photo catalytic behavior analysis*

We studied the influence of grain and particle size on structural and optical properties of $BiFeO_3$ nanoparticles and resulting photo catalytic activity. The smaller particle size in most cases would give rise to a larger surface area and stronger adsorption capacity, which was usually beneficial for photocatalytic activity. Unexpectedly, the photocatalytic activity was found to decrease while the expected surface reactions were decreased by increasing particle size [16]. Optical features, namely the Urbach energy and low energy bands, in the absorption spectra were substantially changed. We argued that these optical modifications related to defects and local distortions were mainly affected at the skin layer that was inherent to oxides like BFO. By reducing the particle size of BFO nanoparticles the skin layer thus altered, which in turns changed the photocatalytic properties. The optical absorption of the Ba and Mn doped BFO nanoparticle was characterized by measuring the UV–Vis diffuse reflectance absorption spectra of samples. Results were presented in samples doped with Ba and Mn have strongly enhanced absorption in visible light, especially 10 % Ba and 10%Mn doped sample, which may be attributed to the induced defects and phase structure changed. But for 15% Mn the effect was gradually decreased due to formation of excessive oxygen vacancy formation that was related to increase the size of BFO nanoparticle size [17].

*3.5 Band gap analysis*

The UV–vis absorption spectrum of as-synthesized BFO nanoparticles was shown in Figure 4. This was seen from the spectrum that BFO nanoparticles had strong absorption in UV–vis region 300–550 nm and slight absorption in 550–700 nm. This absorption characteristic can be used in the photo-catalytic activity in decomposition of organic compounds. The absorption curve gave an absorption edge at 510 nm. The inset of photocatalytic activity showed $(\alpha h\upsilon)^2$–$h\upsilon$ plot for the direct band gap calculation. For any semiconductor, optical absorption near band edge obeys following equation,

$$(\alpha h\upsilon) \propto (h\upsilon - Eg)^n \;\ldots\ldots\ldots\ldots\ldots\ldots (1)$$

where $\alpha$, $h$, $\upsilon$ and $E_g$ are the optical absorption coefficient, Plank's constant, light frequency and band gap energy, respectively. For the direct band gap, the variable '$n$' in the equation is equal to 1/2. As BFO was a direct band gap semiconductor, the extrapolated line intercept at horizontal axis gave the value of direct energy band gap as 1.665 eV with the increase of Mn content. The band gap value was comparable with the earlier reported values. Also, the band gap of BFO at bulk level was reported to be 2.6 eV, which decreased to 1.665 eV for nanoparticles due to the size effect. To investigate the influence of Ba and Mn doping on the optical absorption of BFO, the UV-vis spectra of pure BBFO and Mn-doped BBFO samples were measured at room temperature, as shown in Figure 4.The absorption band edge of BBFO nanoparticles appeared, indicating that BBFO could respond to visible light for photocatalytic reaction. Compared to pure BFO, the Ba and Mn co-doped BFO samples exhibited enhanced absorption capability especially in the visible light region, and the absorption intensity became gradually stronger as increasing the Mn dopant content.[18] Moreover, the band gap could be calculated from the plot of the Kubelka-Munk function ($(\alpha h\nu)^2$ vs photon energy ($h\nu$)) for the direct band gap semiconductor, as presented in the inset of Figure 4. The band gaps were estimated to be 1.725, 1.7, 1.665, 1.69 eV for the pure BBFO, Mn 5%-BFO, Mn 10%-BFO and Mn 15%-BBFO samples, respectively. With the increase of the Mn dopant content, the band gap of doped samples was gradually decreased, in turn leading to a higher absorption capability. Undoubtedly, the enhanced absorption property of Ba and Mn co-doped BFO would probably improve the photocatalytic activity of BFO, as shown in Figure 4.

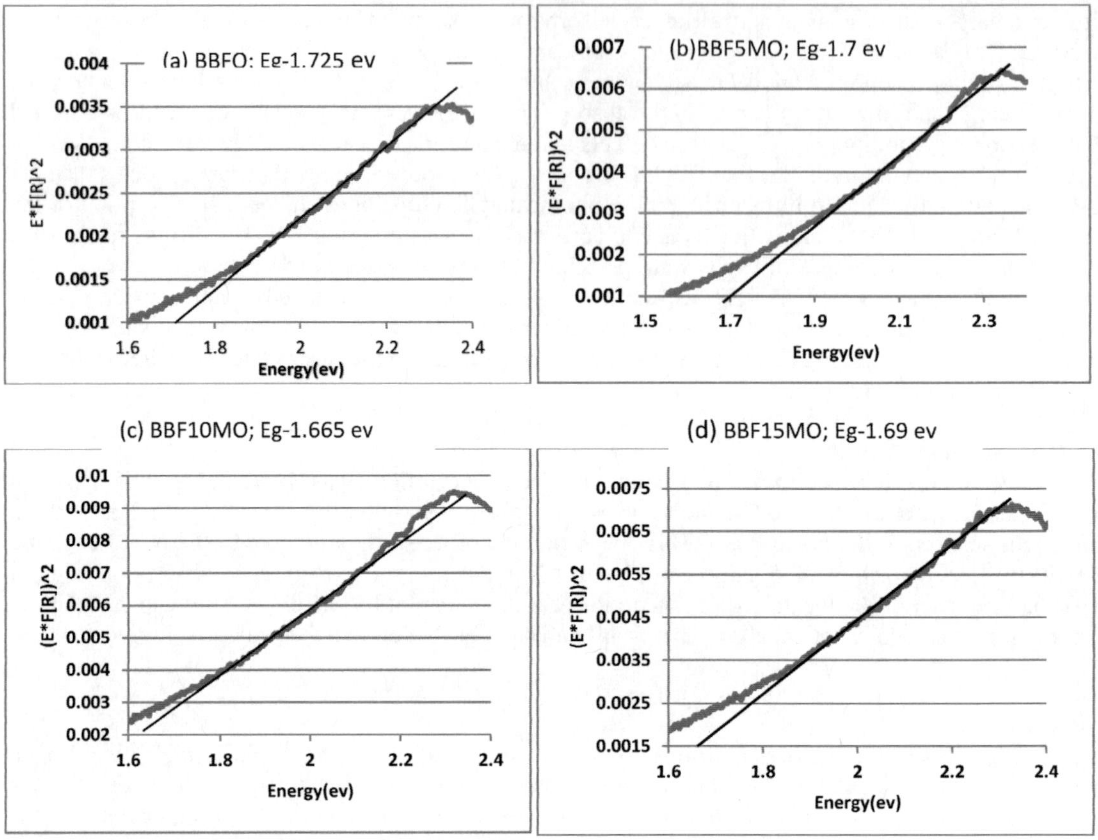


**Figure 4:** Photo-catalytic response of (a) 0%, (b) 5%, (c) 10% and (d) 15% Mn doped BBFO.

### *4. Conclusion*

Ba-Mn co-doped $BiFe0_3$ nanoparticles were successfully synthesized by modified sol-gel method. The magnetic property value of pure BFO was greatly increased due to the size effects and (Ba, Mn) ions doping. Satisfying ferromagnetic performance was achieved due to the suppression of helical order at room temperature. The grain size decreased dramatically with increasing doping concentration for $Bi_{1-x}Ba_xFe_{1-y}Mn_yO_3$ up to Mn content 10%. The grain size of the Ba-Mn doped BFO samples are found to be in the range of 100-120 nm. Absorption increased with the increases of Mn content up to 10% but decreased at 15% due to increase of the size of the particle. Band-gap analysis also agreed with the absorption phenomenon and it was minimum for $Bi_{0.9}Ba_{0.1}Fe_{0.9}Mn_{0.10}O_3$ sample and it increased further for increasing the Mn content due to increase of the grain size.

### 5. Reference

[1] Kuo, C Y, Hu, Z, Yang, J C, Liao, S C, Huang, Y L, Vasudevan, R K, Okatan, M B, Jesse, S,Kalinin, S V, Li, L and Liu, H J, 2016. Single-domain multiferroic BiFeO 3 films. *Nature communications*, 7, p 12712

[2] Wu, R, 2006 Chapter 2 Fundamental properties of magnetic nanostructures: a survey *Contemporary Concepts of Condensed Matter Science Nanomagnetism: Ultrathin Films, Multilayers and Nanostructures*, p 29

[3] Chauhan, S, Kumar, M, Chhoker, S, Katyal, S C, Singh, H, Jewariya, M and Yadav, K L, 2012 Multiferroic, magnetoelectric and optical properties of Mn doped BiFeO3 nanoparticles *Solid State Communications* 152(6), p 525

[4] Kumar, P and Kar M, 2014 Effect of structural transition on magnetic and optical properties of Ca and Ti co-substituted BiFeO3 ceramics *Journal of Alloys and Compounds*, 584, p 566

[5] Zhang, N, Chen, D, Niu, F, Wang, S, Qin, L and Huang, Y, 2016 Enhanced visible light photocatalytic activity of Gd-doped BiFeO 3 nanoparticles and mechanism insight *Scientific reports*, 6, p 26467

[6] Basu, S, Hossain, S K M, Chakravorty, D and Pal, M, 2011 Enhanced magnetic properties in hydrothermally synthesized Mn-doped BiFeO3 nanoparticles *Current Applied Physics*, 11(4), p 976

[7] Fiebig, M, Lottermoser, T, Meier, D and Trassin, M, 2016 The evolution of multiferroics *Nature Reviews Materials*, 1(8), p 16046

[8] Alikin, D O, Turygin, A P, Walker, J, Bencan, A, Malic, B, Rojac, T, Shur, V Y and Kholkin, A L, 2017 The effect of phase assemblages, grain boundaries and domain structure on the local switching behavior of rare-earth modified bismuth ferrite ceramics *Acta Materialia*, 125, p 265

[9] Srinivas, V, Raghavender, A T and Kumar, K V, 2016 Structural and magnetic properties of Mn doped BiFeO3 nanomaterials Physics *Research International*, p 2016

[10] Gao, X, Rodriguez, B J, Liu, L, Birajdar, B, Pantel, D, Ziese, M, Alexe, M and Hesse, D, 2010 Microstructure and properties of well-ordered multiferroic Pb (Zr, Ti) O3/CoFe2O4 nanocomposites. *Acs Nano*, 4(2), p 1099

[11] Khomskii, D, 2009 Trend: Classifying multiferroics: Mechanisms and effects. *Physics*, 2, p.20.

[12] Layek, S, Verma, H C and Garg, A, 2015 Enhancement in magnetic properties of Ba-doped BiFeO3 ceramics by mechanical activation *Journal of Alloys and Compounds*, 651, p 294

[13] Huang, J Z, Wang, Y, Lin, Y, Li, M and Nan, C W, 2009 Effect of Mn doping on electric and magnetic properties of BiFeO 3 thin films by chemical solution deposition *Journal of Applied Physics*, 106(6), p 063911

[14] Rout, J and Choudhary, R N P, 2016 Structural transformation and multiferroic properties of Ba–Mn co-doped BiFeO3 *Physics Letters A*, 380(1-2), p 288

[15] Das, R and Mandal, K, 2012 Magnetic, ferroelectric and magnetoelectric properties of Ba-doped BiFeO3 *Journal of Magnetism and Magnetic Materials*, 324(11), p 1913

[16] Wang, H C, Lin, Y H , Feng, Y N and Shen, Y, 2013 Photocatalytic behaviors observed in Ba and Mn doped BiFeO 3 nanofibers *Journal of Electroceramics*, 31(1-2), p 271

[17] Catalan, G and Scott, J F, 2009 Physics and applications of bismuth ferrite *Advanced Materials*, 21(24), p 2463

[18] Kolte, J, Daryapurkar, A S, Agarwal, M, Gulwade, D D and Gopalan, P, 2016 Effect of substrate temperature on the structural and electrical properties of La and Mn co-doped BiFeO3 thin films *Thin Solid Films*, 619, p 308

## Acknowledgments

We acknowledge financial support from Department of materials and metallurgical engineering, Bangladesh University of Engineering and Technology (BUET). We are very much thankful to some of our seniors who have given me some important advices. We are also grateful to the laboratory assistants for their support.

IOP Conf. Series: Materials Science and Engineering **438** (2018) 012013 doi:10.1088/1757-899X/438/1/012013

# Study of the Structural and Magnetic Properties of Li Substituted Cu-Mn Mixed Ferrites

**M A M Quraishi and A K M Akther Hossain**

Physics Department, Bangladesh University of Engineering and Technology, Dhaka-1000, Bangladesh

**Abstract:** A very fast auto-combustion reaction technique was applied for the preparation of a series of $Li_xCu_{0.12}Mn_{0.88-2x}Fe_{2+x}O_4$ ferrites in a wide range of composition (with x = 0.00, 0.10, 0.20, 0.30, 0.40 and 0.44). Disc and toroid shaped samples were prepared and sintered at various temperatures, in namely, 1200, 1250, 1300 °C in air for 1 hour. The spinel structure of all these samples were confirmed by X-ray diffraction and grain size estimation obtained from the optical microscope. It has been found that the lattice constant increases linearly with increasing Li contents. The grain size increases with increasing Li content and sintering temperature in $Li_xCu_{0.12}Mn_{0.88-2x}Fe_{2+x}O_4$. The magnetic properties of these compositions are characterized by high frequency (10 KHz-120 MHz) magnetic permeability measurements. The real part of initial permeability, $\mu_i^/$ increases with increasing Li content. It has also been observed that $\mu_i^/$ for each sample increases with sintering temperatures because of uniform grain growth. The relative quality factor, Q increases with increasing sintering temperature. Among all the samples, the highest value of Q-factor has been observed for x = 0.40 sintered at 1300 °C.

## 1. Introduction

Ferrites have been considered as extremely vital electronic materials for the greater part a century. It turns into the attractive research field for many researchers and advancement in this field is as yet going ahead, from basic as well as application perspective. The spinel ferrites are compound of iron oxides and transition metal oxides [1]. The physical properties of spinel ferrites are controlled by the preparation conditions, chemical compositions, sintering temperatures and amount of substitution [2]. In particular, lithium ferrite or mixed lithium ferrites have high potential for microwave applications, especially as a replacement of garnets, because of their low cost, squareness of the hysteresis loop, high Neel temperature ($T_N$), high resistivity, low magnetic and dielectric losses [3]. The Li-based ferrite is a standout amongst the most adaptable attractive materials for common use having numerous applications in both low and high frequency devices. It plays a crucial role in various technological applications, for example, microwave device, power transformer, rod -antennas, read-write heads for fast digital tapes [4, 5].

Recently, in our laboratory Hossain et. al [6] have studied the effect of Li substitution on the magnetic properties of $Li_xMg_{0.40}Ni_{0.60-2x}Fe_{2+x}O_4$ ferrites, prepared by the standard solid state reaction technique. It was observed that, the studied spinel ferrites are in ferrimagnetic state at room temperature and have a low saturation field. The real part of initial permeability ($\mu_i^/$) and relative quality factor (Q-factor), both found to be maximum at x = 0.25 for 1100 °C. On the other hand, Hoque et. al. [7] have worked on Cu substituted Li-ferrite and reported that, the structural properties are strongly dependent on the sintering temperature due to the Jahn-Teller distortion of $Cu^{2+}$. An increase in initial permeability has been observed with the increase in Cu content but the resonance frequency shifts towards the lower frequency. Many researchers have investigated Li-Zn-Cu-Mn [8], Li-Cu-Co-Zn [4], Li-Mg-Mn [9], Li-Ni-Zn [10] and Li-Mn-Zn [11], ferrites. To the best of our knowledge, no published literature is available by other researchers on Li substituted $Li_xCu_{0.12}Mn_{0.88-2x}Fe_{2+x}O_4$ prepared by combustion technique. Therefore, there has been a growing eagerness on Li-substituted Cu-Mn ferrite for microwave applications and high permeability with low magnetic loss. In the present work, the $Li_xCu_{0.12}Mn_{0.88-2x}Fe_{2+x}O_4$ has been prepared through auto-combustion technique and sintered at various temperatures. The structural and magnetic properties have been studied as a function of compositions and temperatures.

## 2. Experimental

The Polycrystalline $Li_xCu_{0.12}Mn_{0.88-2x}Fe_{2+x}O_4$ (x = 0, 0.10, 0.20, 0.30, 0.40 and 0.44) ferrites were prepared by auto-combustion technique. The analytical grade of $LiNO_3$, $MnCl_2.4H_2O$, $Cu(NO_3)_2.3H_2O$ and $Fe(NO_3)_3.9H_2O$ were weighed according to the stoichiometric amount and dissolved in ethanol. The mixture was placed in a magnetic heating stirrer at 80 ºC, followed by an ignition, the combustion takes place within a few seconds and fine powders were precipitated. These powders were crushed and ground thoroughly. The fine powders of various compositions were then calcined at 900 °C for 5 h for the final formation of $Li_xCu_{0.12}Mn_{0.88-2x}Fe_{2+x}O_4$ ferrites. Then the fine powders are granulated using Polyvinyl Alcohol (PVA) as a binder and pressed into disk and toroid-shaped samples. The samples were sintered at various temperatures, in namely, 1200, 1250, and 1300 ºC in air for 1 hour.

## 3. Results and Discussions

### *3.1 Structural Properties*

X-ray diffraction was carried out with an X-ray diffractometer (Model: D8 Advance, Bruker AXS) for each sample. The X-ray diffraction (XRD) was performed to verify the formation of spinel structure of various $Li_xCu_{0.12}Mn_{0.88-2x}Fe_{2+x}O_4$ ferrites, in which $Mn^{2+}$ is replaced by $Li^+$ and $Fe^{3+}$. The XRD patterns of these $Li^+$ substituted $Li_xCu_{0.12}Mn_{0.88-2x}Fe_{2+x}O_4$ (with x = 0.00, 0.10, 0.20, 0.30, 0.40 and 0.44) ferrites sintered at 1200 °C in air for 1 h are shown in figure 1.

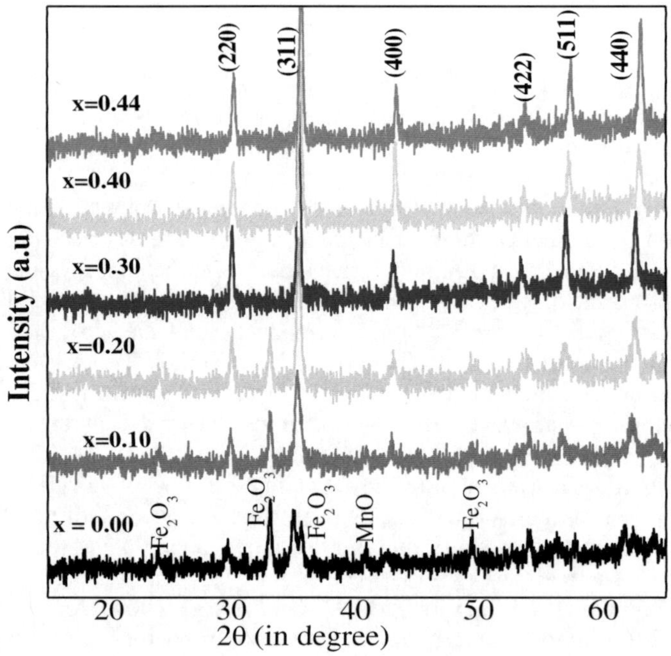


**Figure 1.** The X-ray diffraction patterns for $Li_xCu_{0.12}Mn_{0.88-2x}Fe_{2+x}O_4$

The results indicated that these materials have a formation of cubic spinel structure for each composition. The values of lattice constant obtained from each plane are with the help of Nelson-Riley function [12]. Figure 2 shows that the lattice constant, $a_0$ as a function of lithium contents for various $Li_xCu_{0.12}Mn_{0.88-2x}Fe_{2+x}O_4$ sintered at 1200 °C.

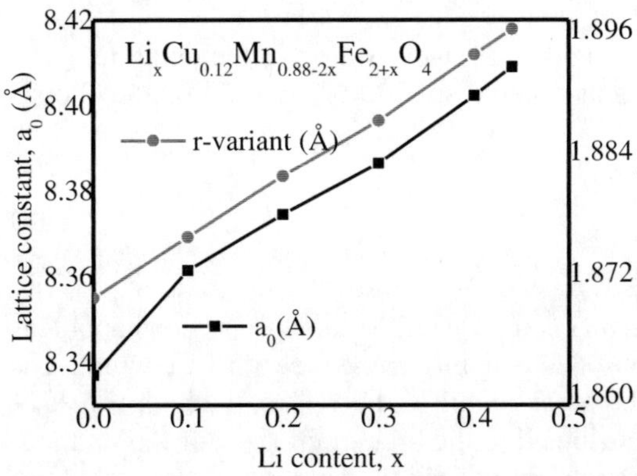


**Figure 2.** Variation of lattice constant with Li contents for $Li_xCu_{0.12}Mn_{0.88-2x}Fe_{2+x}O_4$ sintered at 1200 ºC.

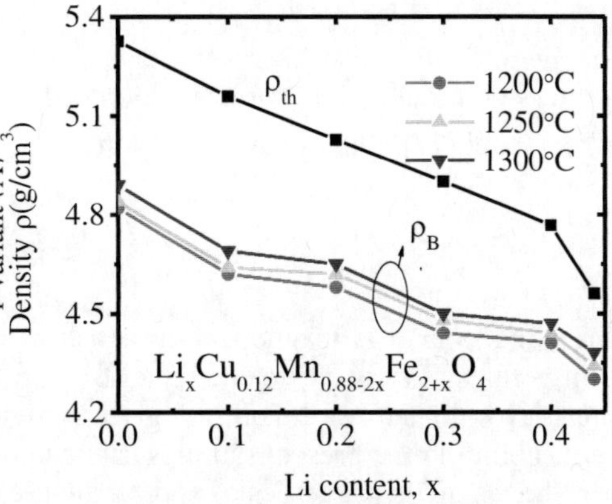


**Figure 3.** Variation of X-ray density and bulk density with Li contents, x for $Li_xCu_{0.12}Mn_{0.88-2x}Fe_{2+x}O_4$

It is noticed that the lattice constant increases with increasing of $Li^+$ content in $Li_xCu_{0.12}Mn_{0.88-2x}Fe_{2+x}O_4$ ferrites. This increase of lattice constant can be attributed to the ionic size differences since the unit cell has to expand when substituted by ions with large ionic size. The ionic radius of $Li^+$ (0.68Å) is greater than that of $Mn^{2+}$ (0.67 Å) [13-14]. When the larger $Li^+$ and $Fe^{3+}$ ions enter the lattice, the unit cell expands while preserving the overall cubic symmetry. On the other hand, the bulk density $\rho_B$ of each composition of $Li_xCu_{0.12}Mn_{0.88-2x}Fe_{2+x}O_4$ ferrites sintered at different temperatures was measured. The theoretical density, $\rho_{th}$ of each sample was calculated and it is noticed that both $\rho_{th}$ and $\rho_B$ decrease with increasing Li substitution in $Li_xCu_{0.12}Mn_{0.88-2x}Fe_{2+x}O_4$ for constant sintering temperature. This phenomenon could be explained in terms of the atomic weight. The atomic weight of Mn (54.94 amu) is greater than combined atomic weight of the Li (6.941 amu) and Fe (55.845 amu) [13].

### *3.2 Magnetic properties*

The optical micrographs have been taken by using optical microscope (Model: NMM-800TRF) of $Li_xCu_{0.12}Mn_{0.88-2x}Fe_{2+x}O_4$ for x= 0.40 are shown in figure 4 sintered at 1200, 1250 and 1300˚C, respectively. Average grain sizes (D) of all samples are determined from optical micrographs by linear intercept technique [14]. The D is significantly dependent on Li substitution. The D increases with increasing Li substitution for fixed sintering temperature. It is also observed that D increases with increasing sintering temperatures. The magnetic property, in particular, magnetic permeability of polycrystalline ferrite is related to two different type magnetizing mechanisms: spin rotation and domain wall motion [15-16], which can be described as, $\mu_i' = 1 + \chi_w + \chi_{spin}$, where $\chi_w$ is the domain wall susceptibility, $\chi_{spin}$ is intrinsic rotational susceptibility. $\chi_w$ and $\chi_{spin}$ may be written as:

$$\chi_w = 3\pi M_s^2 D / 4\gamma \tag{1}$$

and

$$\chi_{spin} = 2\pi M_s^2 / K_u \tag{2}$$

where $M_s$ is the saturation magnetization, $K_u$ is the total anisotropy, D is the grain diameter, and $\gamma$ the domain wall energy. The initial magnetic permeability for each sample was measured by using Wayne Kerr Impedance Analyzer (Model: 6500B). The initial magnetic permeability, $\mu_i'$ increases with increasing $Li^+$ content for different sintering temperatures (1200, 1250 and 1300 ˚C) in air as shown in figure 5. It is also observed that $\mu_i^/$ increases with increasing sintering temperature. The figure 6 shows that the $\mu_i^/$ and D as a function of Li content sintered at 1200 °C. According the Globus and Duplex model [17], the $\mu_i^/$ can be written as

$$\mu_i^/ = \frac{M_s^2 D}{\sqrt{K}} \tag{3}$$

where $M_s$ is the saturation magnetization and K is the magneto crystalline anisotropy constant. This increase in permeability is expected, because grain size of all samples increase with Li content. The large grains favor domain wall mobility, giving rise to high permeability. Therefore, the high values of initial permeability in the present samples can also be attributed to the high grain sizes of the samples. Moreover, the magnetic properties of soft ferrite are strongly influenced by it's composition, additives and microstructures of the material. Among all these factors, the microstructures have great influence on magnetic properties. It is generally believed that larger the grain sizes, the higher the magnetic initial permeability. The increasing value of $\mu_i^/$ with the increase of sintering temperature is due to the lower porosity for samples sintered at higher sintering temperature. As sintering temperature increases pores and voids are reduced with increasing sintering temperature.

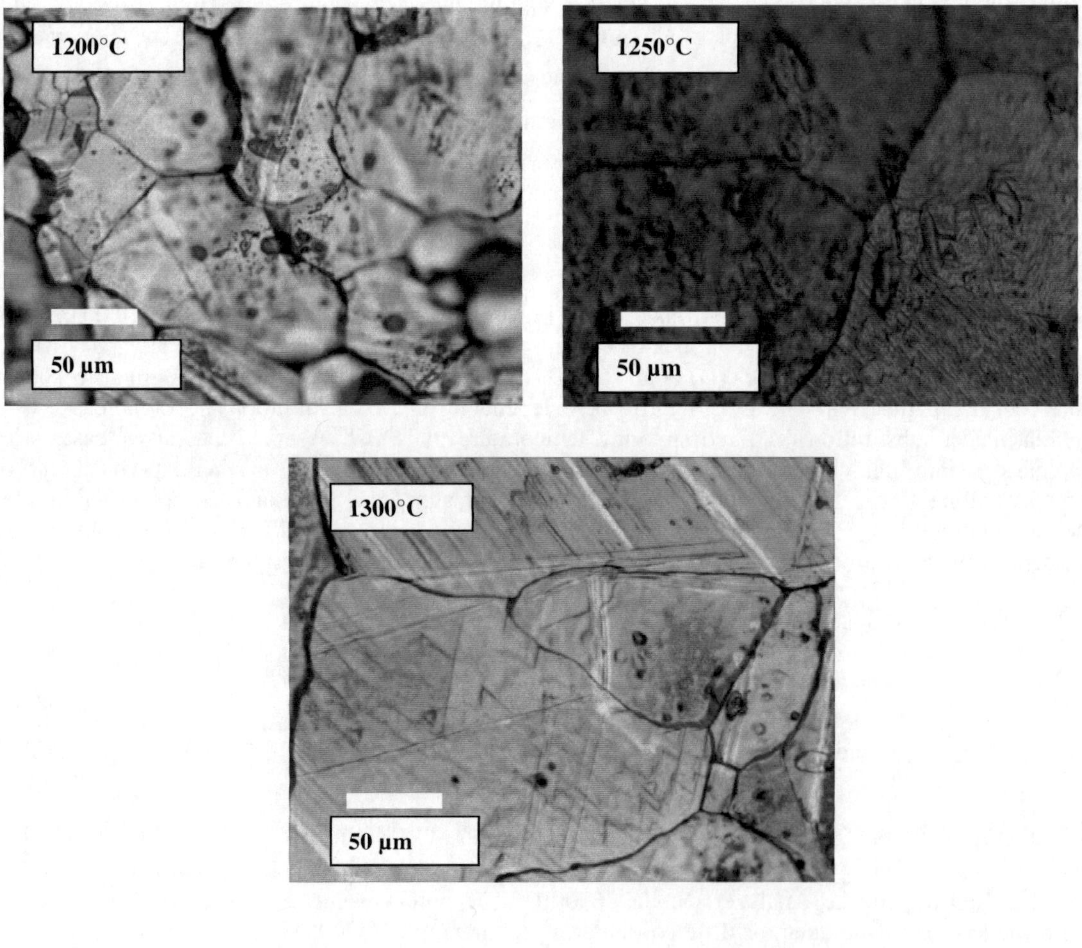


**Figure 4.** The optical micrographs of $Li_{0.4}Cu_{0.12}Mn_{0.08}Fe_{2.40}O_4$ sintered at different temperatures

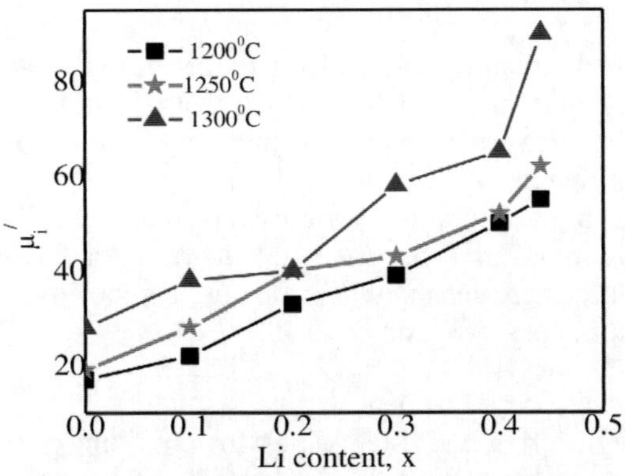


**Figure 5:** The initial magnetic permeability, $\mu_i'$ with lithium content for $Li_xCu_{0.12}Mn_{0.88-2x}Fe_{2+x}O_4$ sintered at 1200, 1250 and 1300 °C in air

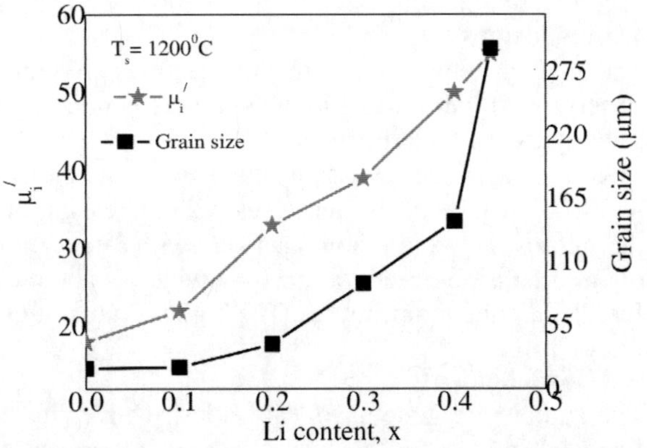


**Figure 6:** The initial magnetic permeability, $\mu_i'$ and grain size, D as a function of Li content sintered at 1200 °C.

For functional application the relative quality factor (Q-factor) is frequently used as a measure of performance. The relative quality factor versus frequency plots of all the samples sintered at 1200, 1250 and 1300 °C are shown in figure 7. It can be seen that the value of *Q*-factor increases with an increase of frequency and shows a peak around 3 MHz. The Q increases with increasing sintering temperature. Among all the studied samples, highest value of Q-factor (=2165) is observed for $Li_{0.40}Cu_{0.12}Mn_{0.08}Fe_{2.40}$ sintered at 1300 °C.

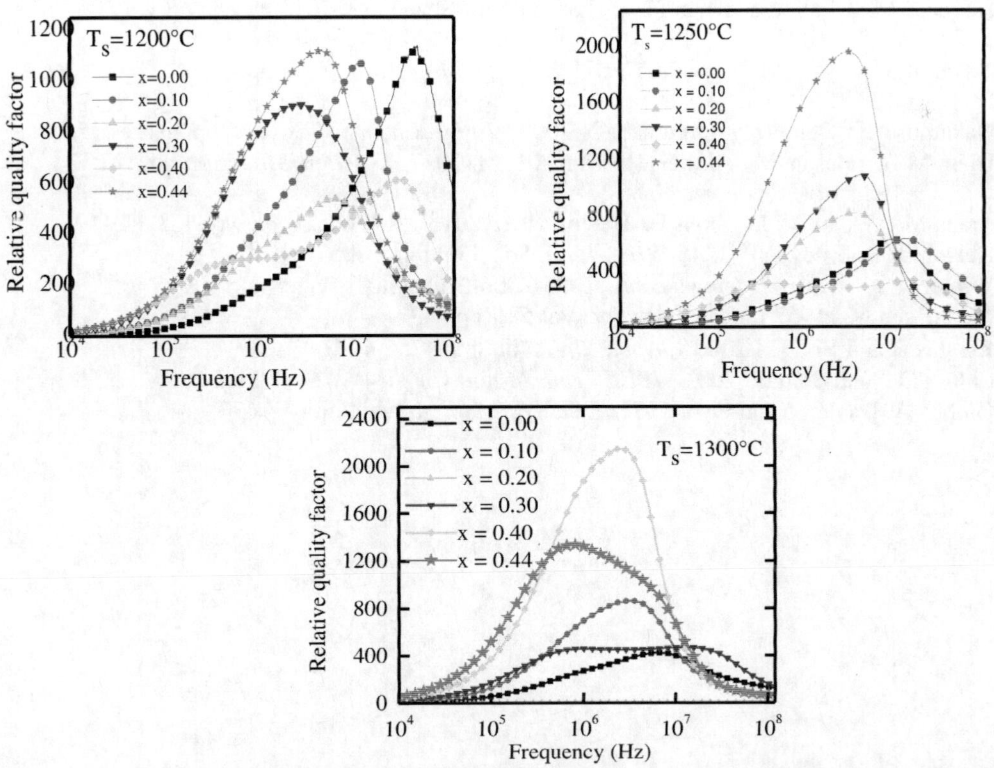


**Figure 7.** The variation of Q-factor with frequency for $Li_xCu_{0.12}Mn_{0.88-2x}Fe_{2+x}O_4$ sintered at 1200, 1250 and 1300 °C

## 4. Conclusions

The XRD patterns confirm that the samples are cubic spinel structure. The lattice parameter increases linearly with increasing Li content. This is due to greater ionic radius of Li ion than that of Mn ion. The study of microstructure shows that grain size increases with increasing Li content in $Li_xCu_{0.12}Mn_{0.88-2x}Fe_{2+x}O_4$ samples. The $\mu_i^/$ increases with increasing Li content for a fixed sintering temperature. The $\mu_i^/$ also increases with increasing of sintering temperatures for all samples because of the microstructure are homogeneous with large grain size and a uniform grain size distribution. The highest $\mu_i^/$ is observed for x = 0.44 which is about three times greater than that of parent composition for all sintering temperatures. The highest value of Q-factor is observed for x = 0.40.

## Acknowledgments

The authors are grateful to Bangladesh University of Engineering and Technology for providing the financial support for this research. The authors are also thankful to Bangladesh Council of Scientific and Industrial Research for providing their laboratory facilities.

.

## 5. References

[1] Omar M A 2000 *Elementary Solid State Physics* (Addison-Wesley) p 453
[2] Bai Y, Zhou J, Gui Z, Yue Z, Li L 2003 *Mater. Sci. Eng. B.* vol 99 p 266
[3] Akther S and Hakim M A 2010 *Mater. Chem. Phys.* vol 120 p 399.
[4] Parvin R, Momin A A, Akther Hossain A K M, 2016 *J. Magn. Magn. Mater.* vol 401 p 760-769.
[5] El-Shabasy M 1997 *J. Magn. Magn. Mater.* vol 172 p 188.
[6] Akther Hossain A K M, Rahman M A, Farhad S F U, Vilquin B, Tanaka H, 2011 *Physica B* vol 406 p 1506.
[7] Hoque S M, Ullah M S, Khan F A, Hakim M A, Shaha D K 2011 *Physica B* vol 406 p 1799-1804,
[8] Nakamura T, Naoe M, Yamada Y 2011 *J. Magn. Magn. Mater* vol 305(1) p 120-126
[9] Nakamura T, Demidzu H, Yamada Y 2008 *J Phys Chem Solids* vol 69 p 2349- 2355
[10] Islam M A, Hasan M, Akther Hossain A K M 2017 *J. Magn. Magn. Mater.* vol 424 p 108-114
[11] Arana M, Galvan V, Jacobo S E, Bercoff P G 2013 *J Alloys Compd* vol 568 p 5-10
[12] Nelson J B, Riley D P 1945 *Proc. Phys. Soc.* London vol 57 p 160
[13] Winter M J *www.webelements.com* 1995-2006 University of Sheffield UK
[14] Mendelson M I 1967 *Ceram J Am Soc.* vol 52(8) p 443
[15] Kang S H and Yoo H I 2000 *J. Appl. Phys* .vol 88 p 4754-4757
[16] Caltun O F and Spinu L 2001 *IEEE Trans.Magn* vol 37(4) p 2353-2355
[17] Globus A, Duplex p and Gyot 1971 *IEEE Trans. Magn.* vol 7 p 617-622

# Effect of Sr and Mn co-doping on multiferroic properties of $BiFeO_3$ via sol-gel chemical process

**Sumaiya Jannat, Md. Rafiqul Islam, Ahmed Sharif**

Department of Materials and Metallurgical Engineering, Bangladesh University of Engineering and Technology, Dhaka 1000, Bangladesh

**Abstract.** In this contribution, Sr and Mn co-doped $Bi_{0.90}Sr_{0.10}Fe_xMn_{1-x}O_3$ (X=0, 5%, 10%, 15%) nanoparticles were prepared by sol-gel (chemical route) method to observe the dominance of co-substitution on their grain size, morphology, UV-vis absorbance and magnetic properties. Co-substitution of Sr and Mn in $BiFeO_3$, leads to structural transformation from rhombohedral (R3c) to orthorhombic ($Pn2_1a$). Field emission scanning electron microscope reveals that upon Mn doping nanoparticles become finer. The M-H loops depict that at 15 mol. % Mn doping the saturation magnetization (Ms) is enhanced significantly. Moreover, the synthesized nanoparticles absorb visible light and their band gaps reduce substantially.

Keywords: $BiFeO_3$ ,XRD, Morphology, Magnetic properties

## 1. Introduction

Multiferroics are the compound of having both ferroelectric and magnetic properties and perhaps $BiFeO_3$ (BFO) is the promising multiferroic material which possesses both of these excellent properties at room temperature. As a result it creates a great field in the application of ferroelectric memories, magnetic tape, nano scale shape memory device, spintronics, and memory device etc. The room temperature structure of BFO is rhombohedral that belongs to point group R3c. It is the only few multiferroics that has Curie (830 °C/1103 K) and Neel (370 °C/643 K) temperatures far above room temperature. [1, 2]

BFO exhibits magneto electric coupling at room temperature and such coupling is strongly influenced by proper doping. However, BFO has some drawbacks in practical application. These are low stability, high leakage current and weak magnetic properties. The rare earth materials like Sr, Ba are doped in the substitution of A-site (Bi) and make the structure more stable, retain the non-Centro symmetry. The transition metal like La, Ni, Ti, Mn, Cr etc are doped in the substitution of B-site (Fe) that minimize the valance fluctuation of $Fe^{3+}$ and compensate the oxygen vacancies [3] . The aim of this synthesis is to investigate the effects of $Sr^{2+}$ and $Mn^{2+}$ doping on the structure, morphology, magnetic properties and uv-vis absorbance.

IOP Conf. Series: Materials Science and Engineering **438** (2018) 012014 doi:10.1088/1757-899X/438/1/012014

## 2. Experimental

Sr and Mn co-doped bismuth ferrites with the composition $Bi_{0.9}Sr_{0.1}Fe_{1-x}Mn_xO3$, where x = 0 ( BSFMO-0), x=0.05 (BSFMO-5), x=0.10 (BSFMO-10) and x=0.15 (BSFMO-15), nanoparticles were synthesized employing soft chemical sol-gel process at low temperature. In this process the raw materials (Bi $(NO_3)_3$ $.5H_2O$; Fe $(NO_3)_3.9H_2O$; Sr $(NO_3)_2$) were weighted respectively in 1:0.95:0.05; 1:0.88:0.05; 1:0.81:0.05; 1:0.76:0.05 ratios. The powders were taken in a beaker and 250 mL de-ionized water was added to make a solution. The solution was heated under stirring on a hot plate at 300 rpm at 200 °C for ½ hr and then required amount of citric acid and ethylene glycol were added to the solution. The solution was heated continuously at 75-80 °C and stirred at 400 rpm until the gel was formed. A vigorous reaction took place to burn the gel. When the reaction was over, the residue was dried in an oven at 120 °C for 18-20 hours. The fluffy power was obtained by crushing the dried substrates in a mortar. To obtain a crystallized sample the fine powders was further annealed at 500 °C for two hours.

The samples were characterized by X-ray diffraction (XRD using $CuK_\alpha$ radiation model 3040-X'Pert PRO, Philip), field effect scanning electron microscope (FESEM: JEOL JSM 7600F) and energy dispersive X-ray spectroscopy (EDS). Magnetic characterization was performed by using vibrating sample magnetometer (Micro sense easy VSM software version 9.13) & Uv-vis spectroscopy (PerkinElmer UV win lab data Processor and viewer version 1.1.00).

## 3. Results and discussion

The XRD pattern of the BSFMO-0, BSFMO-5, BSFMO-10 and BSFMO-15 powder samples at RT confirms the formation of crystalline BFO phase (figure-1). The diffraction was performed by using Bragg angles between 20°-90°. Here, the A-site substitute Sr is fixed concentration that is 10%. Pure BFO shows separated diffraction peaks of the (104) & (110) plane. The peak position of (104) plane of pure BFO is situated at 31.72°. But in this case from Fig. (1) It can be seen that the merged peak has a maximum intensity around 31.98 for BSFMO-0 ceramic i.e. the peak shifted to the right with respect to pure BFO. It occurs due to the ionic radius of $Sr^{2+}$ (1.18 Å) is larger than that of $Bi^{3+}$ (1.17Å) as a result the lattice parameter of BSFMO was increased. [4] Adding $Mn^{2+}$ (5%, 10%, 15%) in substitution of $Fe^{3+}$ shifts the peak to the right because the ionic radius of $Mn^{2+}$(70 pm) on the other hand $Fe^{3+}$ is 60 pm. The R3c to $P_n2_1$ a phase transition is found to increase with increasing Mn doping and single $P_n2_1$ a phase is obtained for BSFMO-15

IOP Conf. Series: Materials Science and Engineering **438** (2018) 012014 doi:10.1088/1757-899X/438/1/012014

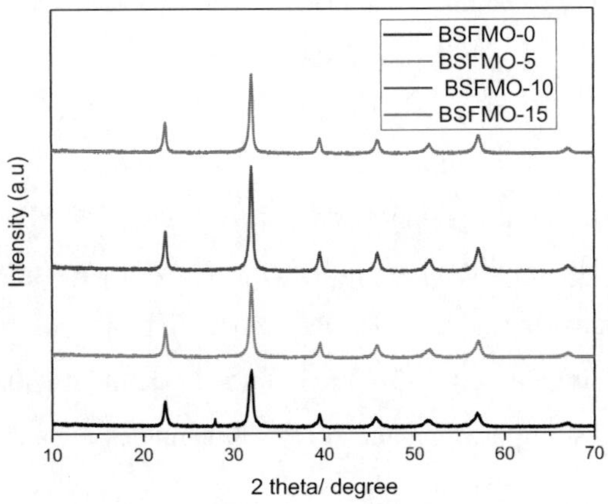


Figure 1. XRD patterns BSFMO-(0, 5, 10 and 15) nanoparticles at room temperature

It is wonder that difference between ionic radii of substituted ions could play a role to modify the crystal structure of BFO.

In BSFMO -0 the surface was irregular, larger grain size and the amount of surface porosity was higher in comparison with other BSFMO ceramics (figure-2).

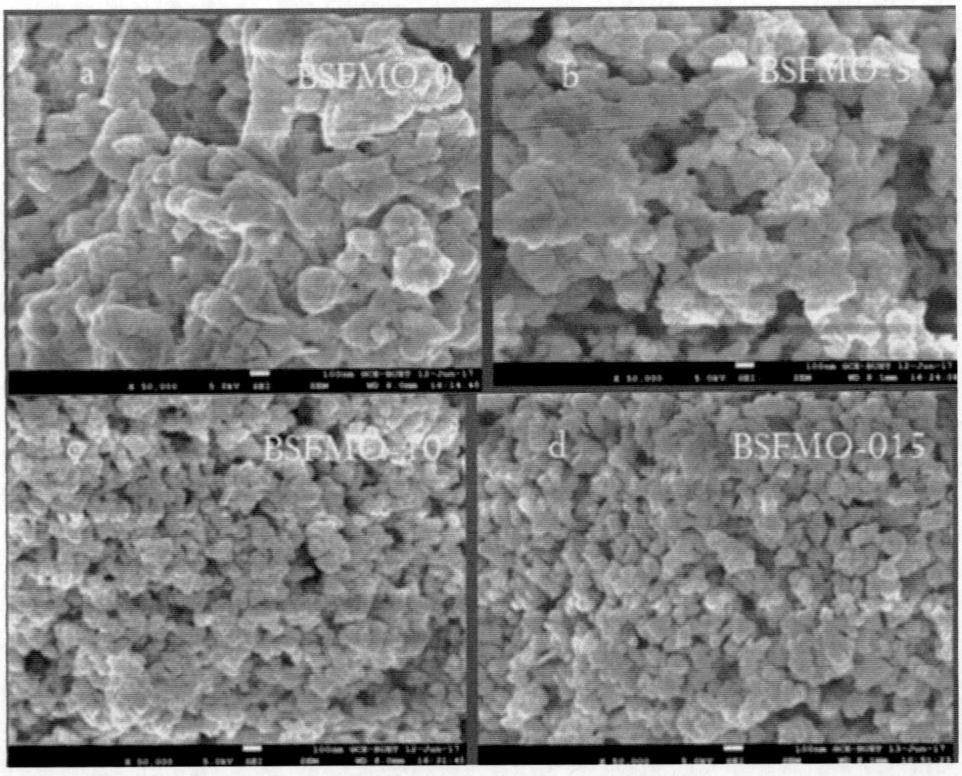

Figure 2. FESEM microstructure of BSFMO nanoparticles (a) BSFMO-0, (b) BSFMO-5, (c) BSFMO-10, (d) BSFMO-15

The higher amount of porosities and rougher surface might arise from the creation of oxygen vacancies in the structure due to hetero-valent substitution of $Bi^{3+}$ with $Sr^{2+}$ in BFO.

The SEM morphology suggests that the grain size and particle size of Sr-Mn co-doped BFO nanoparticles decreases with increasing doping concentrations in $Bi_{0.90}Sr_{0.10}Fe_xMn_{1-x}O_3$ (X=0, 5%, 10%, 15%) . It clarifies that Sr and Mn both acts as a growth inhibitor. The differences in ionic radii between A-site $Bi^{3+}$ (1.03 Å) to $Sr^{2+}$ (1.18 Å) ion and B-site $Fe^{3+}$ (0.645 Å) to multi valance $Mn^{2+}$ (0.83), $Mn^{3+}$ (0.645 Å), $Mn^{4+}$ (0.53 Å) ions lead to a lattice distortion which hinder crystallite nucleation and thereby reduces particle size [5].

When larger ionic radius $Sr^{2+}$ was doped in A-site it modified the lattice structure of the BFO and suppresses the spiral spin structure. Due to formation of oxygen vacancies the bond angle of $Fe^{3+}$-O-$Fe^{3+}$ could be changed and induce canting of the $Fe^{3+}$ions spin. The effect of altering the $Fe^{3+}$-O-$Fe^{3+}$bond angle, rearranged lattice structure with larger ionic replacement and ferromagnetic pinning could produce a net magnetization in the BSFMO-0 structure (figure-3).

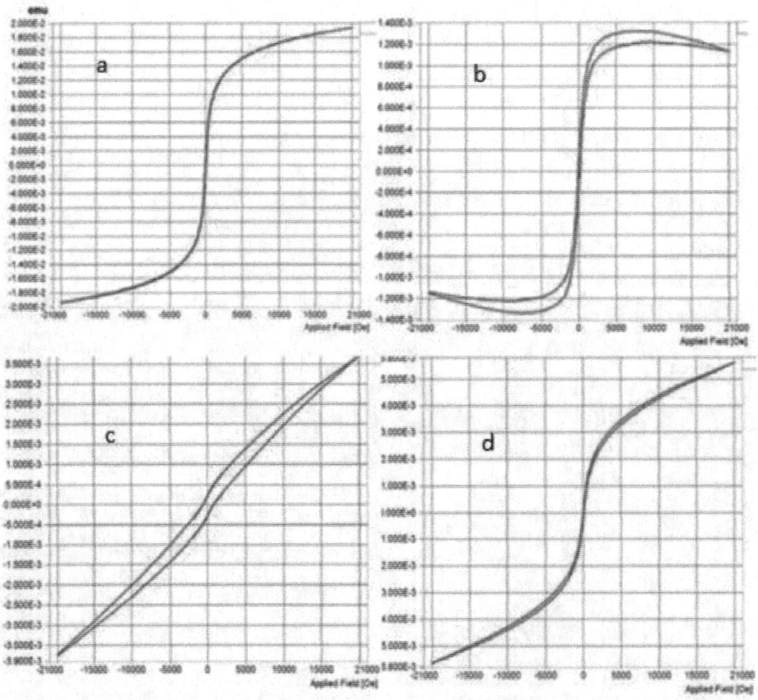


Figure 3. The four figures depict the magnetization vs. applied magnetic field (M _ H) loops for (a) BSFO, (b) BSFMO-5, (c) BSFMO-10 (d) BSFMO-15 nanoparticles at room temperature.

Addition of $Mn^{2+}$ increases the Hc values and at the last sample it decreases (table-1). $Mn^{2+}$ Magnetic, its magnetic moment of Bohr magneton $\mu_B = 5$ .Mn generally stay at multivalent states in BFO and the substitution of $Fe^{3+}$ ([Ar] $3d^5$, $\mu_B = 5$) by $Mn^{2+}$([Ar] $3d^5$, $\mu_B = 5$), $Mn^{3+}$ ([Ar] $3d^4$, $\mu_B = 4$) and $Mn^{4+}$([Ar] $3d^3$, $\mu_B = 3$) ion may cause incomplete compensation of antiferromagnetic SMSS and onset net magnetization. Furthermore, it was mentioned earlier that Sr and Mn co-doping in BFO has significantly reduced particle size of the synthesized nanoparticles. The suppression of spiral modulated spin structure and contribution of uncompensated spins at the particle surface increases sharply with decreasing particle size of BFO nanoparticles and may lead to the enhanced magnetization with increasing doping concentration.

**Table -1**: Magnetic parameters of BSFMO-(0, 5, 10, and 15) ceramics at room temperature

| Samples | $H_c$ (Oe) | $M_r$ (emu/g) *(e-6) | $M_s$ (emu/g) *(e-3) |
|---|---|---|---|
| BSFMO-0 | 43.50 | 948.154 | 10.39 |
| BSFMO-5 | 105.17 | 152.3 | 1.140 |
| BSFMO-10 | 695.61 | 251.46 | 3.72 |
| BSFMO-15 | 41.173 | 118.342 | 5.63 |

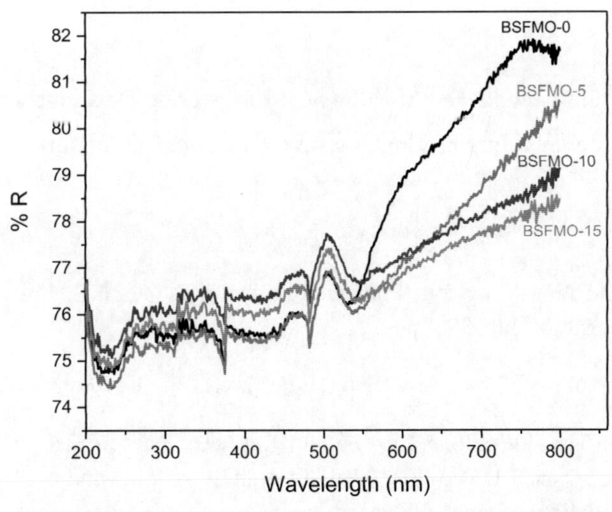


Figure 4. UV-Vis absorption spectra of BSFMO- (0, 5, 10, 15)

BFO is a fascinating optical material which can be used in the application of optoelectronic device. Optical band gap is actually the band gap with energy equals to the energy of visible light. For the pure BFO nanoparticles shows a blue emission around 454 nm and optical band gap is 2.30 eV [6]. But figure 4.

demonstrates that 10% Sr and Mn co-doping has really reduced the optical band gap due to reduced degree of hybridization associated with a stable electronic configuration of $Mn^{2+}$ ( $3d^5$ $4s^0$ half occupation of d orbital ) and therefore green emission around 517 nm is seen . Band gap reduces from 2.30 eV to 2.17 eV (green spectrum). The luminescence center as $Mn^{2+}$ ion creates a amount of electron -hole pairs recombination and emits more photons. [7]

## 4. Conclusion

Structural, magnetic and electrical properties of Sr and Mn co-doped BFO via sol-gel were investigated. sol-gel process has great control over property than other conventional solid state or critical hydrothermal methods. Co-substitution with Sr and Mn leads to structural transformation from rhombohedral (R3c) to orthorhombic ($Pn2_1a$) symmetry which was found to have great effect on optical band gap energy, magnetization and electrical conductivity of BFO. In XRD analysis secondary phase was detected though their effect was hardly observed in their properties. Both $Mn^{2+}$ and $Sr^{2+}$ both act as growth inhibitor. oxygen vacancy and charge imbalance created by $Sr^{2+}$ was not adjusted by other charge because $Mn^{2+}$is also divalent. Thus that leakage current was increased. The magnetic property values of BFO were greatly enhanced due to the size-confinement effects and (Mn , Sr) ions doping. Satisfying ferromagnetic performance was achieved due to the suppression of helical order at room temperature. The improved electric and magnetic properties of the co-doped BFO may be suitable for modern device applications.

**Acknowledgements**
Author would like to thank Materials and Metallurgical Engineering Department for giving support and also thanks to Glass and Ceramic Engineering department of Bangladesh University of Engineering and Technology.

**Reference**

[1]. Gustau Catalan and James F. Scott, Published online: May 4, 2009, Physics andApplications of Bismuth Ferrite Volume 21, pg-2

[2]. Medina Traya Bismuth-Ferrite-review Bismuth Ferrite review Levy on Feb 21, 2014

[3]. Xia Yan,GuoqiangTan, Wenlong Liu, Huijun Ren, Ao Xia Structural, electric and magnetic properties of Dy and Mn co-doped BiFeO3 thin films (7 november,2014) School of Materials Science and Engineering, Shaanxi University of Science & Technology, Xi'an, Shaanxi 710021, China

[4]. Md. Rafiqul Islam, Roisul Hasan Galib , Ahmed Sharif , Mehedi Hasan ,

Md. Abdullah Zubair , Md. Fakhrul Islam ,available 14 july ,2016 Correlation of charge defects and morphology with magnetic and electrical properties of Sr and Ta co-doped BiFeO3

[5]. Mehedi Hasan , M.A. Basith , M.A. Zubair, Md. Sarowar Hossain ,Rubayyat Mahbub , M.A. Hakim , Md. Fakhrul Islam Saturation magnetization and band gap tuning in BiFeO3 nanoparticles via co–substitution of Gd and Mn

[6]. Xuelian Yu, Xiaoqiang AnEnhanced magnetic and optical properties of pure and (Mn, Sr) doped, BiFeO3 nano crystals

[7]. J. F. Ihlefeld, N. J. Podraza, Z. K. Liu,R. C. Rai, X. Xu, T. Heeg Y. B. Chen,J. Li, R. W.Collins,J. L. Musfeldt, X. Q. Pan, J. Schubert,R. Ramesh, and D. G. Schlom, Optical band gap of BiFeO3 grown by molecular-beam epitaxy published online 10 April 2008

# Processing of Ultralight Porous $Al_2O_3$-Ceramics by Biotemplating of Bamboo

**M.A. Rahman and K.M. Shorowordi**

Department of Materials and Metallurgical Engineering, Bangladesh University of Engineering and Technology, Dhaka 1000, Bangladesh

**Abstract**. Biomorphic alumina ceramic with a long, large and oriented growth of alpha alumina grains suitable for the devices such as prosthetic implants, dental implant and control drug release etc. Ultralight porous alumina was synthesized from a carbonaceous preform derived from bamboo using biotemplating technique. Carbonaceous preform (C-preform) of bamboo precursor was prepared by controlled thermal processing in a muffle furnace for pore formation. $Al_2O_3$ was inserted into C-preforms by sol-gel method applying repeated infiltration of low viscous $Al_2O_3$-sol and then dried and sintered at 1500°C in air atmosphere resulting of an ultra-light monophasic $Al_2O_3$ ceramics with replication of the pore structure of the bamboo C-preforms. The microstructure of porous $Al_2O_3$ was characterized by scanning electron microscopy (SEM). Open hierarchical porous structures surrounded by alumina struts web with nominal porosity of 80% inside the biomaterial are visualized which lacking of interconnectivity and low strength.

Keywords: Bamboo precursors, Carbonaceous preform, Ultralight cellular alumina.

## 1. Introduction

The development of highly porous ceramics with a cellular porosity has attracted increasing interest in the recent years for its specific properties such as high surface area, high permeability, high-temperature stability, low weight and low thermal conductivity. Porous ceramics are potential for a variety of applications including filters for molten metals and hot gases, light-weight structural components, electrodes, sensors, bioreactors, catalyst carriers, radiant burners as well as porous implants in the area of biomaterials [1-2].

Artificial porous materials are costly, less available, limited compatibility, specific to Medical/surgical procedure with period of application/usage, relate to characteristics of host with high demand to support in the body natural conditions [3-4]. Compared to synthetic porous materials, natural biomaterials are often available at low cost and causes less health and environmental hazard problems with renewable materials and increased concerns about the environment for people producing the composites as compared to glass/Ceramic materials based composites [5]. The primary reason for the development of biocomposites from natural materials by biotemplating is flexible to the of type/distribution of the reinforcing phases in the composites, surgical techniques, and sterilization methods and the possibility to obtain biocomposites having a wide range of mechanical and biological properties [6].

Biotemplating techniques, in which biological materials are used directly as template structures for high-temperature conversion into technical ceramic materials, overcome this time rate problem [7]. Recently porous ceramics with tailored morphologies and structures synthesized from carbonaceous preform (C-preform) derived by replicating of plants have received wide attention for unique combination of anisotropic structure and properties [8–13]. The retention of parent plant micro-cellular anatomical features and macro-structural integrity in the C-preform plays a very important role in promoting reactivity and producing diverse microstructures bearing on the type of parent biostructure [14]. Previous works on biotemplating was mainly focused on the preparation of biomorphous carbide ceramics, e.g. SiC via a reaction of the biological material derived from biocarbon with different Si-infiltrants such as Si-melt, Si/SiO-gas, Si-containing polymers as well as $SiO_2$-sols [15-17]. Several investigations have been also carried out on the synthesis of biomorphous oxide ceramics. Preparation of $Al_2O_3$- and $ZrO_2$-fibers by oxidizing hydrated cellulose fibers impregnated with solutions of aluminum chloride and zirconium chloride was done earlier [18-19].

Porous ultrahigh alumina ceramics is considered as potential biomaterials and can be produced from Bamboo by biotemplating. Bamboo, the cheapest cellulosic and long fiber can be effective to prepare porous ultrahigh alumina ceramics [20]. In the open literature, due to the certain limitations of artificial materials we want to develop bamboo as biotemplating materials for cellular alumina production is very scarce. Thus the present work studied the fabrication of porous alumina retaining the morphological and macro and micro-structural features of the bamboo precursor.

## 2. Experimental

### *2.1. Conversion into C-preform of Bamboo precursors by thermal processing*

At first sun-dried bamboo after removing the outer hard layer were cut into short lengths (0.02 -0.05 m). The bamboo samples were then oven-dried at 65°C for 24 hours overnight and weight and dimension of this oven dried bamboo were measured. Oven-dried bamboo samples were converted into C-preforms by applying a controlled thermal processing. The bamboo samples were placed in stainless steel close chamber and reduced in an electrically heated muffle furnace under self-generated atmosphere allowing the gases and volatiles to escape through a cold trap where tarry materials condensed. Temperature was raised at different rates as 2 °C/min up to 400°C, and then 1°C/min up to 800 °C and 2 °C/min up to 900°C. The samples were held for 1hour at this pick temperature. The active C-preforms derived from bamboo were recovered after cooling the furnace and weight of the preforms were measured.

### *2.2. Alumina sol infiltration into C-preform and drying*

The alumina sol was prepared by mixing alumina powder (Fluka) by making a 5% weight of aqueous solution. The bamboo C-preform samples, taken in a small beaker, were placed in a vacuum glass chamber fitted with a dropping funnel containing the alumina sol. When the vacuum reached steady state after a certain period of time, the alumina sol was allowed to fall on to the samples drop by drop. The bubbling was appeared during infiltration. At the end of the infiltration, the bubbling was ceased and the vacuum was released. The samples were taken out and were wiped free of any loose sol adhering to the exterior surfaces. Normally eight to ten cycles of infiltration treatment was carried out on sample depending upon weight gained of alumina in preform and sample dimensions. After the infiltration process of each cycle, alumina infiltrated carbon preforms were dried in air and after that in an oven (100-120°C) for 2 hours to 4 hours as per sample dimensions.

### *2.3. Conversion into cellular alumina ceramic by Sintering*

Alumina infiltrated C-preform samples were taken in a crucible. The sample containing crucible was placed in a high temperature furnace for sintering the alumina sol infiltrated oven dried C-preforms and heating was applied. Samples were heated slowly (2°C/minute) in air to sintering temperature of 1500°C. The materials were allowed to stay for 60 minutes at the sintering temperatures for complete oxidation of carbon and to sinter the ceramic. Then the furnace was cooled down to 800°C with a rate of ~2°C/minute followed by furnace cooling to room temperature (Figure 1). Finally cellular ceramics sample was taken from the furnace.

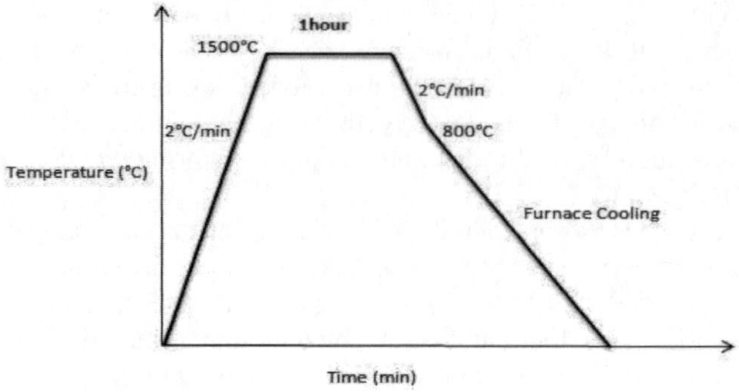


**Figure 1:** Sintering cycle of the Ceramic Alumina sample

*2.4. Characterization of cellular ceramics*

In a FESEM, Electron beam emitted from a field emission electron gun which uses secondary electrons that condense to a fine probe to examine microscopic structure for surface scanning with much higher resolution and three-dimensional appearance of its images with its large depth of field. The signal electrons emitted from the specimen are collected by a detector, amplified, and used to reconstruct an image. Field emission scanning electron microscope (FESEM) was employed to study of the structure of raw bamboo; C-preforms derived from bamboo and cellular alumina ceramics morphologies with a long, large and oriented growth of alumina. This test was carried out to determine/Characterize the pore within cellular ceramic, and to detect the presence of any micro defect in the composite. Images in the FESEM are recorded digitally with ability to generate an image by averaging multiple scans for the same area to reduce background noise in imaging.

## 3. Results and discussion

*3.1. Weight change of bamboo carbon preforms formation and infiltration*

The raw bamboo shorts losses its weight after drying overnight and after C-preform formation. Percentages of this losing weight have presented in (Table 1). Bamboo is composed of cellulose, hemicellulose, lignin, and water. The weight loss occurs after drying of the removal of water in Bamboo and it is around 10%. When the dried samples were heated to 900°C at different heating rate in controlled environment, weight loss occurred drastically due to the removal of the hydrogen, oxygen and other volatile matters from cellulose, hemicellulose, and lignin in bamboo and it is around 90%. Finally a porous carbon preform was obtained with desired bamboo template structure.

**Table 1:** Weight loss of Bamboo shorts after preform formation

| Sample | Weight of Raw sample | Weight after drying (gm.) | Percentage of weight loss after drying | Weight after carbon preform (gm.) | Percentage of weight loss after C-preform |
|---|---|---|---|---|---|
| Bamboo | 1.3050 | 1.1860 | 9.1187% | 0.1136 | 90.42% |

Compare to the previous research, our weight loss percentage is higher 90.42% to ensure porous carbon preform to the weight loss of Jute sample 75.20% [21]. It is found from another research by PK Mandal and R Majumdar that the weight loss of Jute sample and Cane stick is 75.20% and 69.35% respectively [22]

The carbon preform sample derived from bamboo was infiltrated with alumina from alumina sol in several cycles. Infiltration of alumina in carbon preform is a critical step for the fabrication of porous alumina ceramics. After completion of each infiltration cycle, the sample was dried and measured the weight gain (Table 2). It can be seen that the weight gain in grams was increased with increasing infiltration cycle and it is around 70% after the final infiltration cycle. Weight gain of alumina infiltrated was related to drying after each infiltration. Drying temperature and time should be selected in a way that no crack can be formed during drying.

**Table 2:** Weight gain of C-preform sample after alumina Sol infiltration

| Sample | 1st Cycle (gm.) | 2nd Cycle (gm.) | 3rd Cycle (gm.) | 4th Cycle (gm.) | 5th Cycle (gm.) | 6th Cycle (gm.) | 7th Cycle (gm.) | 8th Cycle (gm.) | Percentages of weight gain after final infiltration |
|---|---|---|---|---|---|---|---|---|---|
| Bamboo | 0.1768 | 0.2225 | 0.2630 | 0.3084 | 0.3320 | 0.3570 | 0.3610 | 0.3750 | 69.71% |

PK Mandala, R Majumdar and KK Mukherjee also found that the weight gain percentage of Jute sample after infiltration is 68.1% [21] which is similar to our findings 69.71% to develop a heavy mass of ceramic alumina sample.

### 3.2. Microstructural analysis

Figure 02 shows the physical structure of cross-sectional and longitudinal section of raw dry bamboo. Bamboo contains parallel fibers that are reinforced along the axial direction of the culm which is evidence from the cross-sectional view (Figure 2a). From the longitudinal section, cellular structure is observed in (Fig. 2b)

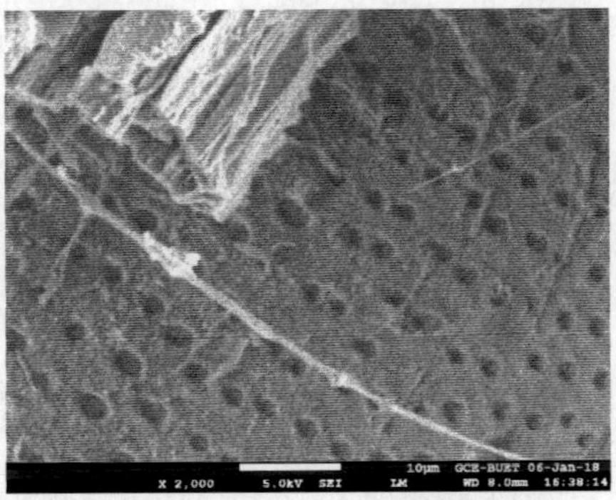


(a)

(b)

**Figure 2:** Showing the microstructure of dry raw bamboo sample (a) cross-section and (b) longitudinal section.

The dried bamboo sample was converted into C-preform and structure of the sample is shown in (figure 03). SEM image illustration for C-preform samples are showing elongated channel pores formed due to the leaving of organic and volatile matters resulted from the heat treatment at 900°C in an inert and self-generated gases. This pours are filled with the accumulation of alumina from alumina sol over cycles of sol-infiltration to fabricate porous alumina.

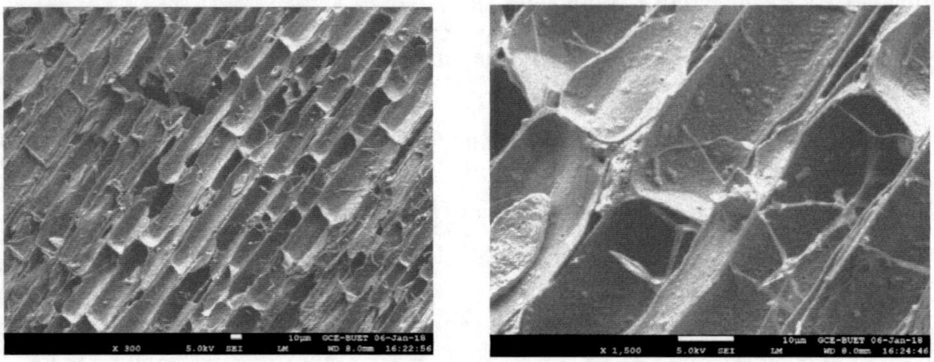

**Figure 3:** SEM image showing the structure of C-preform sample

The final alumina ceramics were fabricated from a thermal C-preform in sintering process at 1500$^0$C, but it was weak to measure the strength. SEM image of the alumina ceramic obtained are shown in (Fig. 04). SEM investigation reveals the porous structure of the alumina ceramics. The porosity oriented within its fiber where alumina particles gathered in this porosity and after sintering porous structure of alumina has formed. The effective properties of porous ceramics are not a simple function of porosity but depend on all features of the microstructure including pores between solid spheres, isolated (closed) spherical pores etc. Bamboo has microstructure of large number of open pore shown in (figure 3). Alumina phase also exhibited round pore. Although larger pore can be seen, the cellular morphology and structure with a longitudinal disposition of carbon preform, their orientation and interconnectivity are not perfectly maintained in the fabricated alumina ceramics.

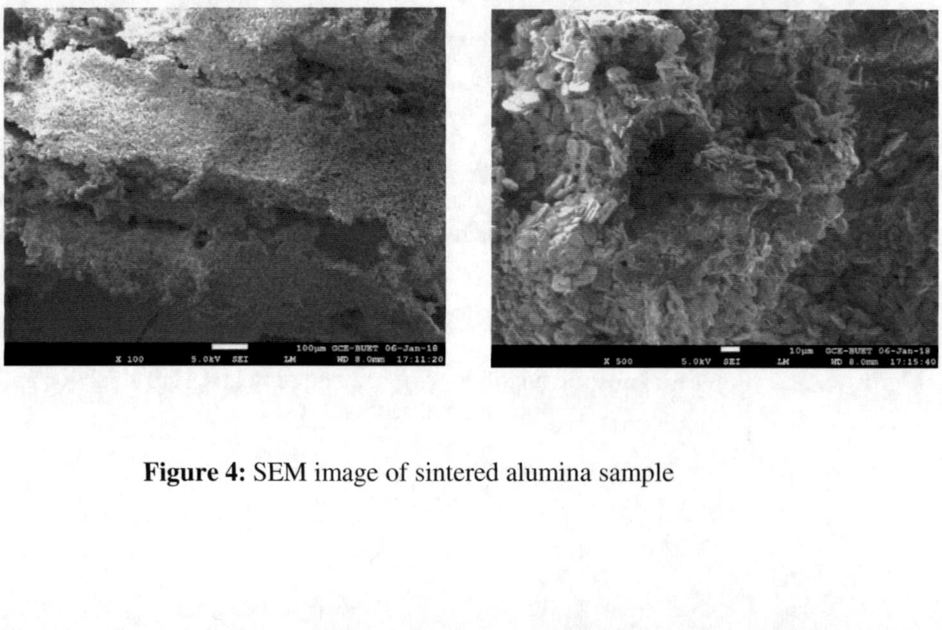

**Figure 4:** SEM image of sintered alumina sample

## 4. Conclusion

The specific aim of this study was to produce porous alumina ceramics in the image of bamboo. Carbon preform from the bamboo was successfully prepared by thermal process. The porous alumina was fabricated from the alumina infiltrated C-preform sintered at 1500$^0$C. Large and oriented growth of alumina grains porous network was found which is likely to be suitable for biomaterial. The pore network of sintered alumina with its strength is important criteria in biomaterial applications and can be obtainable with further study on particle size effect on pore size and shape, infiltration and proper sintering cycle.

## 5. Reference

[01] Brenzy R and Green DJ 1990 the effect of cell size on the mechanical behaviour of cellular materials

[02] Tomasz Wejrzanowski Samih Haj Ibrahimb Jakub Skibinski Karol Cwieka and Krzysztof Jan Kurzydlowski Image Anal Stereol 2017 Appropriate Models For Simulating Open-Porous Materials Vol 36 pp 105-110

[03] Chandramohan and Marimuthu 2011

[04] Schmalz G and Arenholdt-Bindslev D 2016 Chapter 1 Basic Aspects Biocompatibility of Dental Materials Berlin Springer Verlag pp 1–12 ISBN 9783540777823

[05] Jawaid and Abdul Khalil 2011 and Majeed et al. 2013

[06] Ramakrishna et al. 2001 and Vert M Doi Y Hellwich K H Hess M. Hodge P Kubisa P Rinau M Schué F. O. 2012 Terminology for biorelated polymers

[07] Ota T and Takahashi M and Hibi T and Ozawa M and Suzuki S and Hikichi Y 1995 Biomimetic process for producing SiC Wood, ed J Am Ceram Soc

[08] RV Krishnarao and YR Mahajan 1996 Mat. Sci Lett 15 pp 232-235

[09] M Patel and BK Padhi 1990 Mat. Sci 25 pp 1335-1343

[10] T Ota and M Imaeda and H Takase and M Kabayashi and N Kinoshita and H Miyazaki and Y Hikichi 2000 J Am Ceram Soc 83 pp 1521-1523

[11] H Sieber and J Cao and CR Rambo and P Greil 2002 Ceram. Engg. Sci. Proc 23 pp 173-181

[12] CR Rambo and J Cao and H Sieber 2004 Mat. Chem. Phys 87 pp 345-352

[13] C.E. Byrne and DC Nagle 1997 Mat. Res. Innovat pp 137-145

[14] OP Chakrabarti and HS Maiti and R Majumdar 27 July 2004 Indian Patent 0480NF/2002/In

[15] Sieber H and Greil P 1999 Ceramic composites from bioorganic derived materials

[16] Rusnia O and Kiemeier R and Molinero A and Rambo C R and Sieber H 2004 Manufacturing of highly porous SiC-ceramics from Si filled cellulose fibre papers, ed N P Bansal (Cream Trans)

[17] Depczynski W Kazala R Ludwinek K Jedynak K 2016 Modeling and microstructural characterization of sintered metallic porous materials vol 9 No 7 pp 567

[18] Yermolenko IN and Ul yanova and TM Vityaz 1986 IL Reactive ceramic fibers Rev Int Hautes Temper PA Fyodorova

[19] Lueth S Sauter US Bessler WG 2016 An agglomerate model of lithium-ion battery cathodes J Electrochem Soc A 163 vol 2 pp 210–22.

[20] Moore D R and Cervenka A Future requirements in the characterization of continuous fiber-reinforced polymeric composites (IUPAC Technical Report Vol 74 No 4 pp 601- 628

[21] Pijush Kanti Mandala Rabindranath Majumdara Kalyan Kumar Mukherjee Omprakash Chakrabarti and Himadri Sekhar Maiti 2009 An ultralight porous alumina ceramic in the image of jute stem Journal of Ceramic Processing Research Vol 10 No 4 pp. 457-461

[22] PK Mandal R Majumdar KM Mukherjee O Chakrabart HS Maiti 2009 Active cellular preform derived from stems of jute and sticks of cane and their suitability for bulk porous Si/SiC ceramics J Porous Mater Vol No 16 pp 157–163

IOP Conf. Series: Materials Science and Engineering **438** (2018) 012015 doi:10.1088/1757-899X/438/1/012015

**Acknowledgments**
The authors are wishing to acknowledge the Department of Materials and Metallurgical Engineering, BUET for proving the facilities to conduct the research and the department of Glass and Ceramics Engineering to provide the SEM facility.

IOP Conf. Series: Materials Science and Engineering **438** (2018) 012016 doi:10.1088/1757-899X/438/1/012016

# Effects of Gd and Cr co-doping on multiferroic properties of $Bi_{0.9}Gd_{0.1}Fe_{(1-x)}Cr_xO_3$ (x=0–0.08)

**M.N. Hossain, M.A. Matin, M.A. Hakim, M.F. Islam**

Department of Glass and Ceramic Engineering (GCE), Bangladesh University of Engineering and Technology, Dhaka 1000, Bangladesh

**Abstract.** Gd and Cr co-doped $Bi_{0.9}Gd_{0.1}Fe_{(1-x)}Cr_xO_3$ (x=0–0.08) nanoparticles were synthesized via sol-gel method. Precursor salts of bismuth, iron, gadolinium and chromium were used as starting raw materials. Obtained nanoparticles were annealed at various temperatures between 400–700 °C. Optimal properties were found to display at 600 °C. X-ray diffraction (XRD) patterns revealed that the partial substitution of $Bi^{3+}$ by Gd ions and $Fe^{3+}$ by Cr ions in $BiFeO_3$ (BFO) resulted in evolution of phase transition from rhombohedral to orthorhombic structure. The average crystallite size was found to vary from 16 to 68 nm depending on doping level. The field emission scanning electron micrographs demonstrated a distinct morphology of spherical and nano-sized particles. Ferroelectric measurements showed a substantial improvement in polarization in doped BFO under an applied field of ± 30 kV/cm. Magnetic properties measured at room temperature showed enhanced ferromagnetic characteristics with increasing doping level of Cr up to 6 % (at.) compared to that of un-doped BFO. An asymmetric shift both in the field and magnetization axes of M-H loops was also observed.

## *1. Introduction*

$BiFeO_3$ is one of the most promising multiferroic material which exhibits ferroelectric properties in combination with ferromagnetic properties in the same phase. It is an exciting candidate for contemporary, modern and novel applications due to mutual controllability of the electric polarization with a magnetic field and magnetization by an electric field [1]. $BiFeO_3$ has rhombohedrally distorted perovskite $ABO_3$ structure. It is ferroelectric below $T_C$ 829.85 $^0$C and antiferromagnetic below $T_N$ 369.85 $^0$C. Previous research emphasized that impurity free un-doped $BiFeO_3$ is hard to achieve [3]. Various impurity phases include $Bi_2Fe_4O_9$, $Bi_{36}Fe_{24}O_{57}$ and $Bi_{25}FeO_{40}$ [4]. Magnetic ordering of $BiFeO_3$ is of antiferromagnetic type. It has a spiral modulated spin structure (SMSS) with an incommensurate long wavelength period of 62 nm [6]. This spiral spin structure cancels the macroscopic magnetization and prevents the observation of the linear magnetoelectric effect. However, $BiFeO_3$ suffers from current leakage due to the existence of a large number of charge centers caused by oxygen ion vacancies. These problems restrict the use of $BiFeO_3$ for fabrication of multifunctional devices. To overcome these problems, some attempts have already been undertaken such as enhancing magnetism and ferroelectricity in $BiFeO_3$ by the partial substitution of $Bi^{3+}$ by ions such as $Sm^{3+}$,$Nd^{3+}$ etc., and also substitution of $Fe^{3+}$ by ions such as $Cr^{3+}$[5] or simultaneous minor substitution of $Bi^{3+}$ and $Fe^{3+}$ by ions such as $La^{3+}$ and $Mn^{3+}$ or $La^{3+}$ and $Ti^{4+}$ respectively [2]. The partial substitution of $Bi^{3+}$ with ions having higher ionic radii has been found to effectively suppress the spiral spin structure of $BiFeO_3$ giving rise to the appearance of weak ferromagnetism [5]. It was reported that, different kind of doping at A site, specially the single doping of Gd in place of Bi enhances the room temperature magnetization of $BiFeO_3$ as well as improves the phase purity of $BiFeO_3$ [3]. It is also reported that at 200 K due to the substitution of Fe by Cr the area of the magnetization versus magnetic field (M-H) curve increased compared to that of pure $BiFeO_3$ but the M-H loop was still very narrow [5]. Thus, in the present investigation, we studied the effect of co-doping of magnetic Gd and non-magnetic Cr in place of Bi and Fe respectively in $BiFeO_3$. In this context, nominal compositions of $Bi_{0.9}Gd_{0.1}Fe_{(1-x)}Cr_xO_3$ (x=0–0.08) functional multiferroics were synthesized by sol-gel method and their structure, morphology, ferroelectric and magnetic properties were investigated.

## *2. Experimental*

Bismuth nitrate, $Bi(NO_3)_2.5H_2O$ of purity 99% (Merk, India) and iron nitrate, $Fe(NO_3)_3$. $9H_2O$ of purity 98% (Merk, India) were used to synthesize $BiFeO_3$ by sol-gel method. Stoichiometric proportion of precursors were initially dissolved in deionized water. For Gadolinum (Gd) and Chromium (Cr) co-doped $Bi_{0.9}Gd_{0.1}Fe_{(1-x)}Cr_xO_3$ (x = 0–0.08), gadolinium nitrate, $Gd(NO_3)_3.6H_2O$ of purity 99.99% (Sigma-Aldrich Co., Germany) and chromium nitrate, $Cr(NO_3)_3$. $9H_2O$ of purity 99% (Sigma-Aldrich Co., Germany) precursors were used in required stoichiometry with BFO chemistry. When the solution was transparent, citric acid ($C_6H_8O_7$) was added to the solution to complex the metal cations. The solution was then stirred and heated at 70 °C for 3 h to form a transparent blackish-red sol. Next, an appropriate amount of ethylene glycol ($C_2H_6O_2$) was added as a polymerization agent into the solution. Ammonium hydroxide ($NH_4OH$) was added to maintain pH. The resultant solution was then heated at 85 °C to initiate the polymerization reaction and a few minutes later a gel was formed with vigorous boiling and fuming. The gel was dried in a drier to obtain xerogel. The xerogel thus obtained was then ground into powders and annealed at 400–700 °C for 2 h in static air to obtain un-doped BFO and doped BFO nanoparticles. To determine crystallinity and phase(s) of synthesized nanopowders, X-ray diffraction (Empyrean, PANalytical, Netherlands) was employed. Particle morphology was analyzed using field emission scanning electron microscope (JSM 7600F, Japan). Ferroelectric properties were measured using Precision Multiferroic Test System (Radiant Technologies, USA). Magnetic measurements were conducted using a Vibrating Sample Magnetometer (Micro Sense EV9, USA).

## *3. Results and discussion*

Figure 1 shows XRD patterns of the $Bi_{0.9}Gd_{0.1}Fe_{(1-x)}Cr_xO_3$ (x=0–0.08) ceramics sintered at 600 °C indicating the formation of well-crystalline structure. Rietveld refinement of the nanoparticles was carried out by High Score Plus Software and calculated the particle size and lattice strain are presented in Table 1. There are small traces of secondary phase $Bi_2Fe_4O_9$ as marked by * appeared in the un-doped BFO samples. The apparently unavoidable formation of secondary phases during the sol-gel synthesis of un-doped $BiFeO_3$ based materials have been reported earlier [5]. Suppression of the secondary phase in single phase BFO is a major concern because bismuth and/or iron vacancies help to create these types of impurity phases. Interestingly, in this research, second phases (*) $Bi_2Fe_4O_9$ (Fe rich phase) were almost vanished due to doping of $Gd^{3+}$ in BFO. The decrease in the peak intensities for (x = 0–0.08) indicates the incorporation of Cr in $Bi_{0.9}Gd_{0.1}Fe_{(1-x)}Cr_xO_3$ (x = 0–0.08) ceramics [6]. Furthermore, the twin peaks observed at $2\theta$= 32° , 52° , and 57° merges to form a broadened peak which becomes weak and tend to dissolve due to Cr doping. Merging of peak at around $2\theta$ = 32° as Cr content increases suggests a structural change from rhombohedral to orthorhombic and is consistent with that reported earlier [7]. We calculated the average crystallite size with the Scherrer equation shown, $D = k\lambda / \beta \cos\theta$, here $\beta$ is full width at half maximum intensity (FWHM), k is the dimensionless constant with a typical value of 0.9, $\lambda$ is the wavelength of Cu K$\alpha$ radiation. The crystallite size was found to decrease from 68 nm to 16 nm in $Gd^{3+}$ doped BFO. Further addition of doping agent $Cr^{3+}$ in BGFO, the crystallite size increased up to 29 nm for 6% Cr doping.

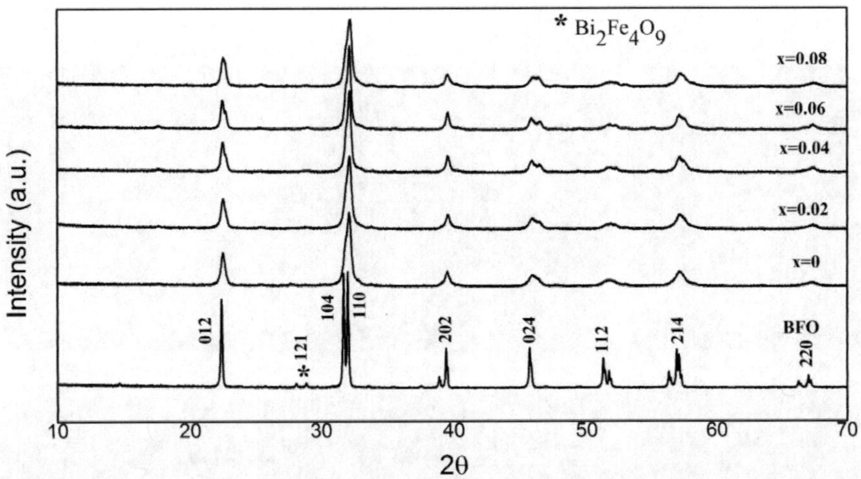


**Figure 1:** X-ray diffraction patterns of un-doped $BiFeO_3$ and doped $BiFeO_3$ ($Bi_{0.9}Gd_{0.1}Fe_{(1-x)}Cr_xO_3$) nanoparticles annealed at temperature 600 °C.

Lattice strain was also calculated using Bragg's law, d = nλ / 2sinθ, here d is lattice spacing and n is the order of reflection. Lattice strain, $\varepsilon = \Delta d/d_1$ where $\Delta d = d_1 - d_2$. A shift in the peak positions was observed with Gd (and/ Cr) doping. As peaks shift to a higher degree (from right to left) the inter planer spacing reduces as a result the lattice strain increases and may shrink the crystal structure. From Table 1 one can see that the lattice strain increases from 0.186% to 0.772% after Gd doping. More interestingly, lattice strain was found to decrease up to 0.433% due to co-doping of Gd and Cr in BFO.

**Table 1.** Refined structural parameters, calculated crystallite sizes (d) along with volume and lattice strain values and % of phases of $BiFeO_3$ and $Bi_{0.90}Gd_{0.10}Fe_{1-x}Cr_xO_3$. (x = 0-0.08).

| Composition [$Bi_{0.90}Gd_{0.10}$ $Fe_{(1-x)}Cr_xO_3$] | a (Å) | c (Å) | R3c % | Pn2₁a % | d [nm] | ε [-] |
|---|---|---|---|---|---|---|
| BFO | 5.57761 | 13.86779 | 100 | - | 68 | 0.186 |
| x=0 | 5.55939 | 13.78815 | 57.2 | 42.8 | 16 | 0.772 |
| x=0.02 | 5.57198 | 13.81291 | 50.1 | 49.8 | 24 | 0.519 |
| x=0.04 | 5.56491 | 13.79472 | 49.7 | 50.3 | 26 | 0.447 |
| x=0.06 | 5.56343 | 13.80535 | 45.4 | 54.6 | 29 | 0.433 |
| x=0.08 | 5.56356 | 13.78329 | 46.5 | 53.5 | 22 | 0.573 |

To investigate the morphology of synthesized nanoparticles, secondary electron microscopy was conducted for all samples. Figure 2(a) shows the morphology of un-doped BFO. Grain shape was not spherical and homogeneous. An average grain size of 185 nm was obtained. SEM micrograph of 10% Gd doped $Bi_{0.9}Gd_{0.1}FeO_3$ sample is presented in Figure 2(b).The average grain size was found to be 62 nm. Figure 2(c) shows the microstructure of the 2 % Cr doped $Bi_{0.9}Gd_{0.1}Fe_{(1-x)}Cr_xO_3$. For this sample, the average grain size was measured to be 53 nm. In addition, the grain size distribution was found to be quite homogeneous. The further increase of the Cr content to 4 % (x =0.04) reduced the average grain size to 50 nm (Figure 2(d)) as well as made the distribution of the grains is more uniform. Notably, increase of the Cr content from x =0.04 to x =0.06 and 0.08 (Figures 2 (e) and (f)) the average grain size becomes 48 nm and 72 nm respectively. Due to the substitution of Cr the average grain size was reduced. Thus, with increasing Cr doping a decreasing trend of the average grain size was observed which is consistent with our findings from XRD results.

(a) $BiFeO_3$ (b) x= 0 (c) x=0.02 (d) x=0.04 (e) x=0.06 (f) x=0.08

**Figure 2**. FESEM micrographs of $BiFeO_3$ and $Bi_{0.9}Gd_{0.1}Fe_{(1-x)}Cr_xO_3$ (x = 0-0.08) nanoparticles: (a) $BiFeO_3$, (b) x= 0, (c) x=0.02, (d) x=0.04, (e) x=0.06, (f) x=0.08 annealed at 600 ºC for 2 hr in air.

However, for 8% Cr doping, the average grain size reduced but due to agglomeration effect it measured to be bigger (breaking the normal trend of decreasing). Previous investigation suggested that the grain growth depends upon the concentration of oxygen vacancies and diffusion rate of the ions [7]. Large number of oxygen vacancies generate in pure $BiFeO_3$ due to highly volatile nature of Bi. Therefore, the significant reduction of the average grain size in Gd and Cr co-doped $BiFeO_3$ samples could be interpreted by the suppression of the formation of oxygen vacancies because of the requirements of the charge compensation. The decreased oxygen vacancies may lead to a lower grain growth rate which is actually a consequence of slower oxygen ion motion [8].

Presented in Figure 3 are P-E hysteresis loops of doped BFO captured at room temperature with an applied field of ± 30 kV. The P-E loop of un-doped BFO is also given as an inset for comparison. Such a shape of P-E loop in un-doped BFO resulted from huge leakage current. Encountered oxygen vacancies may be accounted for this [12]. In contrast, the P-E loops of doped-BFO showed much improved saturation resulting comparatively higher remnant polarization. However, with 10% Gd + 2-6% Cr doping, a gradual improvement in ferroelectric properties was observed in synthesized nanoparticles compared to 10% Gd doped one. With 10% $Gd^{3+}$ doping ionic radii difference may bring about further tilting of the structure resulting an overall improvement in ferroelectric properties. However, we speculate that with the addition of $Cr^{3+}$ dopant at B-site (at the centre of oxygen octahedra) the tilting is in opposite direction to that induced by $Gd^{3+}$ doping. Moreover, the exchange interaction between dissimilar magnetic $Gd^{3+}$ and $Cr^{3+}$ ions could play a role.

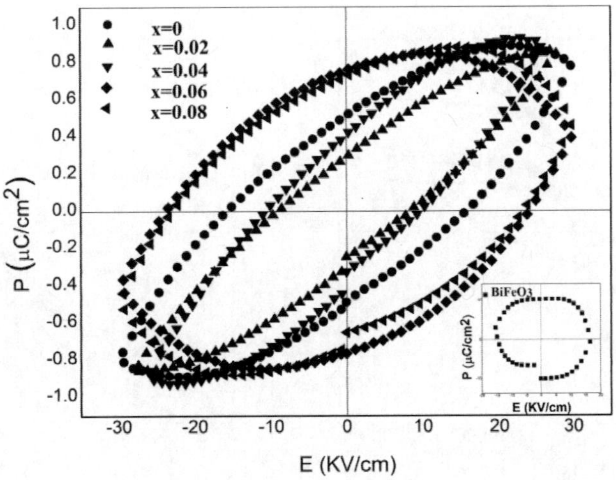


**Figure 3**. P-E hysteresis loops of $BiFeO_3$ and $Bi_{0.9}Gd_{0.1}Fe_{(1-x)}Cr_xO_3$ nanoparticles at applied field ± 30 kV/cm.

From figure 3, we can see that at low applied field the $Bi_{0.9}Gd_{0.1}Fe_{0.98}Cr_{0.02}O_3$ shows a maximum remanent polarization ($P_r$ = 1.21 μC/cm$^2$) compared to others (shown in table 2), but the shape of P-E loop is less elliptical. In the case of the $BiFe_{0.90}Gd_{0.10}O_3$ nanoparticles annealed at 600 °C, having the smallest particle size (16 nm), the P-E loop become more and more typical than other compositions which may be due to their reduced leakage current density [9,12].

**Table 2.** P-E loops and their remanent polarization, coercivity and maximum polarization

| Sample $Bi_{0.9}Gd_{0.1}Fe_{(1-x)}Cr_xO_3$ | Remanent Polarization, $P_r$ (μC/cm$^2$) | Coercivity (KV/cm) | $P_{max}$ (μC/cm$^2$) | Saturation Polarization (KV/cm) | d [nm] |
|---|---|---|---|---|---|
| BFO | - | - | - | 30 | 68 |
| x=0 | 0.16 | 0.69 | 0.25 | 30 | 16 |
| x=0.02 | 1.21 | 0.36 | 3.49 | 28 | 24 |
| x=0.04 | 0.15 | 0.63 | 0.33 | 26 | 26 |
| x=0.06 | 0.41 | 1.64 | 0.22 | 30 | 29 |
| x=0.08 | 0.73 | 1.27 | 0.48 | 30 | 22 |

Figure 4 shows M-H hysteresis curves for un-doped $BiFeO_3$, and (Gd and Cr) co-doped $Bi_{0.9}Gd_{0.1}Fe_{(1-x)}Cr_xO_3$ (x = 0–0.08) ceramics measured at RT with an applied magnetic field of up to ± 20 kOe. All the samples show unsaturated magnetization loops which confirm the basic antiferromagnetic nature of the samples. The un-doped $BiFeO_3$ sample possesses a very narrow hysteresis loop with a very small but non zero remanent magnetization (0.001 emu/g) and a coercive field of 62 Oe at RT. This is due to the fact that pure $BiFeO_3$ is antiferromagnetic (AFM) which does not possess any spontaneous magnetization but has a residual magnetic moment for a canted spin structure.

From Figure 5, a nonzero remanent magnetization and coercive field are observed. The remanent magnetization ($M_r$) is defined as $Mr = |M_{r1}-M_{r2}|/2$ where $M_{r1}$ and $M_{r2}$ are the magnetization with positive and negative points of intersection with H = 0 respectively [2]. The coercive field ($H_c$) is given as $H_c = |H_{c1} - H_{c2}|/2$, where $H_{c1}$ and $H_{c2}$ are the left and right coercive fields respectively [9].

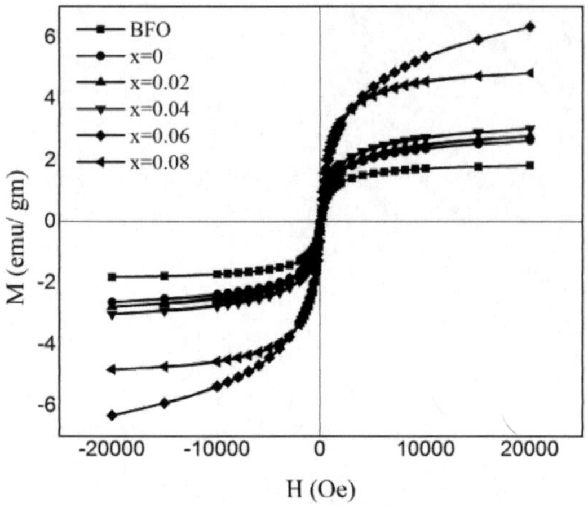


**Figure 4**. M-H hysteresis curves of $BiFeO_3$ and $Bi_{0.9}Gd_{0.1}Fe_{(1-x)}Cr_xO_3$ (x = 0-0.08) nanoparticles measured at RT.

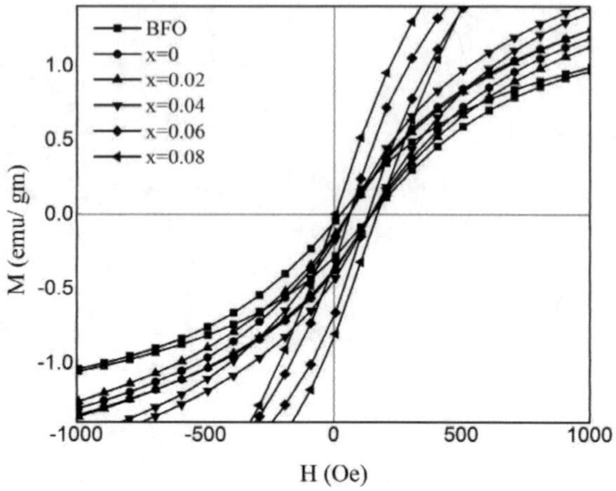


**Figure 5**. An enlarged view of the low-field M-H hysteresis loops of $Bi_{0.9}Gd_{0.1}Fe_{(1-x)}Cr_xO_3$ (x = 0-0.08) samples obtained at RT.

Calculated values of $M_r$ and $H_c$ are presented in Table 3. An increment of the 2% Cr doping concentration in place of Fe in $Bi_{0.9}Gd_{0.1}Fe_{(1-x)}Cr_xO_3$ (x=0–0.08) increased the saturation magnetization although the coercivity decreased. A further increase in Cr doping concentration to 8% dramatically enhanced $M_r$ and $H_c$. Due to 10% Gd doping in $BiFeO_3$ improved magnetic property was observed as reported earlier [7]. This enhancement of the magnetization at RT was attributed to the structural distortion in the perovskite with change in Fe–O–Fe angle. It was expected that this structural distortion could lead to suppression of the spin spiral and hence enhanced the magnetization in synthesized multiferroic system.

**Table 3.** M-H loops and their remanant magnetization, coercivity, saturation magnetization of $BiFe_{1-x}Cr_xO_3$ (x = 0–0.08) ceramics at RT

| Sample $Bi_{0.9}Gd_{0.1}$ $Fe_{(1-x)}Cr_xO_3$ | Remanant magnetization, $M_r$ (emu/g) | Coercivity $H_c$ (Oe) | Saturation magnetization $M_s$ (emu/g) | d [nm] |
|---|---|---|---|---|
| BFO | 0.11 | 62 | 1.9 | 68 |
| x=0 | 0.28 | 45 | 2.9 | 16 |
| x=0.02 | 0.13 | 48 | 3.1 | 24 |
| x=0.04 | 0.16 | 46 | 3.5 | 26 |
| x=0.06 | 0.12 | 36 | 7.1 | 29 |
| x=0.08 | 0.42 | 83 | 5.0 | 22 |

Similar to 10% Gd doped $Bi_{0.9}Gd_{0.1}FeO_3$, the hysteresis loops of Gd and Cr co-doped ceramics were not really saturated at 20 kOe. However, a significant enhancement in remanent magnetization and coercive field was observed with the increase in Cr doping concentration in $Bi_{0.9}Gd_{0.1}Fe_{(1-x)}Cr_xO_3$ (x = 0–0.08) multiferroics. In an investigation, room temperature magnetic properties were carried out for only Cr doped $BiFe_{1-x}Cr_xO_3$ ceramics with varying x up to 0.08 [5]. In that investigation, the highest saturation magnetization of 0.75 emu/g was reported for 6% Cr doped (not co-doped). In the case of presently investigated Gd and Cr co-doped ceramics $Bi_{0.9}Gd_{0.1}Fe_{0.94}Cr_{0.06}O_3$, the saturation magnetization is 7 emu/g, which is nine times higher. The coercive field is also lower for Gd and Cr co-doped samples compared to that of only Cr doped $BiFe_{1-x}Cr_xO_3$ (x = 0–0.08) ceramics. Hence, it is clear that Gd and Cr co-doped $Bi_{0.9}Gd_{0.1}Fe_{(1-x)}Cr_xO_3$ ceramics exhibit enhanced magnetic properties at RT showing weak ferromagnetic (WFM) antiferromagnetism.

The statistical distribution of $Fe^{3+}$ and $Fe^{2+}$ (created due to charge compensation) ions in the octahedral also lead to an increase in magnetization and weak ferromagnetism. However, for 8% Cr doping in $Bi_{0.9}Gd_{0.1}Fe_{(1-x)}Cr_xO_3$, the remanent magnetization and coercive field both again decreased and the antiferromagnetic order typical to $BiFeO_3$ parent phase is likely to be recovered. This might be associated with the structure phase transition from rhombohedral to orthorhombic at $x = 0.08$.

However, beside the enhanced room temperature magnetic properties in Gd and Cr co-doped $Bi_{0.9}Gd_{0.1}Fe_{(1-x)}Cr_xO_3$ ($x = 0$–$0.08$) ceramics, the most intriguing feature observed from the enlarged view (Figure 5) of the low field M-H hysteresis loops is their asymmetric behavior. An asymmetry exhibiting shifts both in the field and magnetization axes can clearly be observed for all the samples. This indicates the presence of exchange bias (EB) effect in it. The presence of EB phenomenon in these multiferroics was assumed to be associated with induced exchange anisotropy at the interface between FM and AFM phases.

## 4. Conclusions

Un-doped BFO and doped $Bi_{0.9}Gd_{0.1}Fe_{(1-x)}Cr_xO_3$ functional multiferroics were successfully synthesized using sol-gel method. XRD patterns shows the coexistence of both the Rhombohedral phase and the Orthorombic phase in the ceramic. The crystallite size (obtained from Rietveld analysis) reduced significantly from 68 nm to 16 nm due to increasing Cr doping level. However, the measured particle size in SEM was found to be substantially higher due mainly to agglomeration effects observed in synthesized nanoparticles (with high surface energy). The ferroelectric property namely remanent polarization and coercivity was enhanced due to co-doping. Due to the combined effects of Gd and Cr, magnetic properties of these ceramics were improved at RT. However, M-H hysteresis loops of co-doped BFO nanaoparticles demonstrate an asymmetric shift towards the field axis. The remanent magnetizations and saturation magnetizations were improved at room temperature but coercivity decreased.

## Acknowledgments

This research is supported by the Department of Glass and Ceramic Engineering, Bangladesh University of Engineering and Technology (BUET), Dhaka, Bangladesh.

## 5. Reference

[1] W. Eerenstein, N. D. Mathur, and J. F. Scott, Nature 442, 759 (2006).
[2] G. Anjum, R. Kumar, S. Mollah, D. K. Shukla, and C. G. Lee, J. App. Phys.107, 103916 (2010).
[3] Q. H. Jiang, C. W. Nan, and Z. J. Shen, J. Am. Ceram. Soc. 89, 2123 (2006).
[4] T. Munoz, J. P. Rivera, A. Monnier, and H. Schmid, Jpn. J. Appl. Phys., Part 1 24, 1051 (1985).
[5] Amit Kumar, K.L. Yadav, Materials Science and Engineering B 176, 227–230 (2011).
[6] R. A. Agarwal, S. S. Ashima, and N. Ahlawat, J. Phys. D: Appl. Phys. 45, 165001 (2012).
[7] M. A. Basith, O. Kurni, M. S. Alam, B. L. Sinha, and B. Ahmmad, Journal of Applied Physics. 115, 024102 (2014).
[8] Z. Chaodan, Y. Jun, Z. Duanming, Y. Bin and Z. Wenli, Integr. Ferroelectr. 94, 31 (2007).
[9] Y. Guo, L. Shi, S. Zhou, W. Liu, and S. Wei, J. Phys. D: Appl. Phys. 46, 175302 (2013).
[10] S. W. Cheong and M. Mostovoy, Nature Mater. 6, 13 (2007).
[11] P. Tirupathi and A. Chandra, J. Alloys Compd. 564, 151 (2013).
[12] A. Ianculescu, F. Prihor, P. Postolache, N. Drgan, and D. Crian, Ferroelectrics 391, 67 (2009).
[13] Chang, C., Mao, D., Int. J. Chem. K., 39, 75, 2007.
[14] R. Ramesh, Nature 461, 1218 (2009).

IOP Conf. Series: Materials Science and Engineering **438** (2018) 012017 doi:10.1088/1757-899X/438/1/012017

# Effects of $CoFe_2O_4$ substitution on magnetic properties of $NiFe_2O_4$ spinel ferrite

**S. A. Ali, M. A. Matin, M. A. Hakim, M. F. Islam**

Department of Glass and Ceramic Engineering, Bangladesh University of Engineering and Technology (BUET), Dhaka-1000, Bangladesh

**Abstract.** Functional magnetic materials of composition $Ni_{1-x}Co_xFe_2O_4$ (x=0, 0.2, 0.4, 0.6, 0.8, 1) were prepared using $NiFe_2O_4$ (soft) and $CoFe_2O_4$ (semi-hard) nanomaterials following solid state synthesis technique. Magnetic properties of synthesized samples were studied employing XRD, FESEM, VSM and DSC techniques. Curie temperature, $T_c$ of all the samples were determined from the temperature dependence of permeability and compared with the peak temperature of DSC thermograph. $T_c$ is found to decrease with increasing $CoFe_2O_4$ substitution which has been ascribed due to weakening of $J_{AB}$ super-exchange interaction between tetrahedral and octahedral sites. The detail hysteresis parameters such as $M_s$, $M_r$, $H_c$ have been determined from M-H loop measured at 80, 200 and 300K. Saturation magnetization, $M_s$ measured at 80K was found much higher compared with data at 300K. $M_s$ is found to increase with decreasing temperature and attained a maximum value corresponding to $CoFe_2O_4$ at any measurement temperature which may be attributed to higher magnetic moment of $CoFe_2O_4$ compared with $NiFe_2O_4$. A gradual magnetic hardening was observed to increase with increasing $CoFe_2O_4$ content with a maximum value of $H_c$ at x=0.8. Initial permeability as a function of frequency has been measured by an impedance analyzer. Phase transition temperature determined from DSC measurements and temperature dependence of initial permeability was found to display a good correlation. In addition, magnetic hardening of soft $NiFe_2O_4$ with increasing amount of $CoFe_2O_4$ substitution is also explained based on the existing theories of magnetism.

## *1. Introduction*

Ferrites are commercially important magnetic materials. Ranging from the very ordinary radio sets to the complicated and exhaustive hardware involved in computers, ferrites have found their way to prove their importance. There are soft and hard magnetic ferrites. Both of these ferrites have enormous practical applications. Soft ferrites have been extensively used for many kinds of magnetic devices such as transformers, inductors, magnetic heads, in resonance circuit for high frequency (ranging from $10^3$ to $10^{11}$Hz) because of their electrical resistivity higher than those of the soft magnetic alloys (ranges from $10^3$ to $10^{11}$ Ω-cm which is up to 15 order of magnitude higher than that of metals like iron), low eddy current losses, high initial permeability, high saturation induction, low hysteresis loss and reduced physical size. Due to these reasons extensive research activities has been carried out during the past 50 years to enhance their efficiency from the application point of view to various technical devices. Cubic spinel ferrite, has two sub-lattices, tetrahedral (A site) and octahedral (B site) in $AB_2O_4$ crystal structure. The important structural, electrical and magnetic properties of spinels, responsible for their application in various fields, are found to depend on the distribution of cations among the sites [1].Therefore the estimation of the cation distribution turns out to be important. Various cations can be placed in A site and B site to tune its magnetic properties. Depending on A site and B site cations it can exhibit various magnetic structures [2]. Both Nickel and Cobalt ferrites have inverse spinel structure in bulk. Some researchers have investigated various properties of bulk Ni-Co ferrites employing co-precipitation, hydrothrermal, sol-gel method but no detail study has been conducted yet to synthesize and characterize nano-power based Ni-Co mixed ferrites [3, 4].

In present work, we studied the structural and magnetic properties of functional magnetic materials of composition $Ni_{1-x}Co_xFe_2O_4$ (x=0, 0.2, 0.4, 0.6, 0.8, 1) prepared using $NiFe_2O_4$ (soft) and $CoFe_2O_4$ (semi-hard) nanomaterials following solid state synthesis technique.

## 2. *Experimental*

Ni-Co ferrites with the composition $Ni_{1-x}Co_xFe_2O_4$ (x=0, 0.2, 0.4, 0.6, 0.8, 1) were synthesized by solid state reaction method. The starting materials for the preparation of the studied compositions were in the form of nano powder of $NiFe_2O_4$ and $CoFe_2O_4$ ferrites of purity 99.99% (Inframat Advanced Materials, USA). Nano powders were mixed (according to their molecular weight) using mortar and pestle and then ball milled for 10 hrs and the slurry was dried and pressed into disc shaped sample. The disc shaped samples were pre-sintered at 850 $^0$C for 4 hrs. Final sintering of the sample was carried out at 1200$^0$C for 4 hrs.

The structure of the sample was determined by the X–ray diffraction method using PHILIPS X'Pert Pro X-ray diffractometer. The microstructural aspects were studied with a field emission scanning electron microscope (JEOL-JSM-7600F). Curie temperature measurement was conducted by using WAYNE KERR impedance analyzer attached with an oven and compared the results obtained from DSC measurement. The magnetic measurement of the sample was carried out by a vibrating sample magnetometer (Model VSM-02, Hirstlab, UK) at 80K, 200K and 300K temperature as a function of applied field.

## 3. Results and discussion

### *3.1. XRD analysis*

The phase identification and lattice parameter determination were performed by XDR technique. XRD patterns of $Ni_{1-x}Co_xFe_2O_4$ samples are shown in Figure 1. All the samples showed a good crystallization with well-defined diffraction lines. The XRD patterns clearly show the formation of spinel structure with identified peaks of (111), (220), (311), (222), (400), (511) and (440) planes. Obtained results are in good agreement with previously reported values [5]. Lattice parameter was determined using Nelson-Riley extrapolation method. The values of the lattice parameter obtained from each reflected planes are plotted against the Nelson-Riley function [6]

$$F(\theta)=\frac{1}{2}\left[\frac{\cos^2\theta}{\sin\theta}+\frac{\cos^2\theta}{\theta}\right] \quad (1)$$

where θ is the Bragg's angle. Experimental data are fitted with a straight line. The accurate values of the lattice parameter were estimated from the extrapolation of this line to F(θ) = 0 or F(θ) = 90$^0$.

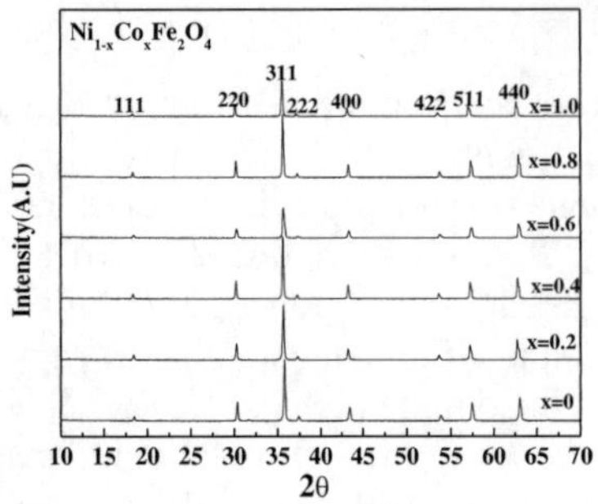


**Figure 1.** XRD pattern of $Ni_{1-x}Co_xFe_2O_4$ ferrites.

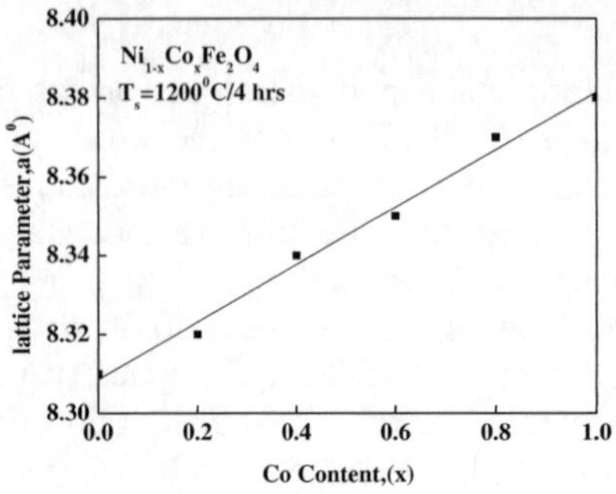


**Figure 2.** Variation of the lattice parameters with Co content in $Ni_{1-x}Co_xFe_2O_4$ ferrites.

The variation of lattice parameter as a function of $Co^{2+}$ content is depicted in Figure 2. It is noticed that the lattice parameter increases with the Co content, obeying Vegard's law [7]. This variation can be attributed to ionic size difference of the component ions. The $Co^{2+}$ (0.745 Å) ions possess a larger ionic radius than $Ni^{2+}$ (0.69 Å) and $Fe^{3+}$ ions (0.645 Å) [8].

Next, the bulk density ($d_B$) was measured by the usual mass and dimensional considerations whereas the X-ray density ($d_X$) of the prepared samples was calculated using the following relation:

$$d_x = \frac{ZM}{Na^3} \qquad (2)$$

where M is the molecular weight of the corresponding composition. N is Avogadro's number, $V = a^3$ is the volume of the cubic unit cell and Z is the number of molecules per unit cell, which is 8 for the spinel structure. The X-ray density is expected to decrease with the Co content because the lattice parameter increases with Co content. Similar result has also been reported in Mg-Mn ferrites [9]. The bulk density is lower than the X-ray density. This may be due to the existence of pores which were formed and developed during the sample preparation or the sintering process. Table 1 shows the variation of bulk density and X-ray density of the samples.

**Table 1.** Variation of lattice parameter, X-ray density, bulk-density, porosity and Curie temperature with cobalt content in $Ni_{1-x}Co_xFe_2O_4$.

| $x^a$ | $a_{th}$ ($A^0$) | $a_{exp}$ ($A^0$) | $\rho_x$ (gm/cm$^3$) | $\rho_b$ (gm/cm$^3$) | P (%) | $T_c$ (C) Impendence Analyzer | $T_c$ (C) DSC |
|---|---|---|---|---|---|---|---|
| 0 | 8.32 | 8.31 | 5.42 | 5.02 | 7.38 | 538 | 556 |
| 0.2 | 8.33 | 8.32 | 5.4 | 4.41 | 18.33 | 527 | 538 |
| 0.4 | 8.35 | 8.34 | 5.36 | 4.34 | 19.02 | 510 | 520 |
| 0.6 | 8.36 | 8.35 | 5.35 | 4.32 | 19.25 | 490 | 510 |
| 0.8 | 8.38 | 8.37 | 5.31 | 4.39 | 17.32 | 460 | 498 |
| 1 | 8.39 | 8.38 | 5.29 | 5.03 | 4.91 | 450 | 485 |

[a] Cobalt (Co) content

$a_{th}$: theoretical lattice parameter, $a_{exp}$: experimental lattice parameter, $\rho_x$: X-ray density, $\rho_b$: bulk density, P: porosity, $T_c$: Curie temperature.

### *3.2. Microstructure*

Magnetic and electrical properties of ferrites highly depend on the microstructure. Between the grain size and porosity of microstructures, the grain size is more important parameter affecting the magnetic properties of a ferrite. Grain growth is closely related to the grain boundary mobility. Recrystallization and grain growth involve the movement of grain boundaries. The grain growth being a result of inter-particle mass transport, appears to be dominated by the bimodal diffusion mechanism of lattice and grain boundary diffusion [10, 11]. Scanning electron micrographs (SEM) of $Ni_{1-x}Co_xFe_2O_4$ sintered at 1200 $^0C$ are shown in Figure 3. As seen from the micrographs, the average dimension of the particles is below 10 μm.

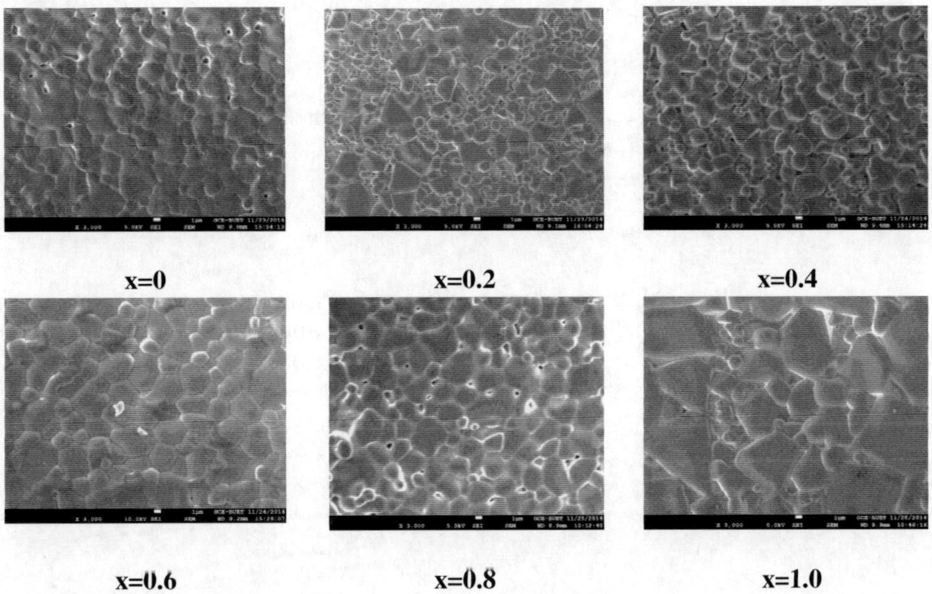


**Figure 3**. SEM micrgraphs of $Ni_{1-x}Co_xFe_2O_4$ samples sintered at 1200 $^0C$.

### *3.3. Temperature dependence of initial permeability*

The magnitude of the initial permeability μ′ of a magnetic material is proportional to the square of the saturation magnetization and inversely proportional to the magnetic magnetocrystalline anisotropy constant. Both quantities vary with temperature, so μ′ is a complicated function of temperature [12]. The variation of μ′ with temperature is a function of the anisotropy constant ($\kappa_1$) and saturation magnetization ($M_s$). Both $\kappa_1$ and $M_s$ usually decrease with increase in temperature. However, the decrease in $\kappa_1$ with temperature is faster than the decrease in $M_s$. When the anisotropy constant reaches zero, μ′ attains its maximum value and then drops off to the maximum value near the Curie point. The thermal variation of the initial permeability μ′ was recorded to determine Curie temperature, $T_C$ for the above investigated compositions of Ni-Co ferrite in the shape of toroid samples which is shown in Figure 4.

The Curie temperature mainly depends upon the strength of the A-B exchange interaction. According to Neel's sub-lattice field model [13], the A-B exchange interaction is stronger than the A-A or B-B interaction. As the concentration of $Co^{2+}$ increases, the amount of $Fe^{3+}$ decreases in the B sites. This reduces the A-B interactions. From figure 4, it can be seen that μ′ increases with increasing temperature, attaining a maximum value just below $T_C$. The maximum value of μ′ just below $T_C$ is a manifestation of the Hopkinson peak attributed to the minimization of magneto-crystalline anisotropy energy with temperature. Beyond this peak, permeability sharply falls to a very low value indicating $T_C$. The sharp fall of the permeability at the Curie point indicates the degree of homogeneity in the sample composition [14]. From the curves of the initial permeability versus temperature, the Hopkinson peak is found which may be attributed to structural transformation from tetragonal to cubic in these samples [15, 16]. Figure 5 shows the phase transition temperature for different composition of $Ni_{1-x}Co_xFe_2O_4$ determined from DSC measurements and Figure 6 shows the variation of Curie temperature with Co content.

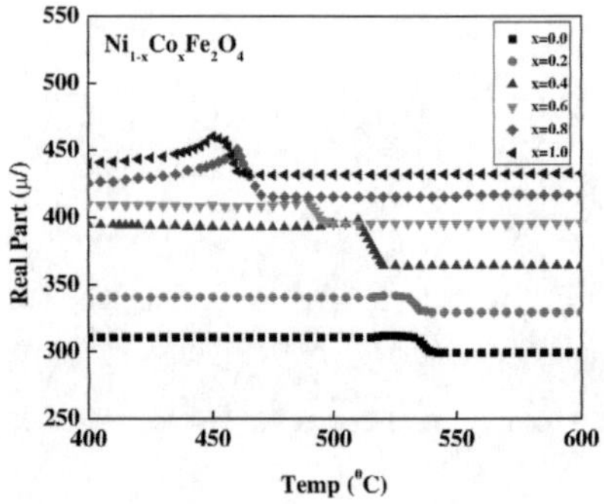


**Figure 4**. Permeability vs Temperature for different composition of $Ni_{1-x}Co_xFe_2O_4$.

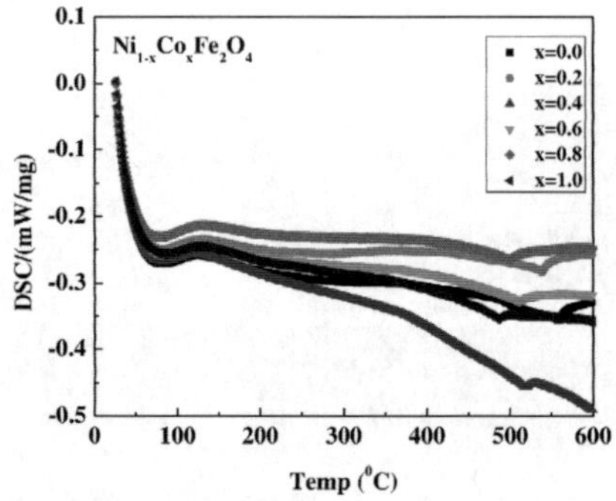


**Figure 5**. Phase transition temperature for different composition of $Ni_{1-x}Co_xFe_2O_4$ determined from DSC

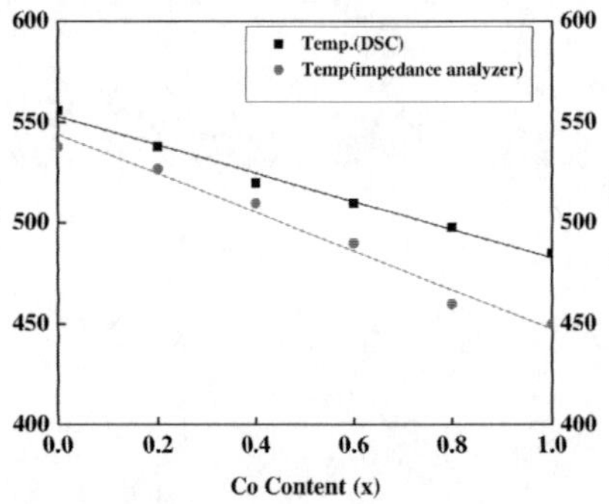


**Figure 6**.Variation of Curie temperature with Co content.

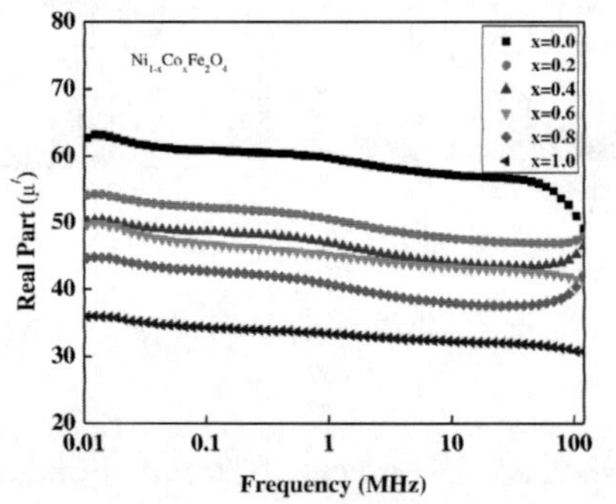


**Figure 7**.Variation of initial permeability with Co content.

### *3.4. Frequency dependence of initial permeability*

Figure 7 shows the variation of the initial permeability spectra as a function of the Co content. It is noticed that the real part of the initial permeability (μ′) decreases with increasing $Co^{2+}$ substitution in $Ni_{1-x}Co_xFe_2O_4$samples sintered at 1200 °C. However the decrease in μ′ with the increase of Co content was not consistent with the variation of microstructures shown in Figure 3. According to the Globus and Duplex model [14], the static permeability and their product are micro-structurally dependent. Ferrites with large grains are preferred to high permeability values because of the contribution of the domain wall motion. A decrease in the grain size results in a decrease in the number of domain walls in each grain. Moreover the decrease in initial permeability is closely related to the densification of the sample. A decrease in the density of ferrites not only results in an increase of the demagnetizing field due to the presence of pores but also reduces the spin rotational contribution, which in turn decreases the permeability [17, 18]. Therefore, the decrease of the initial permeability of Ni-Co ferrites with the increase of the Co content was attributed to the decrease in bulk densities.

### *3.5. M-H loop measurement*

Measurement of a hysteresis loop is a very simple and fast method to characterize ferromagnetic materials. M-H loops at 80K, 200K and 300K temperature for the investigated compositions of $Ni_{1-x}Co_xFe_2O_4$ (x=0.0, 0.2, 0.4, 0.6, 0.8, 1) sintered at 1200$^0$C are depicted in figures 8 to 10.

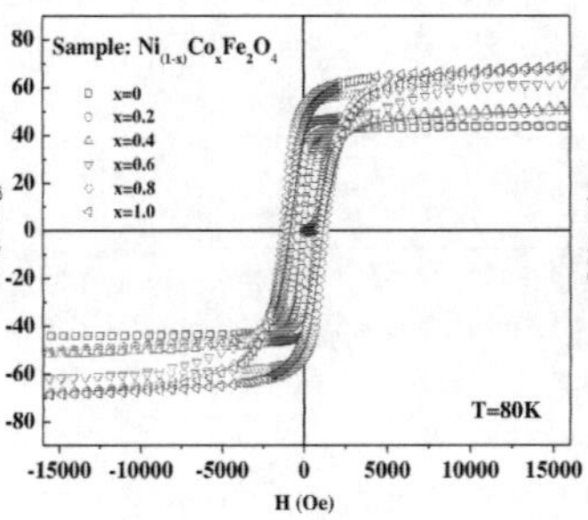


**Figure 8**. The M-H loop for $Ni_{1-x}Co_xFe_2O_4$ at 80K.

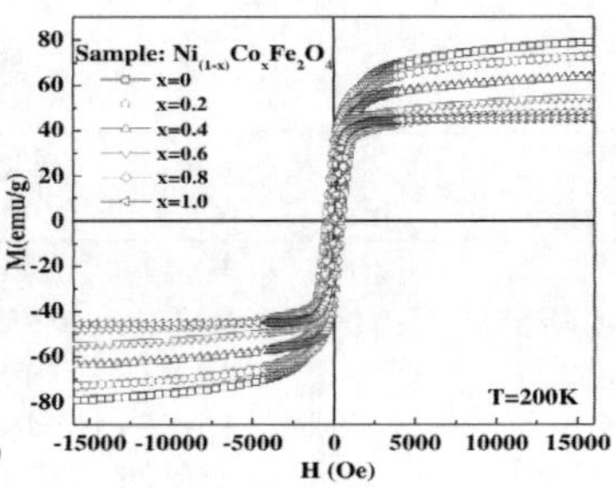


**Figure 9**. The M-H loop for $Ni_{1-x}Co_xFe_2O_4$ at 200K.

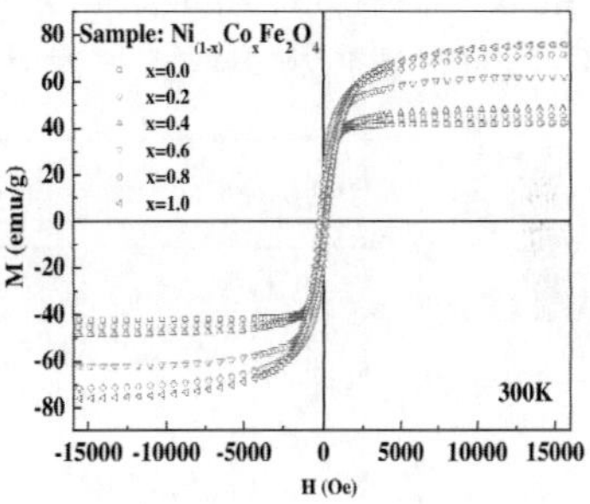


**Figure 10**. The M-H loop for $Ni_{1-x}Co_xFe_2O_4$ at 300K.

The hysteresis curves and their properties such as the saturation magnetization ($M_s$), retentivity ($M_r$), coercivity ($H_c$), and hysteresis loss are extracted from the application point of view. Tables 2-4 summarize hysteresis parameters such as $M_s$, $M_r$, $H_c$ determined from M-H loop measured at 80K, 200K and 300K. Saturation magnetization, $M_s$ measured at 80K was found to be much higher compared with data at 300K. $M_s$ has shown to increase with decreasing temperature and attained a maximum value for $CoFe_2O_4$ at any measurement temperature which may be attributed to higher magnetic moment of $CoFe_2O_4$ compared to $NiFe_2O_4$. A gradual increase in magnetic hardening was observed with increasing $CoFe_2O_4$ content attaining a maximum value of $H_c$ at x=0.8.

**Table 2**. Saturation magnetization ($M_S$), retentivity ($M_r$), coercivity ($H_C$), and hysteresis loss of Ni-Co ferrite samples at 80K.

| Samples | $M_s$ (emu/g) | $M_r$ (emu/g) | $-H_C$ (Oe) | $+H_C$ (Oe) | $\Delta H_C$ |
|---|---|---|---|---|---|
| $NiFe_2O_4$ | 45.58 | 22.55 | -38.23 | 37.54 | 75.77 |
| $Ni_{0.8}Co_{0.2}Fe_2O_4$ | 47.54 | 31.96 | -202.03 | 196.53 | 398.56 |
| $Ni_{0.6}Co_{0.4}Fe_2O_4$ | 54.69 | 33.05 | -233.62 | 234.67 | 468.29 |
| $Ni_{0.4}Co_{0.6}Fe_2O_4$ | 63.71 | 22.25 | -348.28 | 346.57 | 694.85 |
| $Ni_{0.2}Co_{0.4}Fe_2O_4$ | 72.77 | 16.02 | -480.98 | 481.44 | 962.42 |
| $CoFe_2O_4$ | 79.03 | 5.08 | -434.07 | 429.4 | 863.48 |

**Table 3**. Saturation magnetization ($M_S$), retentivity($M_r$), coercivity($H_C$), and hysteresis loss of Ni-Co ferrite samples at 200K.

| Samples | $M_s$ (emu/g) | $M_r$ (emu/g) | $-H_C$ (Oe) | $+H_C$ (Oe) | $\Delta H_C$ |
|---|---|---|---|---|---|
| $NiFe_2O_4$ | 43.95 | 2.51 | 24.75 | 33.57 | 8.81 |
| $Ni_{0.8}Co_{0.2}Fe_2O_4$ | 50.36 | 28.71 | 373.32 | 361.81 | 11.51 |
| $Ni_{0.6}Co_{0.4}Fe_2O_4$ | 51.59 | 42.59 | 774.74 | 771.74 | 3 |
| $Ni_{0.4}Co_{0.6}Fe_2O_4$ | 61.74 | 50.67 | 1047.36 | 1038.79 | 8.57 |
| $Ni_{0.2}Co_{0.4}Fe_2O_4$ | 67.7 | 52.84 | 1267.02 | 1266.42 | 0.059 |
| $CoFe_2O_4$ | 68.67 | 43.73 | 100.63 | 1010.72 | 10.08 |

**Table 4**. Saturation magnetization ($M_S$), retentivity($M_r$), coercivity($H_C$), and hysteresis loss of Ni-Co ferrite samples at 300K.

| Samples | $M_s$ (emu/g) | $M_r$ (emu/g) | $-H_C$ (Oe) | $+H_C$ (Oe) | $\Delta H_C$ |
|---|---|---|---|---|---|
| $NiFe_2O_4$ | 42.16 | 3.16 | 30.41 | 20.82 | 9.58 |
| $Ni_{0.8}Co_{0.2}Fe_2O_4$ | 45.25 | 5.61 | 68.98 | 61.47 | 7.51 |
| $Ni_{0.6}Co_{0.4}Fe_2O_4$ | 48.25 | 7.01 | 82.08 | 73.89 | 8.18 |
| $Ni_{0.4}Co_{0.6}Fe_2O_4$ | 62.27 | 13.18 | 150.58 | 141.63 | 8.95 |
| $Ni_{0.2}Co_{0.4}Fe_2O_4$ | 71.44 | 17.6 | 248.27 | 239.6 | 8.67 |
| $CoFe_2O_4$ | 75.77 | 11.99 | 188.57 | 181.03 | 7.54 |

## 4. Conclusions

The presence of Co ions causes appreciable changes in the physical and magnetic properties of the Ni-Co ferrites. XRD results indicate a single phase of the pure cubic spinel structure. The lattice parameter increases with the increase of Co content. The bulk density and X-ray density are found to decrease linearly with increasing Co content in the mixed Ni-Co ferrite system. The average grain size of $Ni_{1-x}Co_xFe_2O_4$ system increases with addition of Co. The Curie temperature has shown to decrease linearly with increasing Co content. The initial permeability was found to decrease with increasing Co content.

## 5. Reference

[1] Maqsood A Khan, K Rehman M A and Malik M A 2011 *Journal of Alloys and Compound*, Vol. 509 pp 7493–749.
[2] Batoo K M et al. 2013 *Journal of Nanoparticle Research Res* Vol 15 No 2067 pp 1-9
[3] A K M A. Hossain M Seki T Kawai H Tabata 2004 *Journal of Appl. Phys.* Vol 96 (2) No 1273.
[4] A Meenakshisundaram N Gunasekaran V Srinivasan 1982 *Phys. Status Solidi (a)* 69 KIS
[5] C Rath et al 2002 *Journal of Applied Physics* Vol 91 pp 2211-2215.
[6] J B Nelson DP Riley 1945 *Proc Phys Soc (London)* Vol 57 pp 160.
[7] L Vegard 1921 *Z. Phys.* Vol 5 pp 17-26.
[8] R D Shannon 1976 *Acta Crys A* Vol 32 pp 751-767.
[9] Pandit et al 2005 *J. of Mater. Sci.* Vol 40(2) pp 423-428.
[10] T K Gupta R L Coble 1968 *J. Am. Ceram. Soc.* Vol 51 pp 521-525.
[11] Calton et a 2002 *J. Magn. Magn. Matter.* Vol 160 pp 242-245.
[12] Smit J Wign 1959 *HPJ Ferrites (*New York: Wiley).
[13] Neel, L 1948 *Magnetic properties of ferrites. Ann. Phys. Paris* Vol 3 pp 137–198
[14] Globus A Duplex P 1966 *IEEE Trans. Magn.* Vol(3) pp 441–445.
[15] Dawoud HA Shaat SKK 2006 *Islamic University Journal* Vol 14 pp 165–182
[16] Bergestain A Cervinka L 1961 *Czech J. Phys. B.* Vol 11 pp 584–594.

[17] J.J Shorti, S D.Kulkarn, C E Deshpande S K Date 1999 *Mater. Chem. Phys.* Vol 59 pp 1-5.
[18] ST Mahmud et al 2006 *Journal of Magnetism an Magnetic Materials* Vol 305 pp 269-274.

**Acknowledgments**
The authors are indebted to the Department of Glass and Ceramic Engineering, Bangladesh University of Engineering and Technology (BUET) and Material Science Divisions, Atomic Energy Centre, Dhaka-1000, Bangladesh for extending experimental facilities

IOP Conf. Series: Materials Science and Engineering **438** (2018) 012018 doi:10.1088/1757-899X/438/1/012018

# Utilization of Hard Rock Dust in Ceramic Glaze Formulation

**Mst. Shanjida Sultana[1], Aninda Nafis Ahmed[2], Mohammad Nazim Zaman[1], Md. Aminur Rahman[1]**

[1]Institute of Mining, Mineralogy and Metallurgy, Bangladesh Council of Scientific and Industrial Research, (BCSIR), Joypurhat 5900, Bangladesh
[2]Pilot Plant & Process Development Centre,, Bangladesh Council of Scientific and Industrial Research (BCSIR), Dhaka 1205, Bangladesh

**Abstract.** The research focuses on the utilization of hard rock dust for formulating ceramic glazes. The rock dust was collected from Maddhapara Granite Mining area, Dinajpur, Bangladesh. Chemical compositions of raw materials were analyzed by XRF. Rock dust is characterized by high $SiO_2$, $K_2O$, and $Na_2O$ with low content of ferromagnesian element. For Rock dust minor amounts of components may be presents, which will mostly affect the color of the fired product ($Fe_2O_3$,MnO,$TiO_2$,$Cr_2O_3$).Others (MgO,$K_2O$,$Na_2O$) will act as fluxes and may have a strong effect during sintering. Ceramic glazes containing up to 30 wt% rock dust were prepared. The glaze slurries were applied onto ceramic tiles, which were sintered at 1050 to 1100°C with slow and rapid heating rate. Different glaze mixtures containing rock dust were prepared and analyzed for micro hardness and glossiness. Microstructures of the ceramic glaze were observed by FESEM (Field emission scanning electron microscopy). The results indicated that the properties of ceramic glaze were better sintering with slow heating rate and confirmed that dust can be used in glaze formulation. Therefore utilization of hard rock dust in glaze formulation may be new possibility for recycling and conserving natural resources.

## '1. Introduction'

At present, the growth rate of the global production of glazed tiles increased more than other ceramic products with improved aesthetic appearance, gave it a prominent role in the tile market. According to several researches [1-3], uses of natural product have become an important aspect in the ceramic industry, in order to optimize and reduce the consumption of natural resources.

Ceramics industries having a wide variety of process routes. The development of new glass ceramic glazes is limited by some constraints, mainly derived from the heating conditions such as final firing temperature and heating rate. Now most research focus to the development of new glaze products with smoother surfaces and desirable mechanical and chemical properties under similar sintering conditions [4]

There are several reuse and recycling solutions for granite and marble wastes, both at an experimental phase and in practical applications [5-7]. Granite is an igneous rock, with feldspar, quartz and mica as major components and marble is a crystalline metamorphic, shows non-plastic behaviour. Their major constituents are silica ($SiO_2$) and alumina ($Al_2O_3$) with minor quantities of lime (CaO) and alkaline oxides ($Na_2O$, $K_2O$) which acts as fluxing agent. Iron oxide content is also significant, due to the sawing process, and will promote a dark colour of the ceramic products, [8-10].In many studies using granite and marble wastes as raw material in the production of glaze ceramic tiles have been performed[1,6,7]

The hard rock obtained from Maddhapara Granite Mining Company, Bangladesh contains tonalite and granodiorite associated with granodi-oritic gneiss, granite and adamellite [11]. This mining project

produces huge amount of rock dust which is a waste and causing environmental pollution. During the blasting and crushing process, the fine particles can cause more pollution than other forms of dust unless stored properly and further utilized.
The aim of the present work is to utilization of maximum quantities of hard rock dust as raw material in the glaze formulation to develop glazed ceramic tiles.

## '2. Experimental'

### *'2.1 Materials'*

In this study, Rock dust, feldspar, sand, lime stone ($CaCO_3$) and $Li_2O_4B_7$ were used as raw materials glaze composition. Rock dust and feldspar was collected from Maddhapara Granite Mining Company Limited, situated in the district of Dinajpur, Bangladesh. River Sand also used for glaze formulation, which was processed in IMMM (Institute of Mining, Mineralogy and Metallurgy, Bangladesh Council of Scientific and Industrial Research, (BCSIR), Joypurhat laboratory.

**Table-1:** Proximate composition of Raw materials for Ceramic Glaze

| Raw Materials | Wt% |
|---|---|
| Rock dust | 20- 30 |
| Feldspar | 20-25 |
| Sand | 10-15 |
| $CaCO_3$ | 10-15 |
| $Li_2O_4B_7$ | 10-15 |

### *'2.2 Preparation of the Glaze'*

Chemical composition of raw materials analysis of the samples was performed by XRF (X-ray fluorescence). The raw materials were crushed, pulverized and ball milling to powder in form. Prepare a slip with a composition of raw materials (table 1) and adequate water to apply the materials. Pouring the slip on to natural ceramic body and dry the sample about 10 to 15 minutes. Then the specimens were fired to get the final glaze product by heat treatment of 1000-1150 °C temperature in an electric furnace for 30 minutes under heating rate of 5°C/min and 10°C/min (table 2) . Then various properties were analyzed.

**Table 2.**Sample Identification with response to heating temperature and rate

| Sample | Temperature(°C) and rate |
|---|---|
| S1 | 1050, 5°C/min |
| S2 | 1050, 10°C/min |
| S3 | 1100, 5°C/min |
| S4 | 1100, 10°C/min |

IOP Conf. Series: Materials Science and Engineering **438** (2018) 012018 doi:10.1088/1757-899X/438/1/012018

*'2.3 Characterization of the Glaze'*

Glossiness of the prepared glazes was determined by NOVO-GLOSS LITE$^{TM}$ statistical gloss meter. Micro hardness was measured by Vickers indentation using Micro-hardness Tester (HMV-2T, Shimadzu Corporation, Japan). Microstructures of surface of glazes were studied by FESEM-JEOL JSM-7600F.

**'3. Results and discussion'**

*'3.1 Characterization of Raw Materials'*

The chemical compositions of main raw materials analyzed by XRF are shown in Table 3. From 'Table 3' it is evident that rock dust predominantly contains $SiO_2$ with considerable amount of $Al_2O_3$ and minor amounts of components present, which will mostly affect the color of the fired product ($Fe_2O_3$, $TiO_2$).Fluxing agents ($Na_2O$ and $K_2O$) are present in both rock dust and feldspar. Feldspar contains 15.2% $K_2O$, which is excellent for glaze formulation rather than market available feldspar. Sand contains 97.7% $SiO_2$, which have a great influence for crystallization of glaze formulation.

'Figure 1' shows the particle size distribution of raw powder where, grinding a set of natural raw materials into a powder and found the particle size of the powder was less than 70 microns.

**Table 3.**Chemical compositions of the main raw materials.

| Compound | Rock dust | Feldspar | Sand |
|---|---|---|---|
| $SiO_2$ | 53.57 | 63.1 | 97.7 |
| $Al_2O_3$ | 16.51 | 17.2 | 0.21 |
| $Fe_2O_3$ | 9.07 | 0.95 | 1.15 |
| $K_2O$ | 3.15 | 15.2 | 0.13 |
| CaO | 0.76 | 0.95 | 0.16 |
| $Na_2O$ | 3.01 | 1.67 | - |
| MgO | 4.48 | 0.07 | - |
| $TiO_2$ | 0.76 | 0.02 | - |

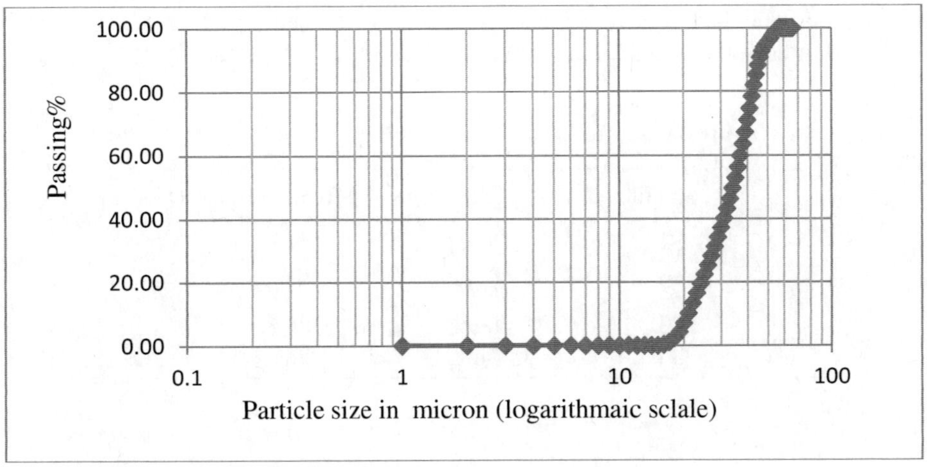


Figure 1: Particle size distribution of the prepared glaze powder.

*'3.2 Characterization of the Prepared Glaze'*

The color of glaze of the ceramic tile can be changed with the help of various pigments, waste products, or natural raw materials containing a lot of colorific oxides. 'Figure 2' show that variety of colors can be obtained by varying the proportion of waste in the composition. S2 samples were found with some bubbles, whereas, S1 found no bubbles which was formulated with slow heating rate at 1050°C.

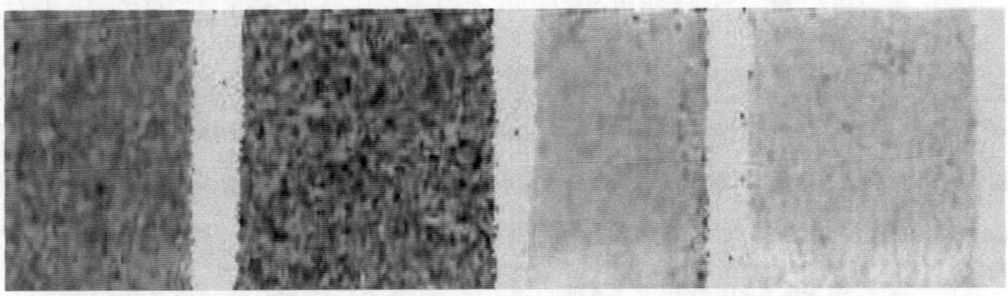

S1 S2 S3 S4

Figure2. Photographs of the developed different glaze ceramic

The prepared glaze coatings obtained were with a smooth surface and had a mat or semi-glossy appearance. Table 4 shows the glossiness and micro hardness of the prepared glaze. The glossiness of S1, S2 and S3 samples were better (46.23, 47.06 and 48.66) than S4 samples. Therefore, slow heating rate might be better for glaze development.

Vickers hardness depends on the type and amount of crystalline phases, composition of residual glassy phases and porosity of the glaze surface. It has been seen that S3 and S4 samples achieved highest micro hardness, however, Vickers hardness of S1 and S2 were better for glazed ceramic tiles.

The glaze coatings that had been obtained on heating temperature 1100°C, 5°C/min (S3) was considerably better than other samples in response to glossiness and hardness.

**Table 4.**Glossiness and Micro hardness of the prepared glaze.

| Sample | Glossiness | Micro hardness (HV) |
|---|---|---|
| S1 | 46.23 | 649 |
| S2 | 47.06 | 621 |
| S3 | 48.66. | 865 |
| S4 | 44.80 | 981 |

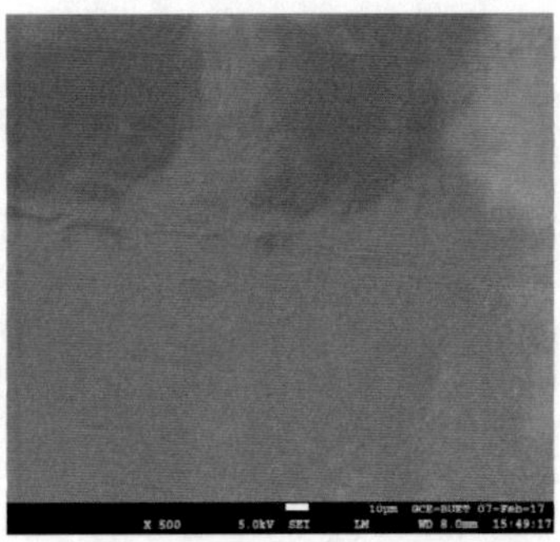

(a)

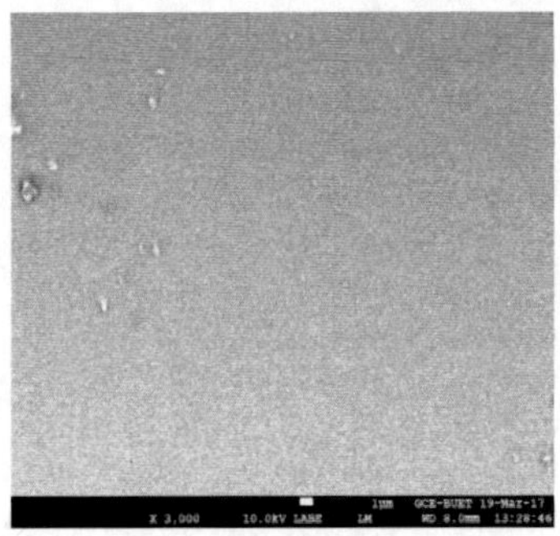

(b)

(c)

(d)

Figure 3. Microstructure of the surface of the glaze for (a) S1,(b)S2,(c)S3 and (d)S4

'Figure 3' show the surface micro structure of the glaze tiles of the prepared samples. The morphology of S1 sample was almost smooth but crack was found and S2, which was formulated at 1050 °C shows smooth surface with very fine crystals.S3 and S4 samples, which developed at 1100 °C temperature show glossy and smooth surface because of the high content of fluxing.

## '4. Conclusion'

- In this research low firing temperature glazes were formulated using hard rock dust, and feldspar that collected from Maddhapara Granite Mining area, Bangladesh.

- The temperature for development of glaze can be reduced to 1050°C which is low for ceramics.
- This low firing temperature glazes were prepared using high percentages (70%) of waste and natural materials.
- The development of glaze with hard rock dust, which is dumping as a waste material will be a new route in ceramic industries as well as production cost of glazed ceramic tiles will be lower and better for environment.

**'5. References'**

[1] Segada˜es, A.M., Carvalho, M.A. and Acchar, W. 2005 Using marble and granite rejects to enhance the processing of clay products, Appl. Clay Sci. 30, 42–52.
[2] Demir, A.2009. Reuse of waste glass in building brick production. Waste Manag. Res. 27,572–577.
[3] Dondi, M. Guarini, G., Raimondo, M. and Zanelli, C. 2009. Recycling, TV waste glass in clay bricks and roof tiles, Waste Manag. 29,1945–1951.
[4] Barbier L, Bonamartini A C, and Lancellotti I 2000 J. Eur. Ceram. Soc. 20,2477-2483.
[5] García-Ten, J., Mallol, G., Bou, E., Silva, G., Fernández, J., Molina, A. and Romera, J. 2003. Recycling marble working wastes in manufacturing ceramic products. II Ceramic wall tile manufacture. Ceramic Forum International, DKG 80(10), 30-32
[6] Saboya, J., Xavier, G.C. and Alexandre, J.2007. The use of powder marble by-product to enhance the properties of brick ceramic. Constr. Build. Mater, 21, 1950-1960
[7 ] Acchar, W., Vieira, F.A. and Hotza, D.2006. Effect of marble and granite sludge in clay materials. Materials Science and Engineering A 419, 306–309
[8] Segadães, A.M. 2006. Use of the diagrams to guide ceramic production from wastes. Advances in Applied Ceramics 105, 46–54(2006)
[9] Scholze, H.1991.Glass–Nature, Structure and Properties, Springer-Verlag, New York, 1991.
[10] Tamsu, N., Bayrak, A.V. and Ozdemir, H. 2013. Effects of Na2O/ K2O ratio on the deformation behavior of the floor tile bodies, Acta Phys. Pol. A 123, 283–284.
[11] Zaman, M.N., Ahmed, S.S. Islam, M.B., Islam, M.S., Ishiga, H. Dozen,K., Daogongand C. And Xiachen, Z.2002. Bangladesh J. Geol., 21, 23–40

**Acknowledgments**

Authors are thankful to BCSIR authority for funding the research .Authors wishing to acknowledge to Department of Glass and Ceramic Engineering, BUET, Bangladesh to help the analysis of glossiness and FESEM.

# Fatigue Properties of Overaged Cast Aluminium-7Silicon-0.3Magnesium Alloy

**R Islam, A Rahman and A K M Bazlur Rashid**

Department of Materials and Metallurgical Engineering, Bangladesh University of Engineering and Technology, Dhaka-1000, Bangladesh

**Abstract.** Over the years, Al–7Si–Mg alloys are used because of their excellent castability, high weight-to-strength ratio, good corrosion resistance and good weldability. In this paper, the fatigue behaviour of Al–7Si–0.3Mg alloys, conforming to A356, has been studied. Specimens of this material were fatigue tested in both the as cast condition, solution treated and aged conditions. The fatigue test was conducted using 3-point fatigue bending setup with the principle of low cycle fatigue failure mechanism. Overaging is found to affect the fatigue behaviour of Al-7Si-0.3Mg alloys. The fatigue properties of as cast alloys showed a three times increase upon solution treatment. Peak ageing condition further improves the fatigue value. Exposure to severe overageing conditions reduced the fatigue properties somewhat from that of the peak values but still maintained higher value than the as-cast conditions, which is constant irrespective of the overageing condition. The improved fatigue properties of heat treated samples can be generally attributed to the spheroidisation of acicular silicon particles.

## *1. Introduction*

Presently the development of light alloys is still dominated by aluminium, with aluminium castings having an important role in the growth of the overall aluminium industry. Among the commercial aluminium castings, Al–7Si–0.3Mg is probably the most widely used alloy mainly because of its excellent stability, specific strength, corrosion resistance, and good tensile and fatigue properties. The popularity of these cast alloys, which are largely used in the automotive, aerospace and defence industries, creates a continuing increase in the demand for components with higher and more consistent mechanical properties. Car wheel rims made from cast aluminium alloys are attractive for several reasons: easy processing and manufacturing, lighter weight leading to lower fuel consumption, and the possibility of recycling the material [1].

Al-7Si-0.3Mg alloys used in automotive applications are exposed to cyclic loadings and fatigue failure. Although considerable progress has been made in understanding the mechanisms of fatigue failure, an accurate fatigue failure prediction is difficult because the different physical processes which prevail to the gradual fatigue damage accumulation during the fatigue life of a metallic component are complex and interrelated [2]. Developing with increasing number of fatigue cycles from atomic to macro-scale damage, the mechanisms entail a host of material, geometric and loading parameters [3]. Cyclic plasticity of engineering polycrystalline alloys is complex and depends on a host of parameters including type of unit cell, value of stacking fault energy, heat treatment, grain size, precipitate geometry and size, distribution and coherence to the matrix etc. [1, 5]. In precipitation hardened alloys, cyclic hardening occurs due to an increase in dislocation density and dislocation-precipitate interactions. Hence, cyclic hardening is highly favoured if the precipitates in the age-hardened alloy are not easily shearable by the dislocations [4].

In the last ten years, there were activities toward improving the understanding of the effect of some intrinsic and extrinsic factors on fatigue crack initiation and propagation in Al-Si-Mg alloy [5]. Extensive research efforts were focused on fatigue behaviour under thermal aging, modification treatment which can improve the mechanical properties of the alloy [6].The influence of microstructure on the fatigue behaviour of Al-Si-Mg alloy was also investigated.

Interactions between the dislocation and precipitates, dislocation and secondary silicon phase are investigated to understand the fatigue behaviour and performance [7]. Several works on the effect of porosity on fatigue of cast Al-Si-Mg have been done so far. Porosity has a catastrophic effect on mechanical properties of aluminium alloys. A 1% volumetric fraction of pores may reduce fatigue life by 50% and strength limit by 20% as compared to the same alloy without pores [8]. The number of cycles needed for crack nucleation depends on loading, ranging from 0% (at start up) to less than 10% of fatigue life [9]. Apart from these, the influence of stress on the crack propagation is also studied. While the time needed for crack initiation decreases with pore size, fatigue life increases when stress decreases and that for a given stress level the number of cycles to failure decreases when amount of pore introduced in the alloy increases [10].

So far, there is very little amount of work done on the influence of overaging on the fatigue behaviour of aluminium alloys. As the cast Al-7Si-Mg alloy being used in the cylinder head, wheel rim, on long time usage the alloy becomes overaged. Therefore, the fatigue behaviour of these alloys in overaged condition is very crucial.

## *2. Experimental*

In the present work, the fatigue response of cast Al-7Si-0.3Mg alloy upon overaging condition was studied. Cylindrical fatigue test samples of 20 mm diameter and 250 mm length were machined from the cylindrical sand cast bars of 22 mm diameter and 280 mm length. The test bars were solution treated at 540±2°C in an air-circulating furnace for 4 hours and then quenched in water of 30±5°C. Then they are refrigerated at -5°C till they were aged at the effective ageing temperature of 170±2°C for 1, 2, 5, 10, 100, 1000, 10000 and 100000 hours. The long ageing times were reduced to a few hours by increasing the ageing temperature using the Arrhenius model [11], which indicates that for every 10°C temperature rise, the ageing time effectively reduces to half. The ageing schedule used for the work is given in Table 1.

Table 1: Ageing schedule used in the work.

| Effective ageing time at 170°C | Actual ageing schedule |
|---|---|
| 1 hr | 1 hr at 170°C |
| 2 hr | 2 hr at 170°C |
| 5 hr | 5 hr at 170°C |
| 10 hr | 1.25 hr at 200°C |
| 100 (98) hr | 3.125 (3) hr at 220°C |
| 1000 (1024) hr | 1.95 (2) hr at 260°C |
| 10000 (10240) hr | 2.44 (2.5) hr at 290°C |
| 100000 (98304) hr | 3.052 (3) hr at 320°C |

The fatigue test was conducted using 3-point fatigue bending setup in the Instron 8801™. The specimens were supported on two lower anvils and the load was applied at the centre of each specimen by a single upper anvil. The fixture had a span length of 18 cm. The parameters maintained throughout

IOP Conf. Series: Materials Science and Engineering **438** (2018) 012019 doi:10.1088/1757-899X/438/1/012019

all the fatigue tests are: Stress ratio = 0.1, Maximum load= 3 kN, Minimum load= 0.3 kN, Average end point = 1.65 kN, Amplitude = 1.35 kN, Frequency= 40 Hz.

The structural morphology of as-cast and heat-treated samples were observed using optical microscopy. Standard metallographic techniques were used to prepare the metallographic specimens.

### *3. Result and Discussion*

The fatigue test of the alloys was done with the principle of low cycle fatigue failure mechanism and as the aluminium is a low strength metal, it has cyclically hardened with repetitive loading and failed in brittle mode [12].

Figure 1 shows the fatigue behaviour of the as-cast and heat-treated samples. It is clear that, the fatigue property is improved during heat treatment. The solution treatment showed almost a three times improvement in fatigue properties over the as-cast samples. The highest fatigue resistance was observed at the peak ageing condition (at about 5 hour ageing time). With over-aging, the fatigue property was found to be decreased somewhat but the fatigue resistance of highly overaged alloys maintained a more-or-less steady value, which is higher than that of the as-cast sample.

This is surprising and contradictory to the results obtained for the strength-ductility behaviour of over-aged aluminium-based aerospace casting alloys [13]. The tensile strengths of these alloys were found be decreased in the overgaed conditions to values which are lower than that obtained in the as-cast condition. Clearly more works are needed to ascertain this issue.

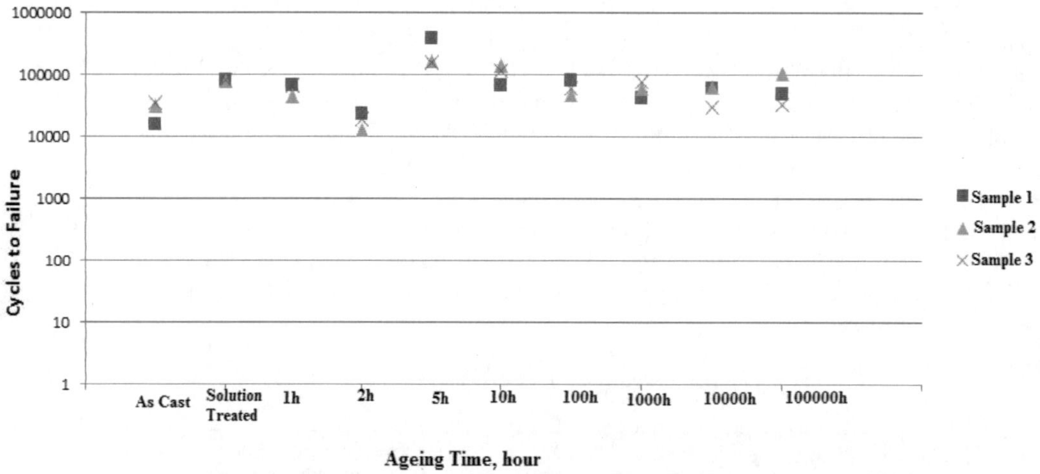


**Figure 1:** Fatigue response to heat treatment.

The size and shape of Si particle plays an important role in affecting the fatigue properties of the alloy. Though the surface of the as-cast samples were generally free from porosity and other defects, the presence of unmodified, coarse acicular silicon particles embedded in the α-Al matrix in the Al-7Si-0.3Mg samples (Fig. 2a) are responsible for the stress concentration in the sample. These additional stress concentrations reduced the fatigue life of as-cast samples.

Upon solution treatment, the silicon particle appeared fibrous and undissolved $Al_5FeSi$, dendritic structured dispersed throughout the structure [14]. Solution heat treatment also causes a substantial degree of spheroidisation and coarsening of silicon eutectics (Fig 2b), thus greatly reducing the stress concentration and improves fatigue life. Furthermore, due to low cycle fatigue (high stress) cracks nucleate by de-cohesion of internal silicon nodules (far from pores) probably located at grain boundaries [15]. So, the fatigue properties of ageing at 170°C for 1, 2, 5 hours are greater than those of as-cast samples.

The maximum fatigue property observed for ageing at 170°C for 5 hours. In this condition the density of the GP zone increases, because of which distortions of the lattice planes are extended to several atomic layers in the matrix. This distortion of the lattice plane causes a hindrance to the dislocation movement. Therefore, the fatigue resistance property of the alloy is further improved for under aged to peak aged conditions. The second reason for the increase in resistance to fatigue fracture behaviour of the alloy is due to the pinning and hindrance of the dislocation movement by $Mg_2Si$ and primary silicon [16].

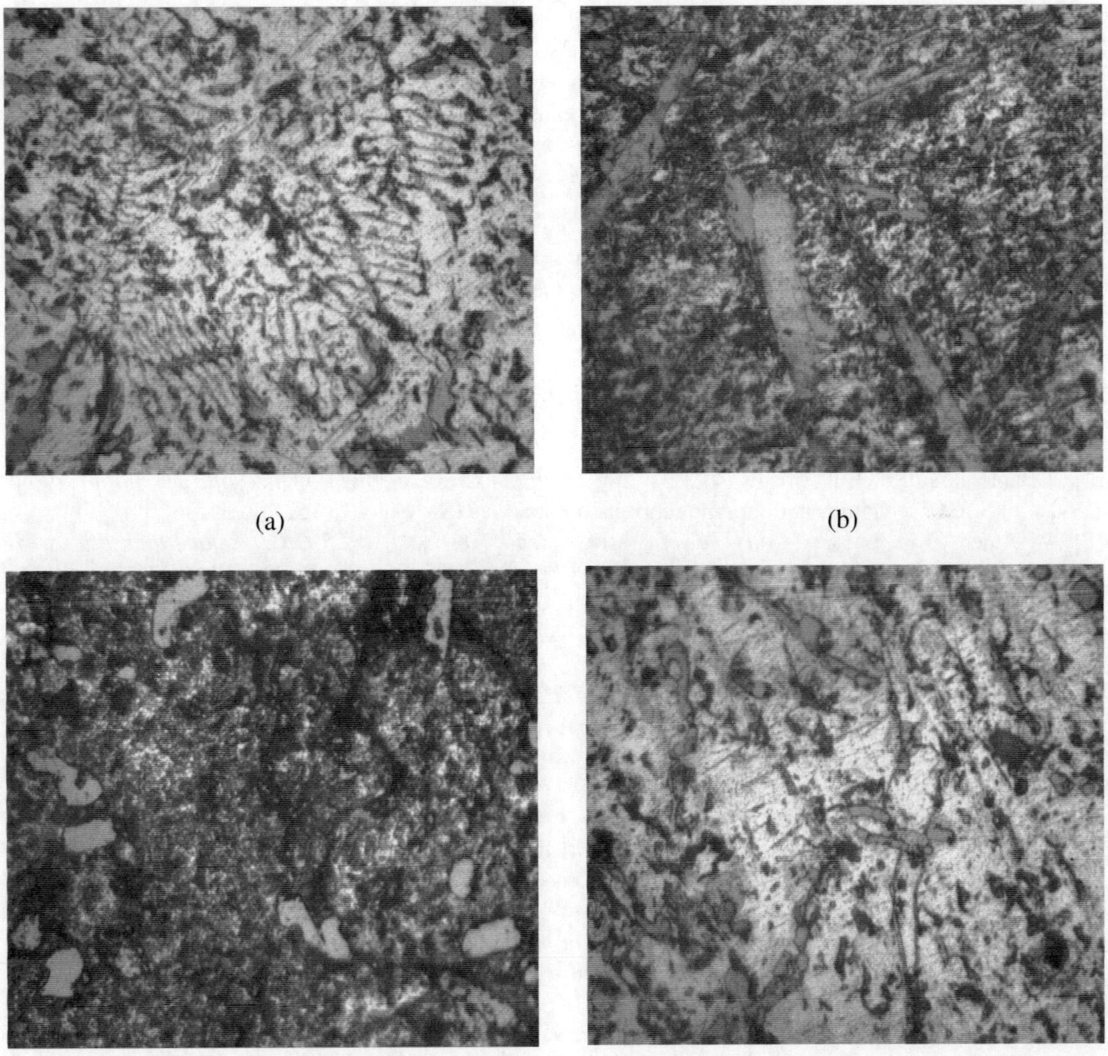

(a) (b)

(c) (d)

**Figure 2:** Optical micrograph of Al-7Si-0.3Mg alloys. (a) As-cast, (b) Solution treated, (c) Aged at 170°C for 5 hours, and (d) Aged at 170°C for 100000 hours (x500).

Upon over-aging, the fatigue properties of the samples are decreased but maintain a steady value. A decrease in the number of cycles to fail in these samples when over-aged for a longer period at higher temperatures could be attributed to further spheroidisation of silicon particles, coalescence of precipitates into larger particles and annealing of the defects in the material. This will cause fewer obstacles to the movement of dislocations. The almost constant fatigue property of the overaged samples can be attributed to the extensive grain growth and migration of silicon particle to the grain boundaries, thus deflecting the fatigue crack propagation. As the inter-dendritic crack propagation occurs alternatively in matrix and along interfaces of silicon/matrix particles, the crack propagation rate (da/dn) is almost constant.

## 4. Conclusion

Overaging is found to affect the fatigue behaviour of Al-7Si-0.3Mg alloys. The fatigue properties of cast alloys showed a three times increase upon solution treatment. Peak ageing condition further improves the fatigue value. Exposure to severe overageing conditions reduced the fatigue properties somewhat from that of the peak values but still maintained higher value than the as-cast conditions, which is constant irrespective of the overageing condition. The improved fatigue properties of heat treated samples can be generally attributed to the spheroidization of acicular silicon particles.

## 5. References

[1] Atxaga G, Pelayo A, Irisarri A M 2001 *Effect of microstructure on fatigue behaviour of cast Al–7Si–Mg alloy* (Materials Science and Technology vol. 17) pp 446-450

[2] Andrei G, Dima D, Anderi L 1997 *Lightweight materials for aircraft applications* (Material characterisation vol. 35 no. 1) pp 41-67

[3] Pantelakis Sp G, Petroyiannis P V, Bouzakis K D, Mirisidis I 2007 *Surface hardness increase of 2024 aluminium alloy subjected to cyclic loading* (Theoretical and Applied Fracture Mechanics vol. 48 no. 1) pp 68-81

[4] Pantelakisa Sp G, Kermanidis Al Th 2008 *Fatigue Crack Growth of Aircraft Aluminium Alloys* (Solid Mechanics and its Applications book series (SMIA) vol. 152) pp 345-366

[5] Feltner C L, Laird C 1967 *Cyclic stress-strain response of F.C.C. metals and alloys—II Dislocation structures and mechanisms* (Acta Metallurgica vol. 15 no. 10) pp 1633-1653

[6] May A, Belouchrani M A, Taharboucht S, Boudras A 2010 *Influence of heat treatment on the fatigue behaviour of two aluminium alloys 2024 and 2024 plated* (Procedia Engineering vol. 2 no. 1) pp 1795-1804

[7] Wang Q Y, Kawagoishi N, Chen Q 2006 *Fatigue and fracture behaviour of structural Al-alloys up to very long life regimes* (International Journal of Fatigue vol. 28 no. 11) pp 1572-1576

[8] Zoroufi M, Fatemi A 2006 *Experimental durability assessment and life prediction of vehicle suspension components: A case study of steering knuckles* (Institution of Mechanical Engineers vol. 220 no. Issue 11)

[9] Jiang X-s, He G-q, Liu B, Fan S-j, Zhu M-h 2011 *Microstructure-based analysis of fatigue behaviour of Al-Si-Mg alloy* (Transactions of Nonferrous Metals Society of China vol. 21 no. 3) pp 443-448

[10] Odegard, Jack A, Pedersen K 1994 *Fatigue properties of an A356 (AlSi7Mg) Aluminium alloy for automotive applications- fatigue life prediction* (SAE Technical Paper International Congress & Exposition)

[11] Rashid A K M B 1994 *Relation Between Time and Temperature of Ageing*, (Journal of the Institution of Engineers', Bangladesh, vol. 22 no. 1) pp.7-11.

[12] Buffière J-Y, Savelli S, Jouneau P H, Maire E, Fougères R 2001 *Experimental study of porosity and its relation to fatigue mechanisms of model Al-Si7-Mg0.3 cast Al alloys* (Materials Science and Engineering: A vol. 316 no. 1-2) pp 115-126

[13] Rashid A K M B and Campbell J 1999 *Strength-Ductility Relationship of Overaged High Strength Aerospace Aluminium Casting Alloys*, (Proceedings of the 6th Asian Foundry Congress, 23-26 January 1999, Calcutta, India) pp 211-217.

[14] Sjölander E 2010 T*he heat treatment of Al–Si–Cu–Mg casting alloys (*Journal of Materials Processing Technology vol. 210 Issue 10) pp 1249-1259

[15] Wang Q G, Davidson C J 2001 *Solidification and precipitation behaviour of Al-Si-Mg casting alloys* (Journal of Materials Science vol. 36 Issue 3) pp 739-750

[16] Ernst F 2004 *Precipitation Hardening of Al-Si-Mg* (Materials Laboratory lll) pp. 325-366.

**Acknowledgments**
We acknowledge financial and technical support from Department of Materials and Metallurgical Engineering, Bangladesh University of Engineering and Technology (BUET).

IOP Conf. Series: Materials Science and Engineering **438** (2018) 012020 doi:10.1088/1757-899X/438/1/012020

# Effect of Boron Addition on the Wear Properties of As-Cast AZ91 Magnesium Alloy

**N Akram, K M Shorowordi, S Roy and A Mahmud**

Department of Materials and Metallurgical Engineering, Bangladesh University of Engineering and Technology, Dhaka 1000, Bangladesh

**Abstract.** AZ91, AZ91-0.5B and AZ91-1.0B magnesium alloys are produced in an inert atmosphere to investigate the influence of B on microstructure and properties of AZ91 alloy. Microstructural study is carried out using optical microscopy. Microhardness measurements are conducted by Vickers hardness tester. Using a pin-on-disk type apparatus, wear tests are conducted under dry sliding condition to assess the beneficial effect of boron on the wear properties of AZ91 alloy in relation to the resulted hardness and microstructures. The worn surface and wear particles morphology are also investigated. It is observed from microstructural investigation that the grains become finer with the addition of boron. The hardness is also found to increase with the addition of boron in AZ91 alloy, however maximum hardness is achieved with the addition 0.5 % B addition. The wear resistance is found to be better in the boron added alloy and maximum resistance is found with the addition of 0.5 % B. Wear mechanisms are also investigated.

## *1. Introduction*

Magnesium alloys have shown greater demand in structural applications for its lightweight and higher strength-to-weight ratio compared to many other structural metals (such as steel, aluminium) [1,2]. These properties of magnesium alloys have attracted the attention in the automotive sector and aerospace industries to reduce fuel consumption and emissions. However, even after the recent developments of magnesium alloys, they are still not used as widely as aluminium alloys because of its poor formability arising from their h.c.p. crystal structure, inadequate mechanical properties, high susceptibility to hot tearing during permanent mould casting and relatively low corrosion resistance [3].

AZ91 is a widely used commercial magnesium alloy containing 9% aluminium and 1% Zn alloy but its mechanical properties are relatively low [4]. Grain refinement and secondary phase formed in this alloy can enhance the mechanical properties by the formation of fine equiaxed grains at the expense of dendrites. The studies on the grain refinement of magnesium based by melt superheating, rapid cooling, $FeCl_3$ addition, carbon inoculation, melt agitation have been reported [5]. It is found that the addition of inoculation is an effective method of refining the Mg based alloys. For the improvement of mechanical properties minor addition of alloying elements have also been studied such as Pb, Sr, RE, Th, Si, B and Ca [6-8]. Addition of some of these elements were found to be effective in improving the mechanical properties by refining the magnesium alloy with the formation of second phases. Boron is thought to be one of the potential element in improving the properties. Generally boron as in the form of master alloys, is added to Al–Si alloys for grain refinement [9]. Suresh et. al. have shown that small amount of boron addition in the form of Al-4B master alloy can cause grain refinement which can lead to improved mechanical properties [2].

IOP Conf. Series: Materials Science and Engineering **438** (2018) 012020 doi:10.1088/1757-899X/438/1/012020

Despite these studies, literatures on effective addition of boron in magnesium alloys is not enough to use boron at commercial scale. The present attempt has been taken to investigate the effect of small addition of boron on the properties of AZ91 alloy.

## *2. Experimental*

Three different Mg alloys AZ91, AZ91-0.5B (0.5% Boron), AZ91-1B (1% Boron) were prepared through casting process using pure ingots of magnesium, aluminum and zinc, and Al-B master alloy (92%Al-8%B) as raw materials. Melting of the starting metals was done in an induction furnace under Ar atmosphere. The melt was then poured into the steel mold and this was also done in Ar protecting atmosphere.

Small pieces of sample were cut from each cast ingot. Samples were prepared by following standard metallographic technique to investigate the microstructure of cast magnesium alloys using optical microscope. Samples were finally etched in acetic-picral reagent (5 ml acetic acid; 10 ml distilled water; 100 ml ethanol and 6 gm picric acid). The micro-hardness of AZ91, AZ91-0.5B and AZ91-1B samples were measured by a Vickers hardness tester under 1 kg load for the dwell time of 10 seconds. For each of the samples, at least ten measurements were taken from different locations of the sample.

For study of the wear properties of magnesium alloys, pins of 5 mm diameter and 12 mm length were machined out from the cast AZ91 alloy plates. The end surface of the pin specimen was polished using a grit paper 1200, cleaned in distilled water and then dried in acetone. A pin-on-disc type setup was used for the wear test in which AZ91 alloy pin was pressed against a horizontal rotating disc made from the hardened steel (HRC 53). Tests were conducted for a total duration of 30 minutes in an ambient air of relative humidity 64±5%and temperature 25±2°C. Three contact loads, viz. 5, 10 and 15 N were used, while the linear speed was kept constant at value of 0.75 $m.s^{-1}$. After the test, the specimens were cleaned in water, dried in acetone and weighed and wear rates ($gm.m^{-1}$) of the specimens were calculated from weight loss measurements. An average of at least six tests for each sliding condition was taken for wear rate measurements.

## *3. Results and discussion*

Figure 1 displays optical micrographs of as cast AZ91, AZ91-0.5B and AZ91-1B. It is seen that white phases have encircled the grains in all the as cast samples. It is reported that the white phase is intermetallic compounds of $\beta$-$Mg_{17}Al_{12}$ phase while the grain is of $\alpha$-Mg [10]. This intermetallic is formed during solidification of liquid metal in permanent mold. Figure 1 also reveals that the grains are almost same in size in AZ91 and AZ91-1B. The grain size of AZ91-0.5B alloy is observed finer than the other two Mg alloys. This indicates in the present study that the 0.5 weight percent boron addition is effective in grain refinement of AZ91, while 1.0 percent boron addition has almost no effect in grain refinement.

Suresh et. al. also studied the addition of boron in AZ91 and found that small addition of boron refines the grains [2]. This is attributed to the presence of $AlB_2$ particles. They reported that the particles are formed at early stage of liquid Mg alloy in the mold and acts as nucleation site for the solidification of liquid melt. The fine particles also have pinning effect which resist the growth of grain during solidification. Thus, it is thought that in the present case, grain size become finer for the addition of 0.5 weight percent boron due to the synergistic effects of fine $AlB_2$ particles early stage of nucleation and its pinning effect during solidification. In previous study, boron addition was limited to in the narrow range of 0.008 to 0.04 weight percentage. They also reported that no further grain refinement was observed beyond 0.032 weight percentage of boron addition [2]. On the contrary, Figure 1(b) shows clear refinement of grains at 0.5 weight percent of boron addition. Joshi et. al. have added boron through Al-5B master alloy and have found that 1%wt boron addition does indeed refine grains significantly. They attribute that the refinement occurs due to Al addition through the master alloy rather than $AlB_2$ particles. Also grain refinement was not very significant at any amount higher than 1 weight percent boron addition [11]. Present study reveals that 1% weight addition of boron does refine the grain. However, 0.5% B addition is found to be more effective in grain refinement (compare between Figure 1a and Figure 1b).

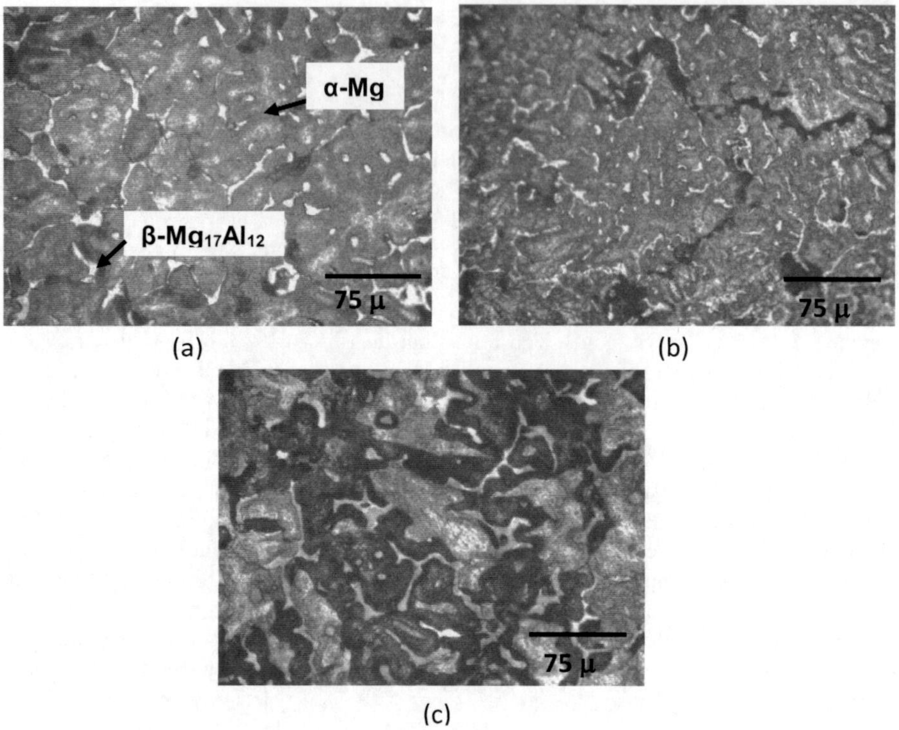


**Figure 1.** Optical micrograph showing the structure of as cast (a) AZ91 (b) AZ91-0.5B and (c) AZ91-1B.

Figure 2 shows a new phase in AZ91-0.5B at high magnifications which is revealed as black phase in boron added sample in Figure 1. This phase is not present in the AZ91 and AZ91-1%B. This phase seems to have a laminar structure and may be stronger than the α-Mg phase. Further study is required to identify the phase.

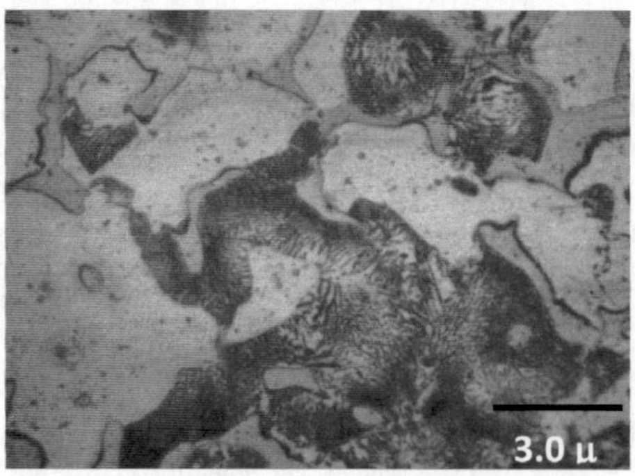


**Figure 2.** Showing a phase of laminar structure in as cast AZ91-0.5B alloy

The hardness of AZ91, AZ91-0.5B and AZ91-1B is shown in Figure 3. It can be seen from the figure that the average hardness of born added alloys AZ91-0.5B and AZ91-1B is higher than the AZ91 alloy. The hardness of 0.5 weight percent boron added sample is higher than the 1.0 weight percent boron added sample. The lowest hardness is found in AZ91 alloy due to its coarse grain structure and absence of second phase particles or laminar structure like phase which are found in the 0.5% boron added sample. The highest hardness is observed in AZ91-0.5B alloy mainly for its fine grained structure, although $AlB_2$ and laminar structure may also have increased the hardness. The hardness of AZ91-1.0B is higher than AZ91 alloy despite having almost same grain size, possibly due to the presence of $AlB_2$ and laminar phase (See Figure 1). The lower hardness of AZ91-1.0B compared to 0.5% B addition can be attributed mainly to the increase of grain size when 1% B is added. The laminar phase found in AZ91-0.5B may also be responsible for its high hardness compared to AZ91-1.0B but cannot be said for certain without further investigation on that phase.

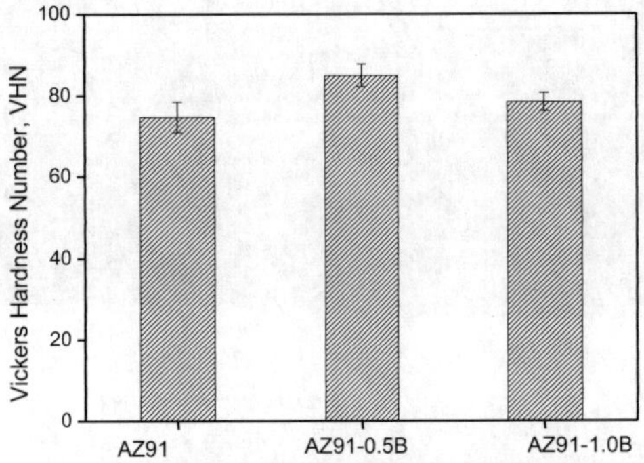


**Figure 3.** Showing the comparison of hardness of AZ91, AZ91-0.5B and AZ91-1B alloy

Figure 4 shows the variation of wear rate of AZ91, AZ91-0.5B and AZ91-1B with normal load. The wear rates of three alloys are found to increase as the normal load increases. The wear rate of AZ91-0.5B is seen to be lower than that of AZ91 and AZ91-1B. The AZ91 alloy shows the higher wear rate than the AZ91-1B. The lowest wear rate found in AZ91-0.5B is due to the higher hardness as compared to both the AZ91and AZ91-1B alloys. The wear properties of materials is dependent on the strength of materials,

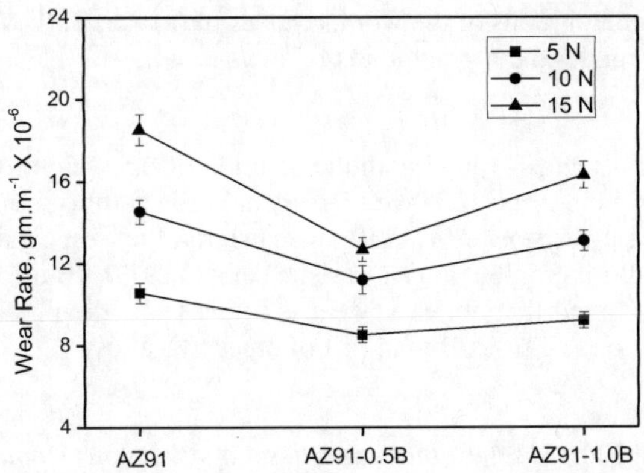


**Figure 4.** Wear rate of AZ91, AZ91-0.5B and AZ91-1B under different normal loads.

IOP Conf. Series: Materials Science and Engineering **438** (2018) 012020 doi:10.1088/1757-899X/438/1/012020

mainly hardness. In this study three alloy shows the wear properties according to their hardness property i.e. AZ91 alloy being of the lowest hardness shows the highest wear rate. The difference in wear rates between the alloys also becomes more pronounced as the load increases.

The morphology of the worn surface of AZ91, AZ91-0.5B and AZ91-1B samples is shown in Figure 5 tested at normal load of 5 N. Parallel ploughing marks along the sliding direction are visible on the worn surfaces of three samples. The ploughing marks on the worn surface and the positive correlation with wear rate suggest that abrasive wear mainly operates under the present experimental conditions. The abrasive wear is found to occur in lesser extent in AZ91-0.5B compared to the AZ91 and AZ91-1B as the ploughing marks are less deep.

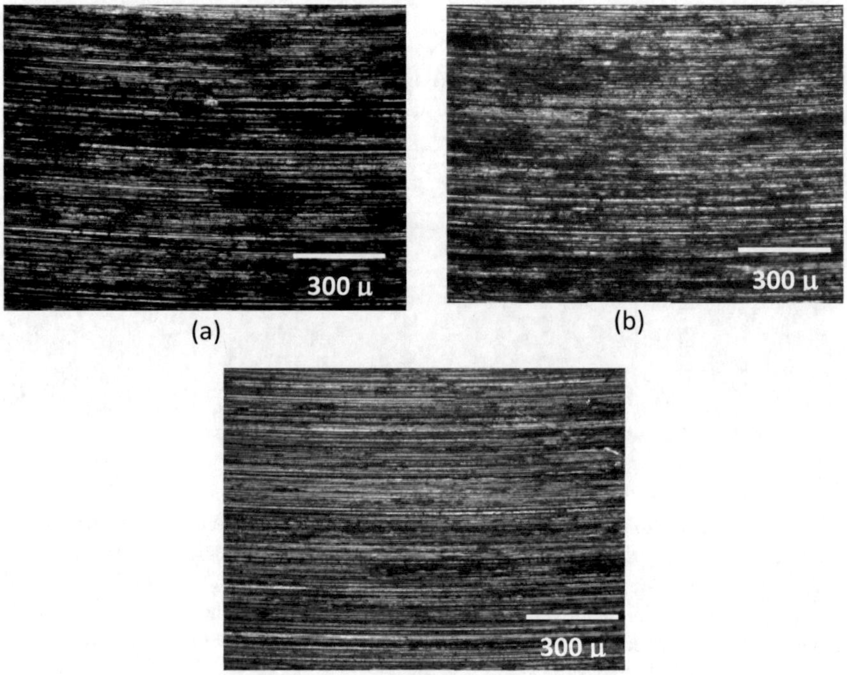

(c)

**Figure 5.** Optical micrographs of the worn surfaces of (a) AZ91, (b) AZ91-0.5B and (c) AZ91-1B tested under the normal load of 5 N.

## ***Conclusion:***

In the present study, grain refinement is found with the addition of 0.5 weight percent boron AZ91 alloy, while less refinement is observed in the sample with 1 percent boron addition. Boron addition significantly increases the hardness and wear resistance of AZ91 magnesium alloy, especially in the case of 0.5% B addition. The $AlB_2$ particles also increase the wear resistance of AZ91-1.0B as compared to AZ91 despite the same grain size. It is also found that abrasive wear is found to be the predominant wear mechanism which occurs at a lesser extent on AZ91-1.0B than that of other two alloys.

## ***References:***

[1] Mordike B L and Ebert T 2001 *Magnesium Properties - applications – potential* Mater. Sci. Eng., A 302 p 37

[2] Suresh M, Srinivasan A, Ravi K R, Pillai U T S and Pai B C 2009 *Influence of boron addition on the grain refinement and mechanical properties of AZ91 Mg alloy* Materials Science and Engineering A, 525, pp 207–210
[3] Fu P H, Luo A A, Jiang H Y, Peng L M, Yu YD, Zhai C Q and Sachdev A K 2008 *Low-pressure die casting of magnesium alloy AM50: Response to process parameters* J. Mater. Process. Technol. 205 p 224
[4] Eliezer D, Aghion E and Fross F H 1998 *Magnesium Science, technology and applications* Journal of Advanced Performance Materials 5 pp 201-212
[5] St John D H, Qian M, Easton M A, Cao P and Hildebrand Z 2005 *Grain refinement of magnesium alloys* Met. Mater. Trans. A 36A pp 1669–1679
[6] Celotto S *TEM study of continuous precipitation in Mg–9 wt%Al–1 wt%Zn alloy* 2000 Acta Mater. 48 pp 1775–1787
[7] Srinivasan A, Pillai U T S and Pai B C 2007 *Effect of Pb addition on ageing behavior of AZ91 magnesium alloy* Materials Science and Engineering A 452–453 pp 87–92
[8] Kim J -M, Kim K -T and Jung W -J 2002 *Effects of isothermal heating procedure and strontium addition on semisolid forming of AZ91 magnesium alloy* Materials Science and Technology 18:6 pp 698-701
[9] Huiyuan G, Yanxiang L, Xiang C, and Xue W 2005 *Effects of boron on eutectic solidification in hypoeutectic Al–Si alloys* Scr Mater 53 pp 69–73
[10] Zhou Z Z, Yang n W, Chen S H, Yu H, and Xu Z F 2014 *Combined effect of non-equilibrium solidification and thermal annealing on microstructure evolution and hardness behavior of AZ91 magnesium alloy* Physica B 443 pp 35–42
[11] Joshi U and Babu N H 2014 *Effect of Al Addition on Microstructure of AZ91D*. In: Alderman M, Manuel M V, Hort N, Neelameggham N R (eds) Magnesium Technology 2014 Springer Cham pp 209–212

**Acknowledgments**
We are very grateful to Dr. Salim Kaiser of DAERS, BUET for helping us perform wear test. We are also thankful to the department of Materials and Metallurgical Engineering, BUET for the lab facilities, technical expertise and support.

# Fatigue Life Dependency on the Cleanliness and Ductility of Structural Steels

**M.A. Islam[1], R.A. Mizan[1], M.A.S.R. Saadi[2]**

[1]Department of Materials and Metallurgical Engineering, Bangladesh University of Engineering and Technology, Dhaka 1000, Bangladesh
[2]Department of Mechanical and Aerospace Engineering, George Washington University Washington, DC 20052, USA

**Abstract**: Plain concrete is very brittle and weak. It cannot withstand sudden shock, cyclic load and high level of tensile stress. So, concrete is reinforced by various types of structural steels. To reduce the overall weight of the RCC structures, worldwide practice is to use high strength structural steels along with high strength concrete. In this research work low strength high ductility (300 grade) and high strength moderate ductility (500 grade) locally produced steel bars of 20mm diameter were used. They were then characterized by means of chemical composition, microstructure, size and distribution of inclusion particles, tensile and fatigue tests. The fracture surfaces of the failed fatigue samples were observed under optical and scanning electron microscopes. After detail investigation, experimental results suggest that compared to ductility, cleanliness of the structural steel bars are much more important for better fatigue life of the RCC structures.

## 1. Introduction

Steel has been used for many decades in structural and engineering purposes, both in land and marine environments, because of its good combination of mechanical, physical and chemical properties. There are many structures that are continuously under cyclic or fatigue loading conditions as bridges, flyovers, runways, jetties, etc. During earthquake all types of structures also face cyclic loading. There is no doubt that, for any RCC structure to be earthquake resistant the ductility of reinforcing materials used for the structure play a vital role [1-3]. In this regard, the qualities of compressive load bearing materials of the concrete such as cement, gravel, sand, etc. are also very important [4,5]. For a particular earthquake, the frequency, intensity and number of vibration depend on the nature of the earthquake. However, it is well established that the loading effects caused by any earthquake are very similar to that of cyclic or fatigue loading. So, use of structural steels having higher strength or high tensile ductility but low resistance to fatigue is not a good choice for earthquake resistant structures. Because of similar loading behaviours, both mechanically simulated fatigue loading and shaking by earthquake induced cyclic loading actions seem to be similar [1]. On the other hand, higher energy dissipation means better earthquake resistance, which is ultimately controlled by both the level of ductility and strength of the material considered. At the same time, ductility also controls the bendability, which is another key property for the earthquake resistant reinforcing steel bar [6-9]. In steels, there always exist a large number of inclusions, which degrade their fatigue lives [10], because the presence of inclusions deteriorates the bending and rebending properties of the rebars.

In Bangladesh, many structural design engineers believe that low strength steel bars as 300 and 400MPa grades having higher level of tensile ductility are more fatigue resistant, i.e. more earthquake safe. The aim of this research work is to discuss the effects of inclusions and tensile ductility on the total fatigue life of locally produced 300 and 500MPa grades of steel bars.

## 2. Materials and Experimental Procedure

The materials used in this research work were locally produced conventional low (300MPa) and high strength TMT (500MPa) steel bars of diameter 20mm. After collection, they were cut into two different sizes as ½ inch and 12 inch. For metallography and chemical analysis ½ inch length samples were used, whereas, for tensile and fatigue testing 12 inch long samples were used. In order to know the microstructures of the steel bars, metallographic samples were prepared for study under metallurgical microscope following standard procedure. After complete polishing, the faces of each sample were etched by 2% Nital solution. For all cases, metallography was performed on both unetched and etched samples. Using 500kN capacity Universal Testing Machines (Shimadzu, Japan) tensile tests were performed. Fatigue tests were conducted on Servo Hydraulic Fatigue Testing Machine (Instron, USA) of Model No.8801 of 100kN capacity. Here, required load (variable), frequency (70Hz), span length 160mm and stress ratio (0.1) were adjusted for each sample tested under compression-compression type loading mode. The collapsed (half broken) samples were forced for complete separation into two halves. Fracture surfaces were then photographed under optical and scanning electron microscopes to know various fracture features.

## 3. Results and Discussion

*3.1. Chemical compositional analysis*
The average chemical compositions of these steels are shown in Table 1.

**Table 1**. Chemical compositions of the steel bar used.

| Sample ID | C | Si | Mn | P | S | Cu | Ni | Cr |
|---|---|---|---|---|---|---|---|---|
| 300 MPa | 0.21 | 0.44 | 0.93 | 0.064 | 0.081 | 0.134 | 0.052 | 0.151 |
| 500 MPa | 0.18 | 0.24 | 0.94 | 0.041 | 0.033 | 0.203 | 0.031 | 0.084 |

From Table 1, it is clear that both grades of bars are of C-Mn structural steels. Here 500MPa steel bars conform to the ASTM Standard Specifications, however, 300MPa grade bars do not conform because of their higher level of phosphorus and sulfur contents [10]. Here it is be mentioned that these two elements are very detrimental for the fatigue properties of the steel bars as they make the bars brittle and crack sensitive because of their tendency to segregate at grain boundaries [11,12]. The structural steel bars of secondary steel routes are made by melting steel scraps, sponge irons, cast irons, etc and refining the molten steel to the required level. It is of note worthy that in Bangladesh almost all steel companies are using induction furnace melting process, which is really melting furnace rather than used for steel refining by international communities. From this melting route, to some extent, quality steel is possible if high grade steel scraps are used as input materials and the molten steel thus produced is further refined using ladle refining furnace. However, this practice is almost absent here.

*3.2. Metallographic observation*
The photomicrographs of various steel bars in unetched condition are presented in Figures 1 and 2. During the steel melting process, lots of fluxing agents along with other additives are added to the molten steel to adjust its chemical compositions within the prescribed range of the standard. These additives also produce slag that tries to float from inside the molten steel towards its surface. Moreover, oxides, dust particles and other nonmetallic components also produce additional slag and floats towards the surface. If sufficient time is not allowed for the inclusions to float to the top of the

IOP Conf. Series: Materials Science and Engineering **438** (2018) 012021 doi:10.1088/1757-899X/438/1/012021

molten steel, the inclusions will remain as a part of the steel matrix. The usual practice of production of TMT 500MPa grade steel requires refined steel billets. From Figure 1, it is clear that the inclusion level of this steel bar is very low. Moreover, particles sizes of inclusions are also small and they are well distributed. In the case of refining, argon gas is passed from the bottom of the ladle refining furnace, which helps break down the lump of the inclusions and distribute them in the molten steel. Argon gas flow also helps float the slag from inside the molten steel to the surface making overall inclusion content lower.

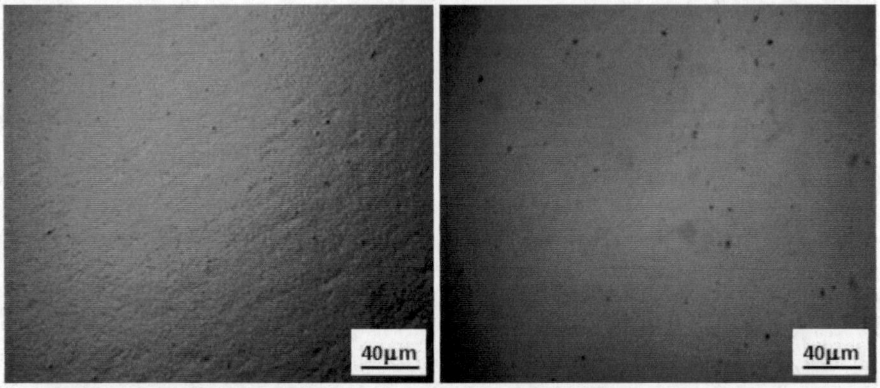


**Figure 1**. Photographs on unetched samples of 500MPa grade steel bars.

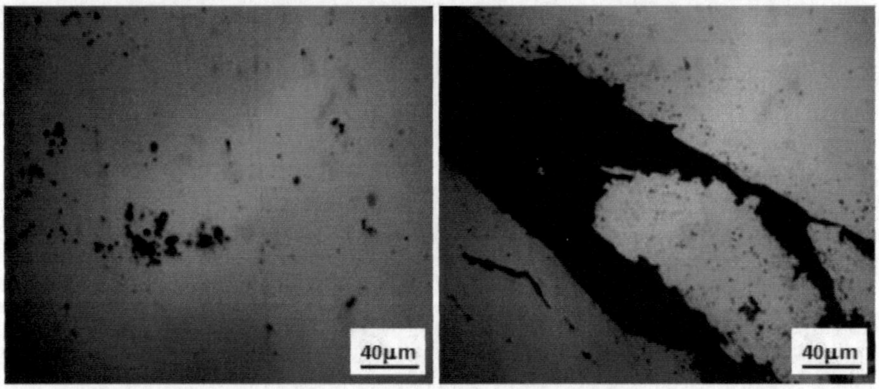


**Figure 2**. Photographs on unetched samples of 300MPa grade steel bars.

Micrographs presented in Figure 2 were taken on the unetched samples of 300MPa grade bar. These micrographs revealed somewhat more inclusion particles, which are also coarser in sizes. This is nothing but the result of insufficient refining and/or argon gas blow. In some samples very long and thick inclusion particles are also present in the steel bars. These types of inclusions are very dangerous for mechanical properties, especially, for cyclic property if they are unfortunately present at the critical location of the loaded steel bars. So, it is a great concern for the safety of any structures also if these bars are used [13-15]. Figure 3 confirmed that the 500MPa grade bar was of TMT process (presence of outer ring, left), whereas, identical surface throughout the whole section indicates that 300MPa bar was made via hot rolling and normalizing process (right side macrograph). Tempered martensitic case and fine grained ferrite-pearlitic core microstructures TMT 500MPa bar are shown in Figure 4.

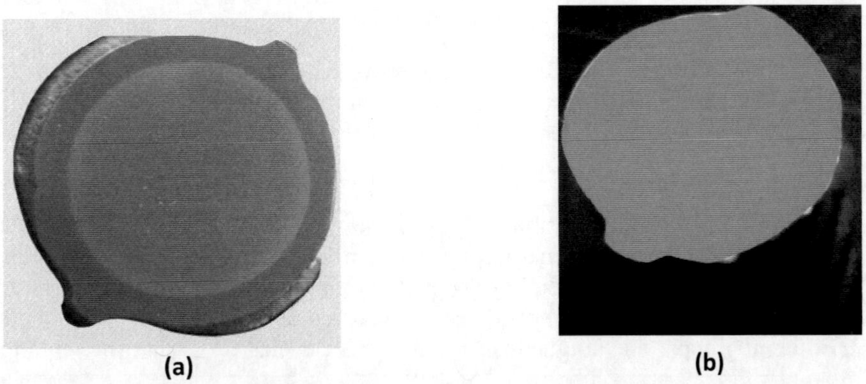


**Figure 3**. Macrophotographs showing etched surfaces of (a) 500MPa and (b) 300MPa bars.

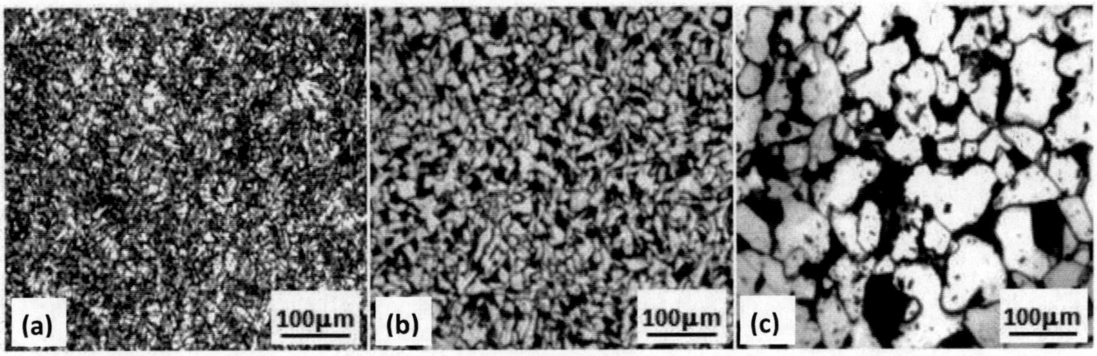


**Figure 4**. (a) Tempered martensites at the case, (b) ferrite-pearlite at the core of 500MPa bar and (c) ferrite-pearlite microstructures of 300MPa bar.

On the other hand, 300MPa bar is usually cooled in the open air after the completion of the hot rolling process. As a result, both the outer layer and inner core became to be composed of similar type of ferrite-pearlite microstructures, Figure 4c. The unetched micrographs revealed a huge amount of inclusion particles in 300 grade steel bar, which was very low for 500MPa bar (Figures 1 and 2). The etched micrographs presented in Figure 4 also confirm the similar observation. For 300 grade bar, especially, in the ferrite grains lots of black dots are visible which indicate the presence of coarse and overall huge amount of inclusion particles in the steel, but in the case of 500 grade bar this is almost absent.

### *3.3. Tensile behaviours*

Following standard procedure tensile tests of steel bars were carried out on computer controlled 500kN capacity Universal Tensile Testing Machine. For all cases, at least, three samples were tested and the average values are presented in Table 2.

**Table 2**. Average tensile properties of the steel bar tested.

| Sample ID | YS MPa | Ave YS MPa | UTS MPa | Ave UTS MPa | %El | Average |
|---|---|---|---|---|---|---|
| | 312 | | 481 | | 17 | |
| 300 MPa | 420 | 364 | 539 | 497 | 24 | 21 |
| | 360 | | 473 | | 22 | |
| | 520 | | 621 | | 16 | |
| 500 MPa | 543 | 541 | 636 | 644 | 15 | 15 |
| | 560 | | 675 | | 13.5 | |

IOP Conf. Series: Materials Science and Engineering **438** (2018) 012021 doi:10.1088/1757-899X/438/1/012021

From Table 2, it is revealed that all steel bars provide the minimum levels of yield strength and ductility to satisfy the standards. Having almost similar chemical compositions, the production technologies made the bars of different grades.

### *3.4. Fatigue behaviours*

Using the three point bend configuration, fatigue crack propagation tests were performed. Both fatigue cracking and crack growth tests to final failure were carried out at room temperature in the laboratory air on the same machine. Here, the total cycles elapsed for fatigue cracking and crack growth up to the final collapse of the as received samples were considered as the total fatigue life of individual test sample in different groups. The fatigue load versus fatigue cycles curves are shown in Figures 5-10.

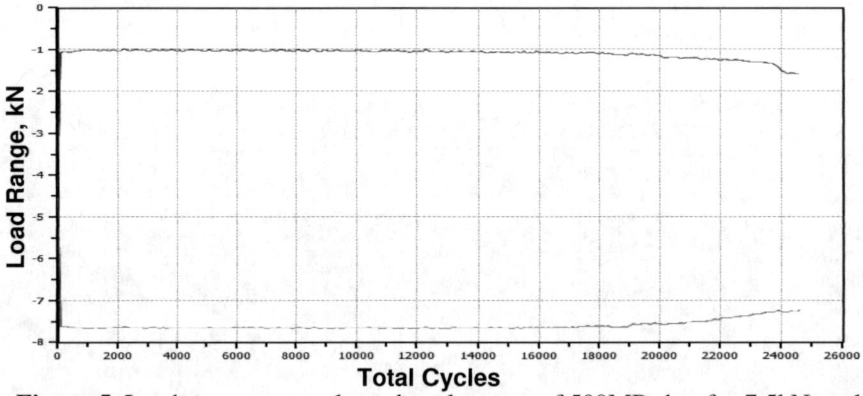


**Figure 5**. Load range versus elapsed cycle curve of 500MPa bar for 7.5kN set load.

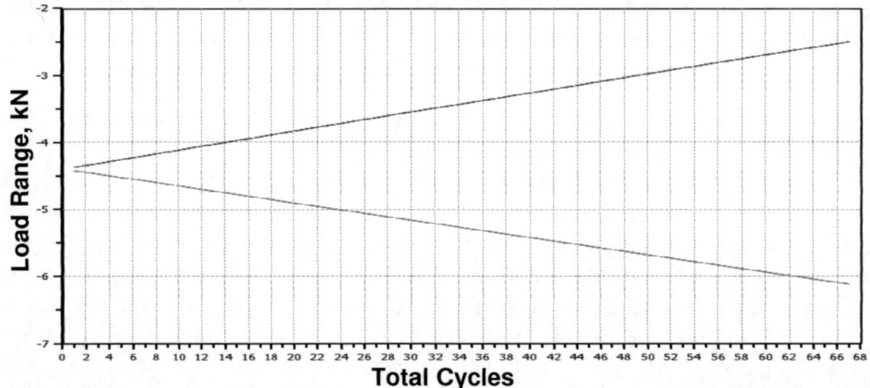


**Figure 6**. Load range versus elapsed cycle curve of 300MPa bar targeted for 7.5kN set load.

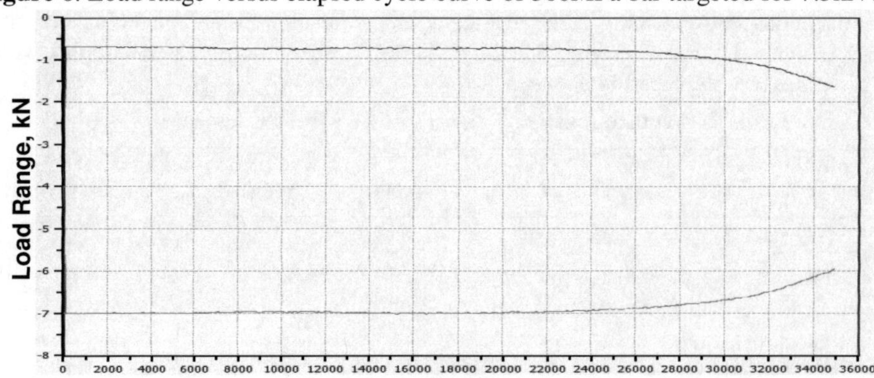


**Figure 7**. Load range versus elapsed cycle curve of 500MPa bar for 7.0kN set load.

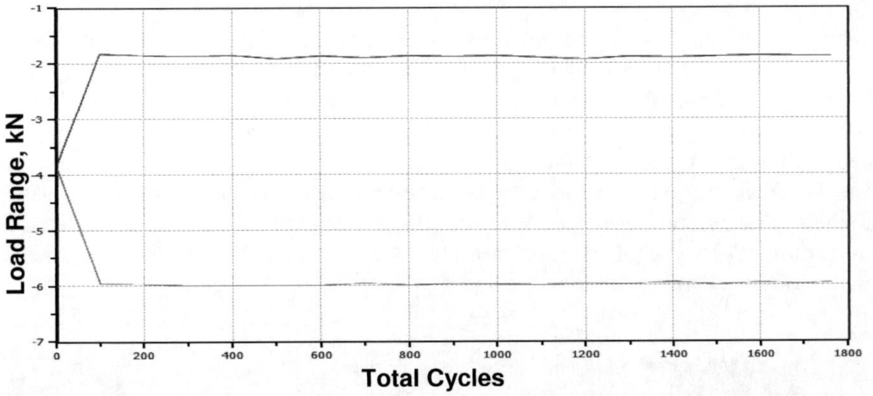


**Figure 8**. Load range versus elapsed cycle curve of 300MPa bar targeted for 7.0kN set load.

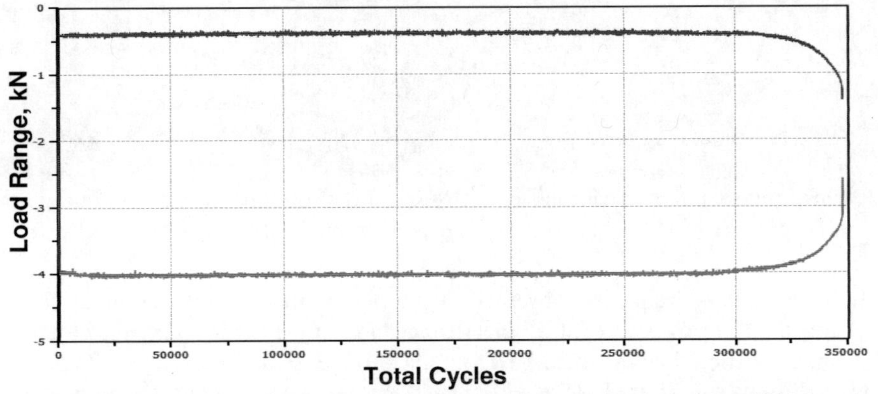


**Figure 9**. Load range versus elapsed cycle curve for 500MPa bar for 4.0kN set load.

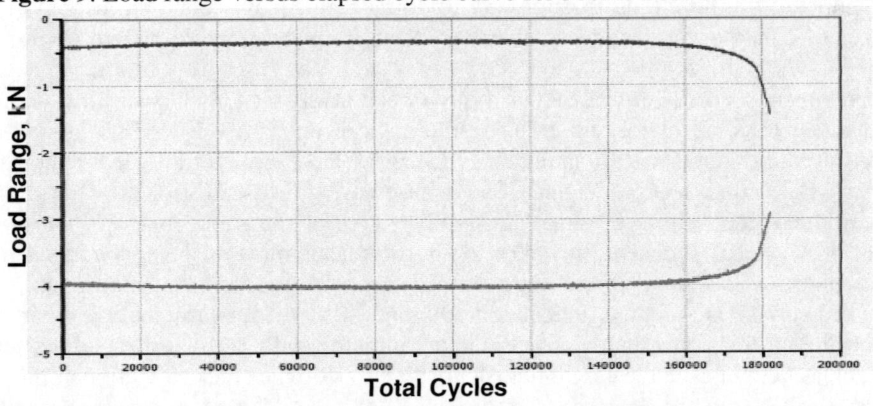


**Figure 10**. Load range versus elapsed cycle curve of 300MPa bar for 4.0kN set load.

Consider the cases at around 7.5kN (375MPa stress range) load, where 500 and 300 bars failed after, respectively 24500 and 66 cycles. At the same time, one very alarming observation is that, for 300MPa bar, the load was not possible to increase beyond 6.0kN (230MPa stress range), although the targeted load was 7.5kN. At reduced applied load of 7kN (321MPa stress range), bars of both grades tolerated higher levels of cycles. However, fatigue life of 300MPa bars is still alarming, which is only around 1800 cycles. The deterioration of fatigue resistance of 300 grade steel bars is mostly related to the inclusion content as well as higher level of trace elements sulphur and phosphorus in this grade of steel bars. Other researchers have also investigated the effects of trace elements and non-metallic inclusions on toughness and fatigue load tolerance and found severe degradation effects [11-13,16-19].

As per their observation, inclusions are detrimental for fatigue properties, although their harmful effects may vary widely. In the case of lower level of applied load, which is experienced for high cycle fatigue, e.g. 400kN (144MPa stress range) load in the present work, the performance of the 300 grade bar is still not good, Figs.9-10.

As mentioned earlier, the presence of inclusion particles is not good for both toughness and fatigue resistance of steels. Whatever may be the nature of the non-metallic inclusions, they deteriorate the fatigue life of steels by accelerating the crack formation and growth, Figure 11. This overall scenario makes an early failure. This type of deterioration increases with increase in the area fraction of the non-metallic inclusions.

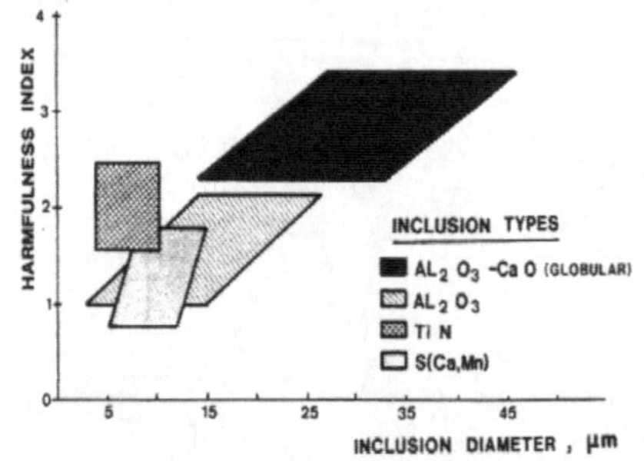


**Figure 11**. Harmfulness index of various inclusions on fatigue life.

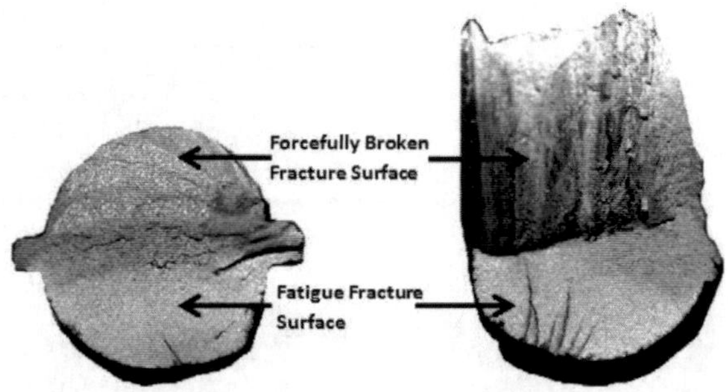


**Figure 12**. Low magnification fatigue fracture surfaces of 500 (left) and 300 (right) steel bars.

Inclusion particles were very frequently observed in the microstructures of 300 grade bars. Not only inside of the bars, they are also present at the surface and sub-surface areas that helped nucleate fatigue crack from the critical loading area, i.e. from the surface. Because of high volume fraction and closely spaced inclusions, along with segregation of trace elements sulphur and phosphorus fatigue cracks were initiated from multiple locations for 300 grade bar, they were then merged quickly that caused the bar to collapse very suddenly, which is clear from the fracture pattern of this steel bar, Figure 12.

Experimental results clearly revealed that both steel bars passed the tensile test requirements, however in fatigue tests, absolutely upsetting results have been observed for 300 grade bars. Now, why is this different behaviour of 300 grade bars? In tensile test, applied fully axial load is uniformly distributed on whole cross section, Figure 13. Because of loading scenario, in a particular section (marked blue dotted line), stress becomes uniformly active at the same time. If inclusion zone is considered as broken/weak ligament, then metallic portion bears most of the applied tensile load. So, inclusions become less sensitive for tensile test. However, in fatigue test bending stress becomes operative, where surface is the most critical zone, Figure 14. If there is any inclusion particle on the surface or subsurface areas, it instantly acts as a crack initiator and/or propagator and the surrounding metal cannot be as helpful as in the case of tensile loading to stop the crack formation or propagation.

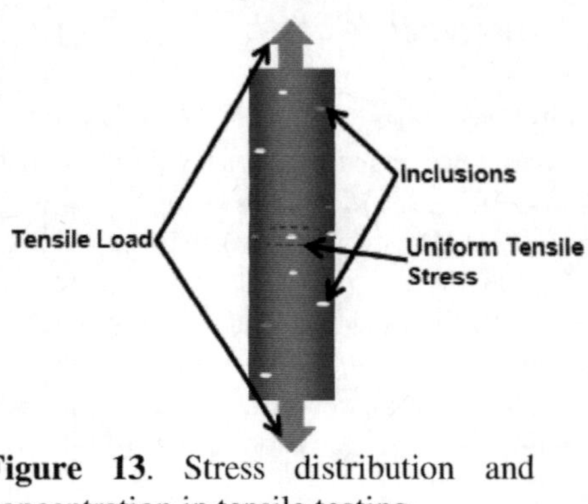


**Figure 13**. Stress distribution and concentration in tensile testing.

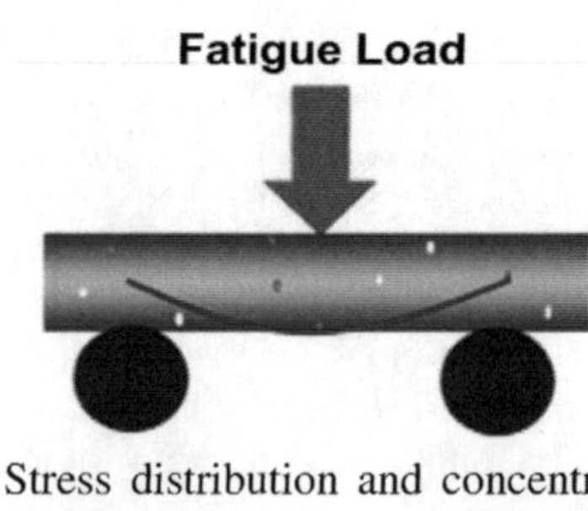


**Figure 14**. Stress distribution and concentration on the steel bar in bending-bending type fatigue testing.

## 4. Conclusions

After detail experimental studies on locally produced 300 and 500MPa grade steel bars, the following conclusion are drawn:

a. The chemical compositions of 300MPa grade bars are not controlled enough to meet the standard specification because of higher level of P and S contents.

b. At higher load, 300MPa bars collapsed almost instantly, whereas 500MPa bars withstand several thousands of cycles. At low load condition, both types of steel bars showed higher fatigue life, however, 500MPa bars showed significantly better performance for all cases.

c. Fractographic observation revealed that the worst fatigue lives and occasionally, the instant collapse of the 300MPa bars are very much related to the densely populated large size inclusion particles at the highly tensile critical location of the surface and also due presence of high percentages trace elements like S and P.

d. It is thought that steel bars with higher tensile ductility provide better toughness and fatigue life. However, present experimental results revealed that the higher ductility obtained via tensile test alone is not sufficient to provide better fatigue life, for which volume, size and distribution of inclusion particles play more vital role.

## References

[1] Sing JG, Bandyopadhyay N and Mohanty ON 2001. *Practical Failure Analysis*, **1** 53.
[2] Otani S 2004. *J. Adv. Concrete Tech.* **2** 3.
[3] Australian/New Zealand Standard for Steel Reinforcing Materials, AS/NZS 4671 2001.
[4] Saleem MA and Ashraf M 2008. *Pak. J. Eng. Appl. Sci.* **2**, 59.
[5] Commonwealth Report 2001. *News Letter* **4** 1.
[6] Beynon D, Jones TB and Fourlaris G 2005. J. Mat. Sci. Techn. **21** 510.
[7] Tomota Y, Tokuda H, Adachi Y, Wakita M, Minakawa N and Moriai A 2004. *Acta Materialia* **52** 5737.
[8] Hulka K 2005. J. Mat. Sci. Forum **473** 91.
[9] Islam MA 2012. *J. Scientific Research* **4** 51.
[10] ASTM Standard Specification: A 706/A 706M – 04.
[11] Islam MA 2007. *JMEPEG, ASM International* **16** 73.
[12] Islam MA, Knott JF and Bowen P 2004. Mat. Sci. Tech. **21** 76.
[13] Kiessling R 1978. *Metals Society*, Part III 43.
[14] Atkinson MJ 2004. *J. Iron Steel Inst.* **195** 64.
[15] Yang ZG, Zhang JM, Li SX, Li, GY, Wang QY, Hui WJ and Weng YQ 2006. *Mater. Sci. Eng. A,* **1–2** 167.
[16] Peaslee KD, Singh V and Lekakh SN 2009. M U Sci. & Tech. and Nucor Steel 1.
[17] Juvonen P 2004. PhD Thesis, Helsinki University of Technology.
[18] Deng H, Li W, Sakai T and Sun Z 2015. *J. Materials* **8**, 8338.

## Acknowledgement

The authors would like to express his gratitude of CASR, BUET to provide necessary funding for this research work. They also indebted to MME Department for ensuring the access to all laboratories and equipments required for the course of the research work.

IOP Conf. Series: Materials Science and Engineering **438** (2018) 012022 doi:10.1088/1757-899X/438/1/012022

# Finite Element Analysis of an Aluminum Alloy Using Progressive Failure Algorithm

***M.F. Hossain[1], K.M. Shorowordi[1], M.S. Islam[2]***

[1]Department of Materials and Metallurgical Engineering, Bangladesh University of Engineering and Technology, Dhaka 1000, Bangladesh
[2]Department of Naval Architecture and Marine Engineering, Bangladesh University Engineering and Technology, Dhaka 1000, Bangladesh

**Abstract.** This study aims to develop a reliable finite element analysis procedure to model the complete fracture of ductile specimens using the progressive degradation of the material stiffness algorithm under tensile load. The ductile specimen in this study is an aluminum alloy. The progressive failure algorithm used here is based on the assumption that the material behaves like a stable progressively fracturing solid. The stiffness reduction is carried out at the reduced integration gauss points of the finite element mesh depending on the mode of failure. A number of material properties are necessary for such simulation to carry out and experimentation of the metal are needed to evaluate these properties. The actual tensile tests data are applied to the finite element simulation. The verified simulation method has a great importance in practical design of structures and metals. A renowned finite element analysis software ABAQUS is used in this study.

## 1. Introduction

Aluminium is the key material in aerospace and automobile industry. Among the aluminium alloys, Al 7075 has high strength to weight ratio and is increasingly applied to manufacture of automobiles and aircrafts [1]. Recent researches shows that these aluminium alloys can attain strength as high as 1 GPa [2]. Having boron in this alloy decreases its grain size and thus improve its strength. The addition of nickel hardens the surface of the alloy and thus increases its overall property [3,4]. A 7075 alloy modified with nickel and boron can be potential to achieve excellent properties which are lucrative for the strengh related applications. Being this highly important material, it is evident that a properly developed simulation model will allow further maneuverability in its application by aiding in complex designs.To understand the material, tensile tests until fracture are often employed. Having a realistic tensile simulation of this material until fracture using damage models means that the simulation results will also, allow researchers to have an in depth understanding of this material. And also, these simulation will enable further ease of process designing such as in various drawing processes [5].

Obtaining reliable experimental results are extremely important to obtain a reliable simulation but the key challenges here are choosing the right damage criterion and the proper meshing of the specimen. Through the use of meshing in a rectangular tensile specimen EL, Amri Abdelouahid et al. showed that different size of meshing shows different results in the simulation [6]. Most of the time using finer mesh means that the result will converge to the real experiments but that also means higher computational time so a meshing should be such that it matches the realistic data but not too much fine that it consumes unnecesssarily high amount of computaional energy. And about the fracture damage models, although there are many damage models, it is reported that the phenomenological damage models (i.e. johnson cook , ductile damage models) are the best to employ when experimental data are at hand since the continuum damage mechanics shows systematically the effects of damage on the mechanical properties of materials and structures as well as the influence of external conditions and damage itself, and thus accurately representing the real situations [7,8]

In this study a simulation of tensile test using a computer program with the aid of real experimental data is done by applying various mesh element sizes and using the ductile damage model as a representative of phenomoligical damage model.

## 2. Methodology

### *2.1 Material Preparation*

The present investigation was carried out on modified 7075 Al alloy with nickel and boron whose composition is shown in Table 1. The material was cast at 850°C in the form of rectangular bar in a metal mold.

**Table 1.** Chemical composition of Al 7075 alloy

| Elements | Weight Percentage | Elements | Weight Percentage |
|---|---|---|---|
| Zinc | 6.0 | Chromium | 0.3 |
| Magnesium | 3.0 | Zirconium | 0.2 |
| Copper | 2.0 | Boron | 0.01 |
| Nickel | 1.0 | Aluminum | Balance |

### *2.2 Heat Treatment*

Two types of heat treatments were applied namely solution heat treatment with rolling and precipitaion hardening. Samples were machined and then Ni and B modified Al 7075 alloy was at first solution treated at 480 °C for 5 hours and later hot rolled at 450 °C to facilitate recovery and recrystallization [9]. After the reduction of about 66 percent in height the material was quenched in cold water. During the reduction the specimen was reduced from 15 mm to 5mm decreasing about 0.66mm at each rolling pass to avoid the material failure during the hot rolling. Then the sample was aged at 120 for 24 hour to impart some precipitation strengthening.

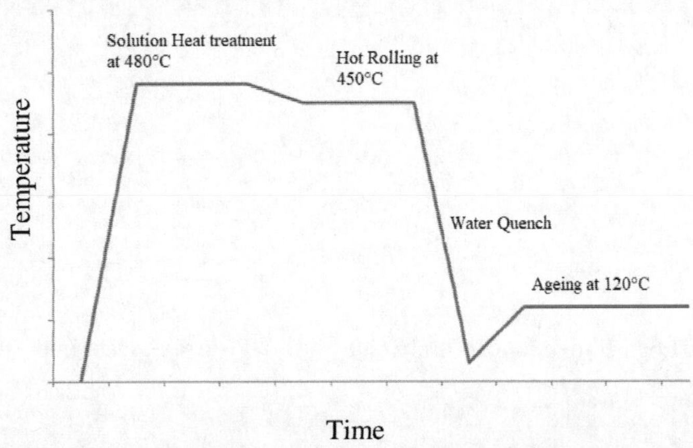


**Figure 1.** Heat treatment cycle of modified Aluminum 7075 alloy

IOP Conf. Series: Materials Science and Engineering **438** (2018) 012022 doi:10.1088/1757-899X/438/1/012022

*2.3 Simulation Data*

*2.3.1 Density, Plastic and Elastic parameters.* To simulate tensile damage and failure we have collected our tensile test data in the load elongation form. This data is later converted into stress and strain form. But to use these data in finite element software we need true stress-true strain data because with the true stress data we can simulate realistic results which can be obtained by the equations below

$$\sigma = S(1+\delta) \tag{1}$$

$$\in = \ln(1+\delta) \tag{2}$$

- $\sigma$ = True stress
- $\in$ = True strain
- S = Engineering stress
- $\delta$ = Engineering strain

From the true stress true strain curve modulus of elasticity is calculated with the equation below:

$$\text{Modulus} = \text{True Stress at Yield / True Total Strain at Yield} \tag{3}$$

We calculated our youngs modulus to be 73 MPa. Densty was calculated 2.4g/cm$^3$ and the poisons ratio was taken from literature which is .33[10]

In order to include plasticity within Abaqus, the stress-strain points past yield, must be input in the form of true stress and logarithmic plastic strain. The logarithmic plastic strain required by Abaqus can be calculated with the equation given below.

$$\in^{plastic} = \ln(1+\delta) - \frac{\sigma}{\text{Youngs Modulus}} \tag{4}$$

Now, this true stress plastic strain data is good to apply in Abaqus as input data.

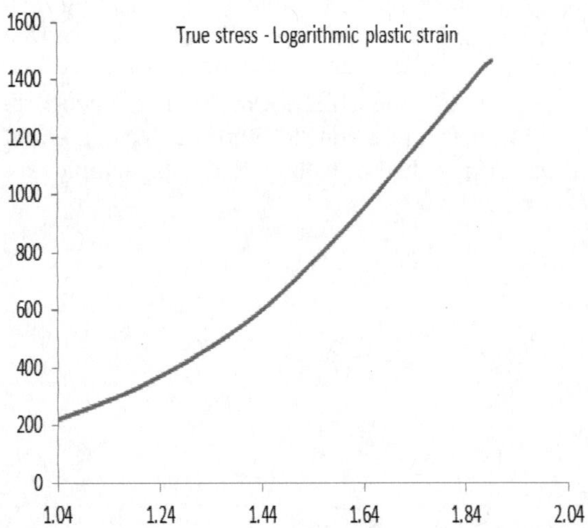


**Figure 2**. True stress-Logarithmic plastic strain data

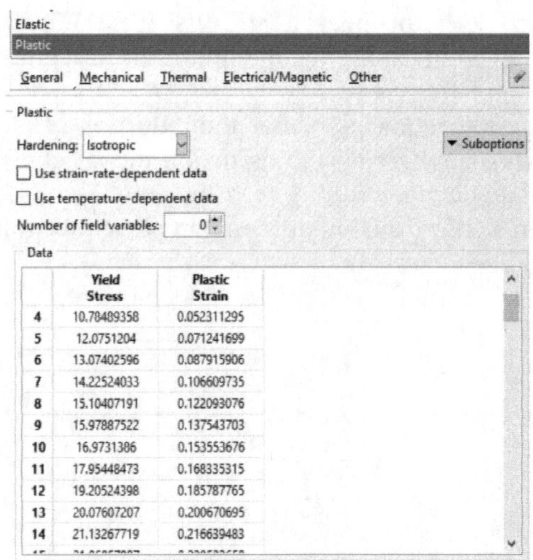


**Figure 3.** Data imported to Abaqus

*2.3.2 Damage Parameters.* In the damage parameters strain rate can readily be obtained from the experiment itself . In our case the strain rate was 0.25mm/s. For stress triaxiality as a practical approach, the stress triaxiality is used to identify the state of stress and is defined as follows:

$$\text{Stress triaxiality} = \frac{\sigma(\text{mean normal stress})}{\sigma(\text{effectice stress})} \tag{5}$$

$$\sigma_{\text{mean normal stress}} = (\sigma_x + \sigma_y + \sigma_z)\,/3 \tag{6}$$

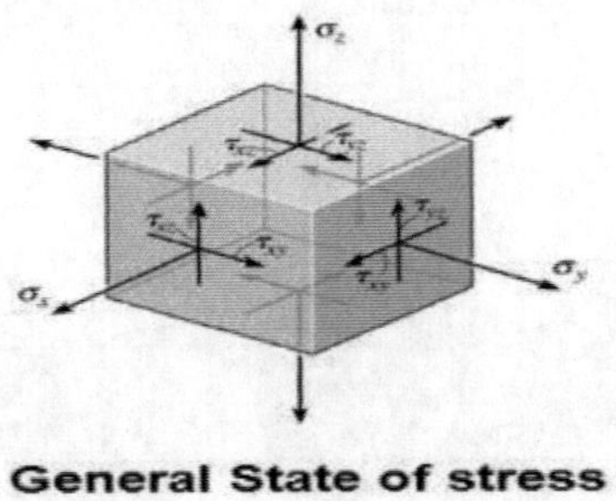


**Figure 4.** State of stresses in various directions

$$\sigma_{\text{effective stress}} = \frac{1}{\sqrt{2}}\sqrt{\left(\sigma_x - \sigma_y\right)^2 + \left(\sigma_y - \sigma_z\right)^2 + (\sigma_z - \sigma_x)^2 + 6\left(\tau_{xy}^2 + \tau_{yz}^2 + \tau_{xz}^2\right)} \tag{7}$$

- $\sigma_x, \sigma_y, \sigma_z$ = stress in three directions.
- $\tau$ = shear stresses.

The unique situation for uniaxial test is that it can be assumed until necking all of the stress state except in y and z directions are zero and also the scalar value of the stresses in y and z direction are equal with having opposite signs. Using the equation 5, 6 and 7 the value of stress triaxiality leads to 0.33. So, we can assumed our stress triaxiality value as 0.33 upto necking[11]. And for fracture strain it is equals to the equivalent plastic strain and under tensile loading, the equivalent plastic strain exactly equals to the tensile plastic strain at rupture from stress-strain diagram. Al 7075 properties are given in Table 2.

**Table 2.** Properties of Al 7075 alloy

| Property | Value |
|---|---|
| Density | 2.4 g/cm$^3$ |
| Youngs modlus | 73000 |
| Poissons ratio | 0.33 |
| Fracture strain | 0.6 |
| Stress triaxiality | 0.33 |
| Strain rate | 0.25 mm/s |
| True stress- plastic strain | Data imported in abaqus in tabular form |

*2.3.3 Specimen geometry*. The test specimen is modelled on Abaqus in accordance with the tensile sample as much as possible. Dimension of the sample is given in Table 3 and the samples are shown in figure 4.

**Table 3.** Dimension of the sample

| Gauge length | Width | Thickness | Fillet radius | Overall length | Length of reduced section | Length of grip section | Width of grip section |
|---|---|---|---|---|---|---|---|
| 25 mm | 15 mm | 4 mm | 6 mm | 80 mm | 32 mm | 22 mm | 7.5 mm |

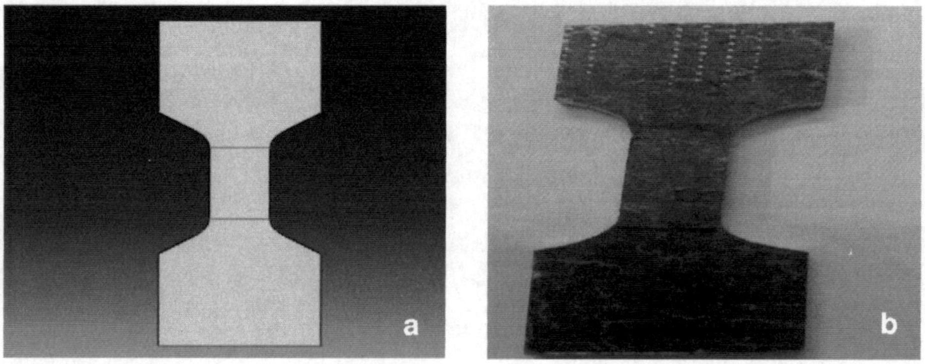


**Figure 5.** Showing the tensile specimen for (a) simulation and (b) real test

*2.3.4 Meshing*. As discussed above meshing is an important part for obtaining proper results in the simulation. Meshing of the sample was done in four different ways with C3D8 8-node linear brick, hourglass control mesh elements. The data obtained by meshing in the simulation of the material are tabulated in Table 4 and meshing is shown in Figure. 5.

**Table 4.** Element size and Simulation run time

| Model No. | Total Number of Elements | Total number of nodes | Mesh size | Simulation run time required (s) |
|---|---|---|---|---|
| Model 1 | 1674 | 2496 | 1.6 | 118.23 |
| Model 2 | 2973 | 4312 | 1.2 | 245.33 |
| Model 3 | 11780 | 14940 | 0.8 | 1795.3 |
| Model 4 | 95790 | 108273 | 0.4 | 38770.8 |

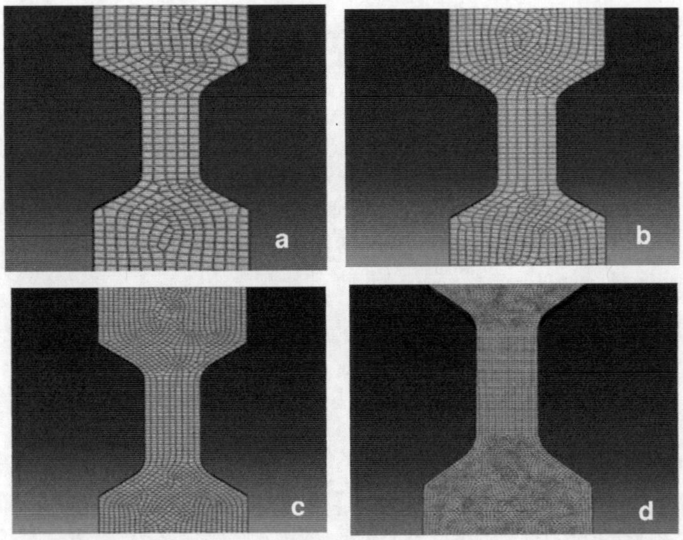


**Figure 6.** Showing different meshing style of the sample 3

## 3. Results and Discussion

The data obtained from the simulations show that as the mesh becomes finer the simulation time increases dramatically. On the other hand with the finer mesh size the simulation starts to converge with the real result. Simulation with 95790 elements showed very good convergence with the real data which is apparent in figure 6(d) whereas the model 1 with the lowest number of elements showed poorest result as evident in figure 6(a). So, it can be concluded that in the simulation particularly in tensile test mesh size plays an important role and with the higher mesh density, we got more realistic result.

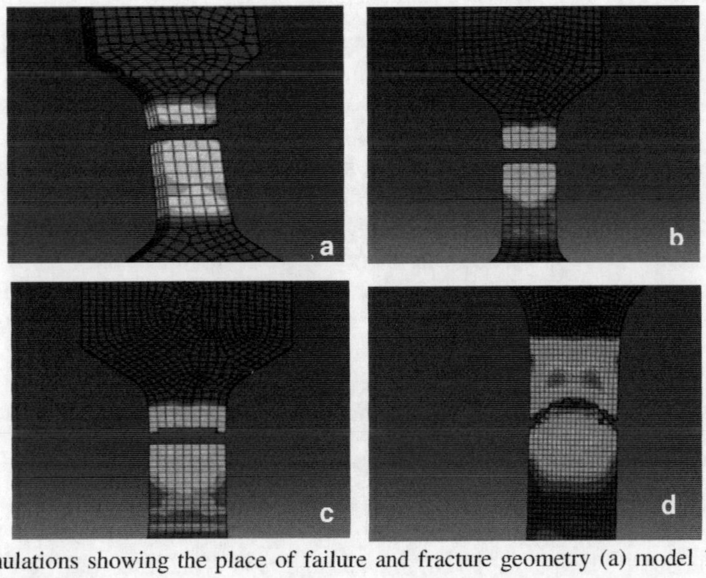


**Figure 7.** Simulations showing the place of failure and fracture geometry (a) model 1, (b) model 2, (c) model 3, (d) model 4.

Also, the place of the fracture in all the simulations was nearly as same as the real experiments. So, it is evident that the simulation is quite accurate in the context of fracture geometry and the place where the fracture occurred. And that is obtained just by inputting values of only a few variables obtained from a tensile test. So, using this simulation results it is possible to predict what may happen in the complex structures that uses this material by using a lot less experimental resources that would otherwise be required.

There is, also, a discrepancy with the real results and simulation in the portion where the crack propagated. The simulation shows how ideally fracture should look like and initially in the real experiment, the crack started to propagate as it should be. But as there are lots of defects in the real sample, the crack had a lot of possible pathways to propagate. So, as evident in the figure 7, the crack path differed at the later portion of the specimen. Nevertheless, from figure 7(c) it can be seen that the final fracture geometry is somewhat similar to the simulation.

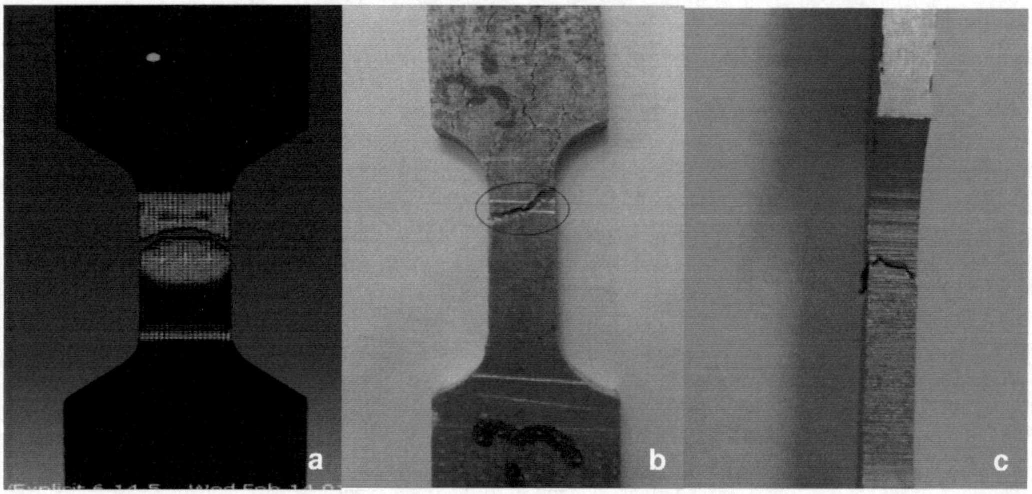


**Figure 8.** Fracture geometries. (a) Simulated fracture geometry (b) Fracture geometry from the front view (c) Fracture geometry from the side view.

## 4. Conclusion

Finite element analyses of uniaxial tensile tests were conducted using ABAQUS software. With the tricky choice of mesh size and the damage model it could quite accurately predict the place of necking and fracture geometry. Although, there were some differences in crack propagation due to the presence of defects in the real sample it can be said that the simulation did almost match the real scenario. Also, it is evident that the numerical accuracy of the simulation increased with the decreasing mesh size but the computing time increased with the decreasing mesh size. Overall, this simulation showed high convergence to the real data so the data from the simulation can be used in the process and structural areas where these materials are used.

## 5. Reference

[1] E.A.StarkeJr 1996 Progress in Aerospace Sciences (Virginia, Charlottesville) vol 32 pp 131-172.

[2] Liddicoat, P., Liao, X., Zhao, Y., Zhu, Y., Murashkin, M., Lavernia, E., Valiev, R. and Ringer, S. 2010 Nanostructural hierarchy increases the strength of aluminium alloys. Nature Communications 1(6), pp.1-7.

[3] Jasim M. Salmanl , Shaymaa Abbas Abd Alsada2 and Khadim F. Al-Sultani 2013 *Research Journal of Material Sciences* vol 1(6) pp 12-17.

[4] Brunelli K, Dabalà M, Martini C. 2006 Surface hardening of Al 7075 alloy by diffusion treatments of electrolytic Ni coatings *Metallurgia Italiana* p37.

[5] Venkateswarlu, G., Davidson, M. and Tagore, G. (2011). Influence of process parameters on the cup drawing of aluminium 7075 sheet *International Journal of Engineering, Science and Technology* vol 2(11).

[6] EL Amri Abdelouahid 2015 22ème *Congrès Français de Mécanique* (Lyon,France)

[7] Lemaitre, J. 1985 *A continuous damage mechanics model for ductile fracture*. J. Eng. Mater. techno. 107 pp 83–89.

[8] Lemaitre, J., 1985 *Coupled elasto-plasticity and damage constitutive equations* Comp. Meth. appl.Mech. Eng. pp 31–49.

[9] Abolhasani, A. Zarei-Hanzaki, A., Abedi, H. and Rokni, M. 2012 The room temperature mechanical properties of hot rolled 7075 aluminum alloy. *Materials & Design*, 34, pp.631-636.

[10] Asm.matweb.com.(2018). *ASM Material Data Sheet*. [online] Available at: http://asm.matweb.com/search/SpecificMaterial.asp?bassnum=MA7075T6.

[11] Mirzajanzadeh M, Canadinc D. A 2016 Microstructure-Sensitive Model for Simulating the Impact Response of a High-Manganese Austenitic Steel *Journal of Engineering Materials and Technology* vol 138(4).

IOP Conf. Series: Materials Science and Engineering **438** (2018) 012023 doi:10.1088/1757-899X/438/1/012023

# Stress, Deformation and Failure Analysis of Parabolic Leaf Spring by Finite Element Analysis with Material Optimization

**Tasmeen Inteeser Khan[1], M I Washif Rahman Omi[2], Md. Jubayer Hossain[3], Md. Merajul Haque[4], Fazlar Rahman[5]**

[1, 2]Department of Mechanical and Production Engineering, Ahsanullah University of Science and Technology, Dhaka-1208, Bangladesh.
[3, 4, 5]Assistant professor, Department of Mechanical and Production Engineering, Ahsanullah University of Science and Technology, Dhaka-1208, Bangladesh.

**Abstract.** A leaf spring is a simple form of spring commonly used for the suspension system in heavy duty vehicles which is originally called laminated or carriage spring. It performs the isolation task in transferring the vibration due to road irregularities to the passenger's body or goods transported on it. The advantage of leaf spring over helical spring is that the ends of the spring have been guided along a definite path as it deflect to act as a structural member in addition to an energy absorbing device. Increasing competition and innovations in the automobile sector, tends to modify the existing products with new or advanced material products. To improve the performance of the suspension system, many modifications have been taken place overtime but the recent innovations imply parabolic leaf spring and application of composite materials, as the composite materials have high strength-to-weight ratio compared to the conventional steels. The present work attempts to analyze the comparison between the conventional steel (AISI1030) leaf spring and the composite (Carbon/Epoxy) leaf spring with respect to static and fatigue analysis. Here static analysis determines the safe stress, deformation and corresponding pay load and also studies the behavior of structures under practical conditions. The research also focuses on fatigue analysis to determine its life cycle and also to observe its fatigue failure characteristics. The model of the leaf spring has been carried out into the CAD software. The analysis of the steel leaf spring has been divided into theoretical, experimental and simulation sections .The simulation of steel leaf spring has been validated with the theoretical and the experimental results. Further simulation has been done for composite leaf spring by FEA software. Finally, the result shows that the composite leaf spring has better load carrying capacity in reduced weight, and better fatigue behavior rather than the conventional steel material for the leaf spring.

## 1. Introduction

The main focus of the present automobile manufacturer has been the weight reduction due to conservation of natural resources, emission gas regulation and economization of energy. The main approach of the weight reduction can be achieved by design optimization, introduction of better material and better manufacturing process [4]. Leaf spring is the simple form of spring commonly for the suspension system of the vehicle. It is one of the potential items for weight reduction in the automobiles as it accounts for ten to twenty percent of the un-sprung weight. A leaf spring can either be attached directly to the frame at both ends or attached directly at one end, usually the front, with the other end attached through a shackle, a short swinging arm [3]. Generally the capacity of absorbing strain energy, makes the suspension system more comfortable. So the strain energy plays a vital role in designing the suspension system. Besides the introduction of composite materials have made it possible to reduce the weight of the leaf spring without any reduction of stiffness and load carrying capacity.

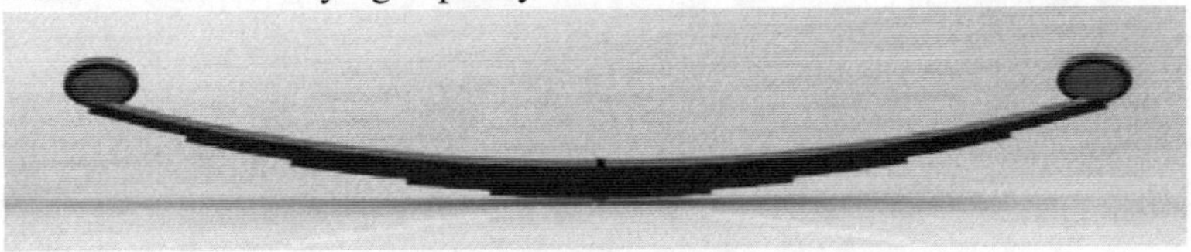

**Figure 1.** Leaf spring.

## 2. Experimental Analysis

**Table 1.** Dimension for the specimen leaf spring

| Parameter | Dimension (mm) |
|---|---|
| Length of the Span | 581 |
| Number of Full length Leaves | 2 |
| Number of Graduated Leaves | 4 |
| Length of the Master Leaf | 596 |
| Thickness of the Leaf | 6 |
| Width of the Leaf | 49.6 |
| Height of Camber Profile | 79 |

The experiment has been done in a Universal Testing machine. Load has been applied on the center of the leaf spring. The range of operating load for this spring is 1800 N-2200 N.

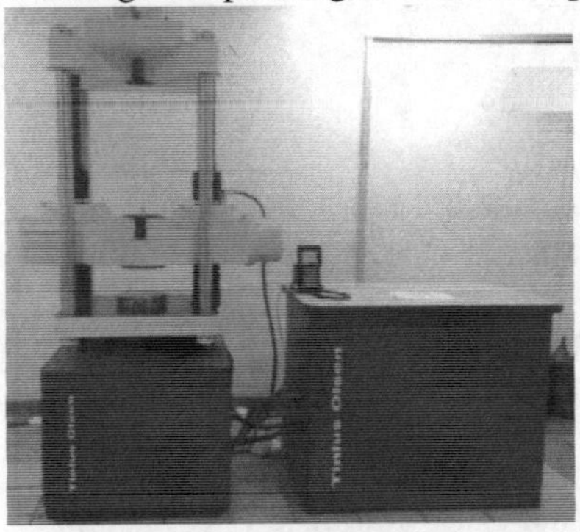

**Figure 2.** UTM machine.

**Figure 3.** Experimental Setup for Leaf spring.

**Table 2.** Deformation of the steel leaf spring

| Load (N) | Deformation (mm) |
|---|---|
| 543 | 2.69 |
| 1033 | 5.12 |
| 1542 | 7.65 |
| 2200 | 10.91 |
| 2544 | 12.59 |

The chemical composition test for the specimen leaf spring has been done. From the result of the test, specimen is a medium carbon steel (Grade AISI 1030).

**Table 3.** Result for chemical composition test

| Element in Leaf spring | % of Element (standard) | % of Element (obtained) |
|---|---|---|
| Carbon | 0.28-0.34 | 0.310 |
| Manganese | 0.60-0.90 | 0.731 |
| Phosphorous | 0.04 (max) | 0.012 |
| Sulfur | 0.05 (max) | 0.012 |

## 3. Analytical Stress and Deformation for Steel Leaf Spring

The bending stress and deformation of the leaf spring is calculated using following equations,
Bending stress for the full length leaves,

$$\sigma_F = \frac{6W_F L}{n_F b t^2} = \frac{6L}{n_F b t^2}\left(\frac{3n_F}{2n_G+3n_F}\right)W = \frac{18WL}{bt^2(2n_G+3n_F)} \tag{1}$$

$$\text{As,}\ \sigma_F = \frac{3}{2}\sigma_G$$

Bending stress for the graduated leaves,

$$\sigma_G = \frac{2}{3}\sigma_F = \frac{2}{3}\times\frac{18WL}{bt^2(2n_G+3n_F)} = \frac{12WL}{bt^2(2n_G+3n_F)} \tag{2}$$

Total bending stress,

$$\sigma = \sigma_F + \sigma_G \tag{3}$$

Deformation,

$$\delta = \frac{12WL^3}{Ebt^3(2n_G+3n_F)} \tag{4}$$

Where, (t = thickness of the leaf, *b*=width of the leaf, *L*= length of the leaf, *E*= modulus of elasticity, *W*= load applied on the spring, $n_F$ = number of full length leaves , $n_G$ = number of graduated leaves)

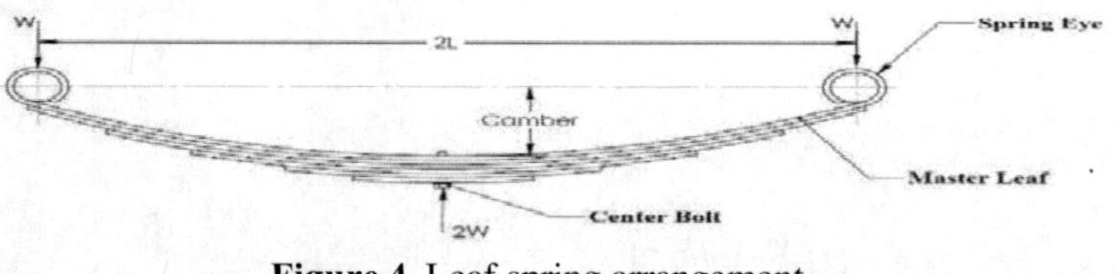


**Figure 4.** Leaf spring arrangement.

Now,
for the operating load 2200 N, *W*= 2200/2 = 1100 N, effective length *L* = 581 mm, width *b* = 49.6 mm, thickness *t* = 6 mm, modulus of elasticity E= 200 GPa, number of full length leaves, $n_F$ = 2, number of graduated leaves, $n_G$= 4
Putting these values in equation (1), stress for full length leaves, $\sigma_F$= 232.90 MPa
And from equation (2), stress for graduated leaves, $\sigma_G$ = 155.26 MPa
So, total bending stress, $\boldsymbol{\sigma} = \boldsymbol{\sigma_F + \sigma_G}$ = 232.90 + 155.26 = 388.16 MPa
And from equation (3), deformation, δ = 10.91 mm.

## 4. FEA Analysis of Leaf spring

The modeling of leaf spring has been done from the dimensions of experimental specimen leaf spring in

IOP Conf. Series: Materials Science and Engineering **438** (2018) 012023 doi:10.1088/1757-899X/438/1/012023

the SOLIDWORKS 2017.

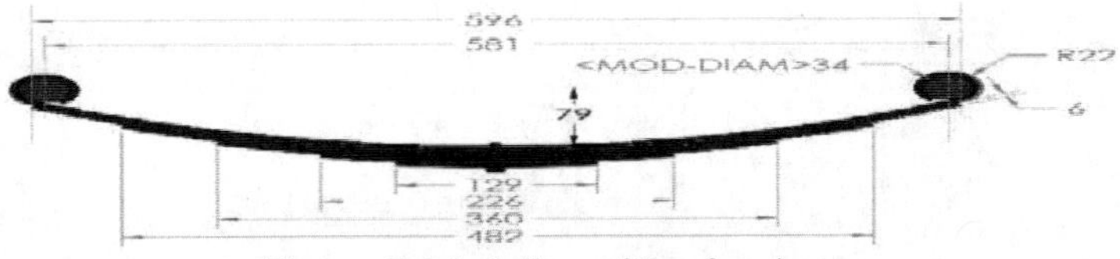


**Figure 5.** Modeling of Leaf spring.

The FEA analysis has been done in simulation software ANSYS workbench 17.1 for the AISI 1030 steel leaf spring and also for the Carbon Epoxy composite leaf spring.

**Table 4.** Material properties of existing leaf spring (AISI 1030).

| Parameter | Value |
|---|---|
| Young's Modulus (E) | 200 GPa |
| Poisson's Ratio | 0.29 |
| Tensile Strength Ultimate | 586 MPa |
| Tensile Strength Yield | 441 MPa |
| Density | 7850 kg/m$^3$ |

**Table 5.** Mechanical properties of Carbon Epoxy.

| Properties | Value |
|---|---|
| Tensile modulus along X-direction (Ex) | 121000 MPa |
| Tensile modulus along Y-direction (Ey) | 8600 MPa |
| Tensile modulus along Z-direction (Ez) | 8600 MPa |
| Shear modulus along XY-direction (Gxy) | 4700 MPa |
| Shear modulus along YZ-direction (Gyz) | 3100 MPa |
| Shear modulus along ZX-direction (Gzx) | 4700 MPa |
| Poisson ratio along XY-direction (PRxy) | 0.27 |
| Poisson ratio along YZ-direction (PRyz) | 0.40 |
| Poisson ratio along ZX-direction (PRzx) | 0.27 |
| Density | 1490 kg/m$^3$ |

After importing the assembly file from SOLIDWORKS 2017 into ANSYS Workbench 17.1, proper boundary conditions have been applied as the one eye as a fixed support and the other eye can translate in the X direction only. The operating load 2200 N has been applied at the center of the spring. The failure analysis has been done with fully reversed load of 2200 N.

The maximum von mises stress has generated in (AISI 1030) steel leaf spring 389.8 MPa.

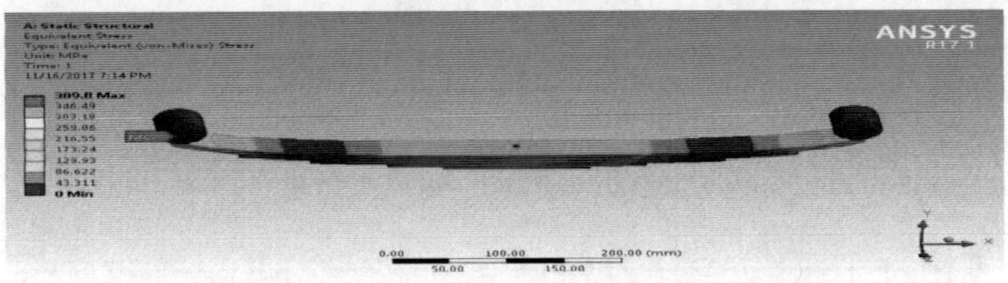


**Figure 6.** Von mises stress in steel leaf spring.

IOP Conf. Series: Materials Science and Engineering **438** (2018) 012023 doi:10.1088/1757-899X/438/1/012023

The maximum deformation has been obtained 8.62 mm.

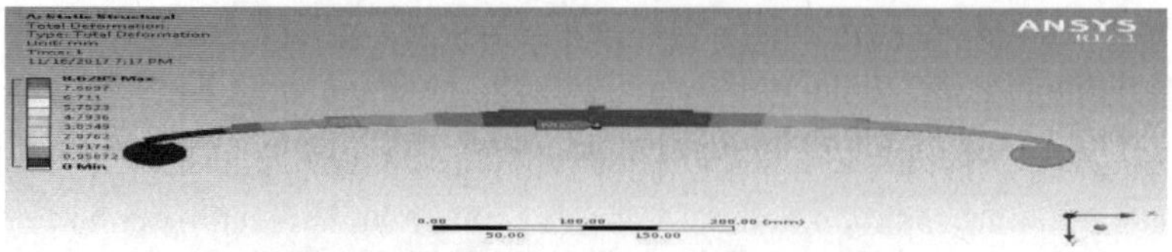


**Figure 7.** Deformation of steel leaf spring.

The maximum strain energy has been obtained 13.96 MJ.

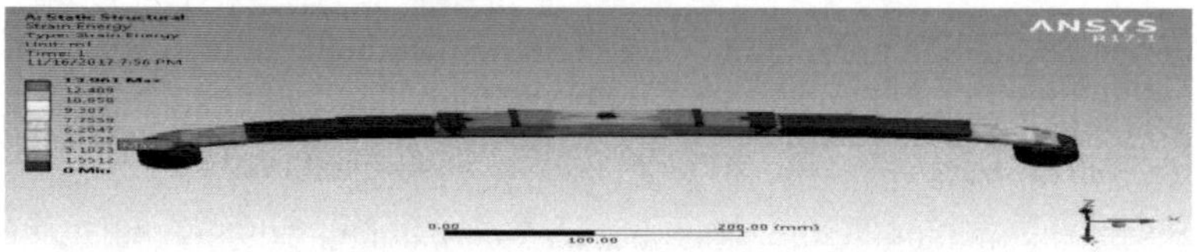


**Figure 8.** Strain energy of steel leaf spring.

The minimum life cycle of the steel leaf spring has been obtained 95966.

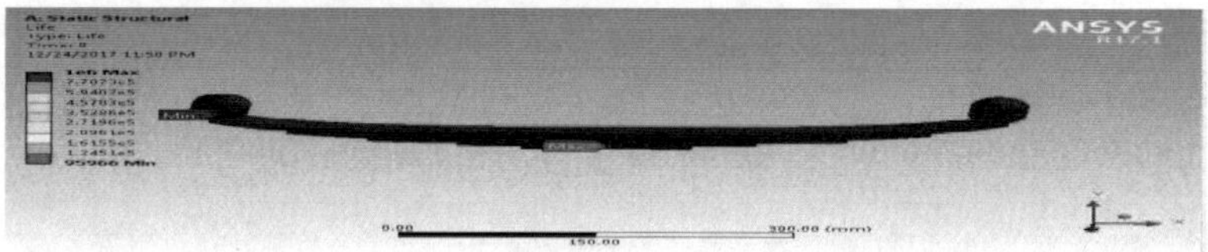


**Figure 9.** Life Cycle of steel leaf spring.

The maximum von mises stress has generated in Carbon Epoxy leaf spring 106.66 MPa. And the maximum deformation has been obtained 2.32 mm.

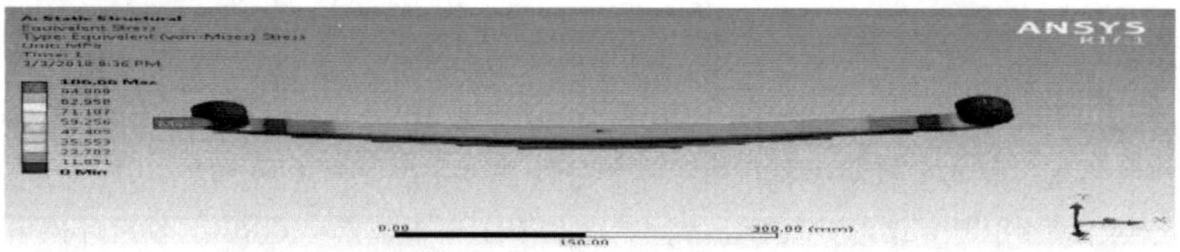


**Figure 10.** Von mises stress in composite leaf spring.

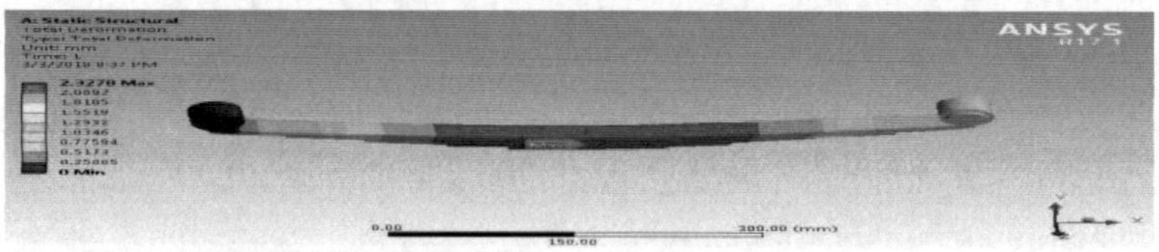


**Figure 11**. Deformation of composite leaf spring.

The maximum strain energy has been obtained 7.56 MJ.

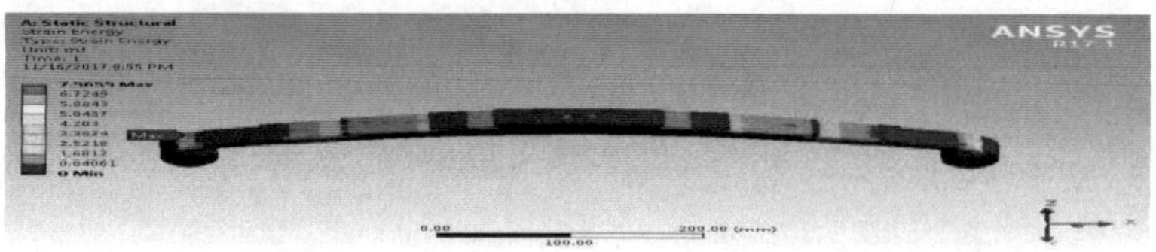


**Figure 12.** Strain energy of composite leaf spring.

The Minimum life cycle of the composite leaf spring has been obtained 10000000.

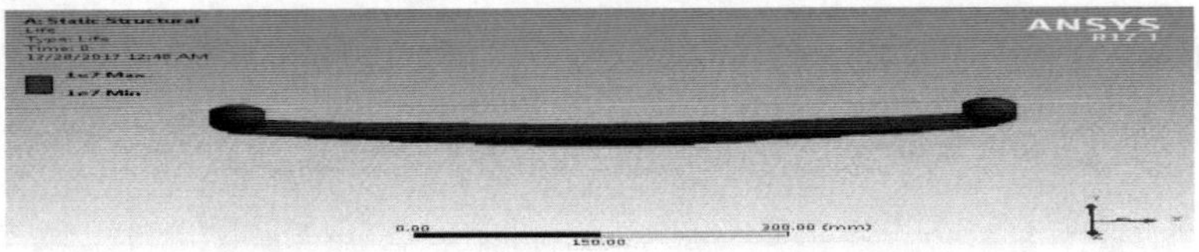


**Figure 13.** Life cycle of composite leaf spring.

## 5. Results and Discussion

The stress and deformation analysis of the Steel leaf spring has been done by analytical, experimental and simulation sections. The validation of the results has been shown in Table 6 and Table 7.

**Table 6.** Validation of Stress.

| Load (N) | Stress (MPa) (Analytical) | Stress (MPa) (Experimental) | Stress (MPa) (Simulation) | Variation (%) (Analytical vs. Experimental) | Variation (%) (Analytical vs. Simulation) | Variation (%) (Experimental vs. Simulation ) |
|---|---|---|---|---|---|---|
| 543 | 95.8 | 84.01 | 96.2 | 12.30 | 0.41 | 14.51 |
| 1033 | 182.26 | 165.67 | 183.03 | 9.10 | 0.42 | 10.47 |
| 1542 | 272.07 | 254.55 | 273.21 | 6.43 | 0.41 | 7.3 |
| 2200 | 388.16 | 390.85 | 389.80 | 0.68 | 0.42 | 0.27 |

**Table 7.** Validation of Deformation.

| Load (N) | Deformation (mm) (Analytical) | Deformation (mm) (Experimental) | Deformation (mm) (Simulation) | Variation (%) (Analytical vs. Experimental ) | Variation (%) (Analytical vs. Simulation) | Variation (%) (Experimental vs. simulation ) |
|---|---|---|---|---|---|---|
| 543 | 2.69 | 2.36 | 2.12 | 12.26 | 21.18 | 10.16 |
| 1033 | 5.12 | 4.66 | 4.05 | 8.98 | 20.89 | 13.09 |
| 1542 | 7.65 | 7.16 | 6.04 | 6.27 | 20.89 | 15.76 |
| 2200 | 10.91 | 11.01 | 8.62 | 0.91 | 20.98 | 21.70 |

**Table 8.** Comparison between Steel and Composite leaf Spring.

| Material | Stress (MPa) | Deformation (mm) | Stiffness (N/mm) | Minimum Life Cycle | Strain Energy (MJ) | Weight (kg) |
|---|---|---|---|---|---|---|
| AISI 1030 Steel | 389.80 | 8.62 | 255.22 | 95966 | 13.96 | 7.01 |
| Carbon Epoxy | 106.66 | 2.32 | 948.27 | 10000000 | 7.56 | 1.14 |

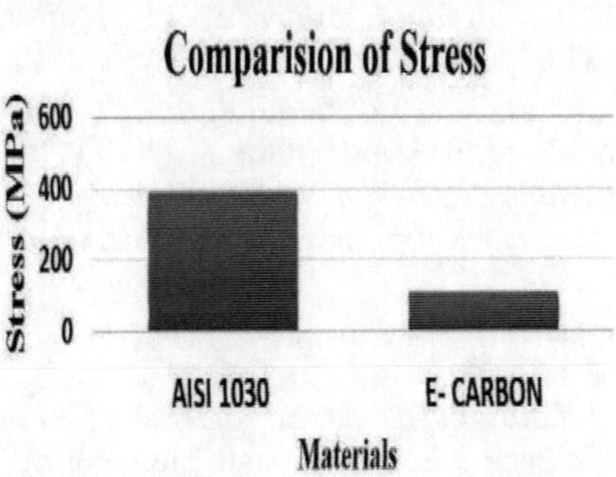


**Figure 14.** Comparison of Stress.

Carbon Epoxy composite leaf spring generates 72.65% less stress compared to steel leaf spring.

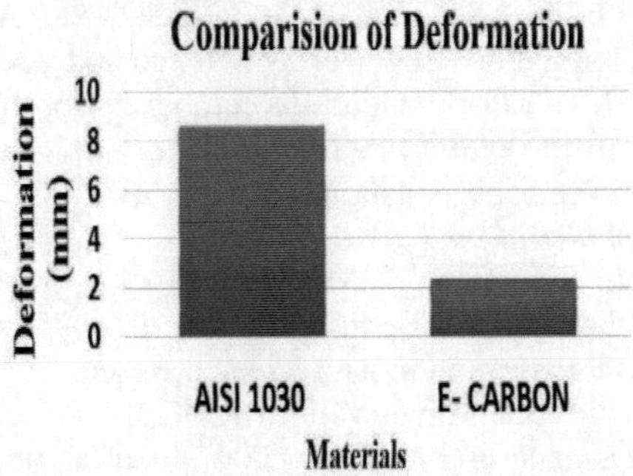


**Figure 15.** Comparison of deformation.

Deformation for the Carbon Epoxy leaf spring is 73% lower than the steel leaf spring.

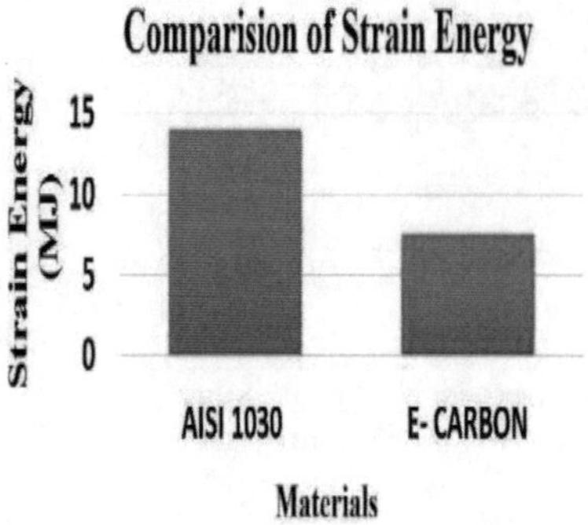


**Figure 16.** Comparison of Strain Energy. Strain energy for the composite leaf spring is 45.8% lesser than the steel leaf spring.

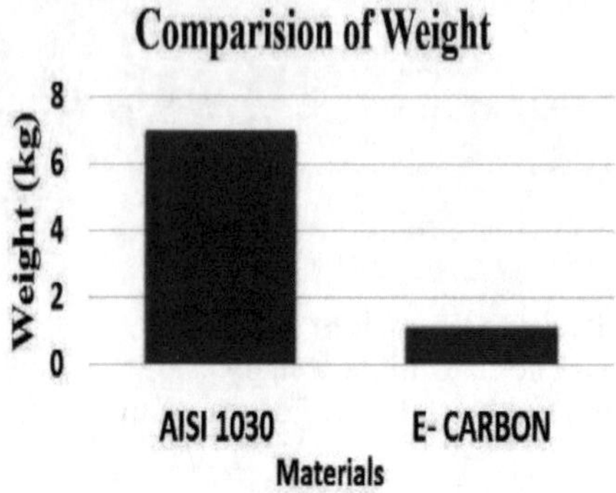


**Figure 17.** Comparison of weight. Weight of composite leaf is 83.73% lesser than the steel leaf spring.

The main objective of this research work is to study the stress, deformation and the failure analysis of an existing steel (AISI1030) leaf spring by analytical, experimental and simulation method, and also to compare it with the composite (Carbon Epoxy) leaf spring. From the static analysis it has been obtained that the maximum stress has generated in the master leaf near the eye of fixed end. For an operating load 2200N, results has been validated by analytical, experimental and simulation process with slight deviation. The stress variation for the analytical, experimental and simulation are 388.16 MPa, 390.85 MPa and 389.80 MPa respectively. And the variation for the deformation are 10.91mm, 11.01 mm and 8.62 mm respectively. The failure analysis of these models have been done by finite element method in the ANSYS 17.1 for fully reversed load of 2200 N. From the fatigue analysis it has been obtained that the maximum probability for the crack initiation is on the master leaf near the eye of fixed end. The life cycle for the composite leaf spring has been obtained $10^7$ cycles and for the steel leaf spring 95966 cycles.

## 6. Conclusion

From this research it can be concluded that with respect to, load carrying capacity and longevity the composite (Carbon Epoxy) leaf spring will ensure a better performance in all aspects compared to AISI 1030 steel leaf spring for the same dimensions and for the same load, in reduced weight.

## 7. References

[1] G Harinath Gowd, E Venugopal Goud, August 2012, *"Static Analysis of Leaf Spring"*, International Journal of Engineering Science and Technology (IJEST), Vol. 4 No.08,

[2] MD Tahsin Ishtiaq, December 2016, *"Optimized Modelling and Fabrication of a Parabolic Leaf Spring with Finite Element Analysis"*, Capstone Design Project (Grand Valley State University).

[3] Syambabu Nutalapati, December 2015, *"Design and Analysis of Leaf Spring by using Composite material for Light Vehicles"* International Journal of Mechanical Engineering and Technology, Vol. 6, Issue 12.

[4] Dev Dutt Dwivedi, V.K. Jain, 2016, *"Design and Analysis of Automobile Leaf Spring using ANSYS"*, Organization India Technical Research, Vol 3, Issue 1.

[5] Trivedi Achyut V, Prof R.M Bhoraniya, May 2015, *"Static and Dynamic Analysis of Automobile Leaf Spring (TATA ACE)"*, International Journal of Science Technology and Engineering, Vol.1, Issue 11.

[6] Makarand B. Shirke, March 2016, *"Static and Fatigue Behavior of Steel and Glass Epoxy Composite Leaf Spring of Light Motor Vehicle"*, International Advanced Research Journal in Science Engineering and Technology, Vol.3, Issue 01.

IOP Conf. Series: Materials Science and Engineering **438** (2018) 012024 doi:10.1088/1757-899X/438/1/012024

# Effect of Fiber Content and Post Stress on Moisture Absorption of Jute Polyester Composite

**S.Z. Ali, M.K. Nahian and M.A. Islam**

Department of Materials and Metallurgical Engineering, Bangladesh University of Engineering and Technology, Dhaka 1000, Bangladesh

**Abstract.** Polymer composites have gained much more attention because of their lightweight, low cost, ease of fabrication, corrosion resistance, wear property etc. In this regard, natural fiber became very popular because of its biodegradability. The moisture absorption of the natural fiber reinforced polymer composites, however, causes many adverse effects on their mechanical properties as well as on long-term performance. This moisture absorbing tendency limits their uses in many engineering applications also. In this research work jute polyester composites containing 5, 10 and 15wt% jute fiber were fabricated in hand lay-up method applying various post-stress conditions and then moisture absorption tests were carried out on prepared samples. It has been found that for unstressed composites, water absorption increased linearly with increasing jute fiber content. Interestingly, with an increase in post-stress, water absorption decreased. In this regard, better results were found for composites containing a higher proportion of jute fibers.
*Keyword:* Jute fiber composite, Polyester, Fiber content, Post-stress, Moisture absorption.

## 1. Introduction

Nowadays, natural fiber composites are emerging as an alternative to replace the glass or other man-made fiber reinforced composites in many engineering applications. Natural fibers as an alternative reinforcement in polymer composites have attracted the attention of many researchers and scientists due to their advantages over conventional glass and carbon fibers [1]. Most composites have strong, stiff fibers in a matrix and in this case, the matrix is weaker and less stiff. The objective is usually to make a component which is strong and stiff, often with a low density [2]. The important driving force for such emergence of utilizing natural fibers is that they are abundant, inexpensive and biodegradable and impose no adverse effects on the environment [3].

One of the main drawbacks of using natural fibers as reinforcement is their hydrophilic nature. Polymeric materials are normally hydrophobic, so there lays an incompatibility between fiber and matrix. Due to the hydrophilic nature of natural fibers, they absorb a significant amount of moisture which has a deleterious effect on the mechanical properties of composites [4]. The influence of water uptake on the mechanical properties of jute fiber reinforced polypropylene composites was studied by A C Karmaker and co-workers [5]. They have observed that the swelling of jute fiber in composite material has a positive effect on the mechanical properties. When the fiber absorbs a small amount of water, the fibers tend to swell slightly. This results in the development of a strong mechanical interlocking between the fiber and the matrix. So at very low water absorption, the natural fiber reinforced polymer composites

IOP Conf. Series: Materials Science and Engineering **438** (2018) 012024 doi:10.1088/1757-899X/438/1/012024

show a slight increase in their mechanical properties. However, in the case, composites which were immersed in cold water for 48 hours, fibers absorb more water and swell considerably. This excess water absorption and the resultant uncontrolled swelling of the fiber leads to matrix cracking and ultimate debonding of the fiber. It has been reported that the water absorption of cellulose fiber reinforced polymer composites depends on the concentration of the fiber [6].

In order to decrease moisture absorption, fiber surface can be modified by employing chemical modifications to improve the adhesion between fiber and matrix [7]. Effect of fiber loading on moisture absorption was also studied by several researchers [8]. Effect of hybridization with glass fibers and chemical modification of fiber on moisture absorption of banana fiber/polyester composite were also studied [9]. The long-term water absorption behavior of composites with different natural fibers and polypropylene were investigated [10]. Fiber surface modification to reduce the moisture absorption of sisal/polyester was also studied [11].

In this research work effect of different fiber content on moisture absorption of continuously aligned jute fiber reinforced polyester composite was studied. Later post stress induced moisture absorption reduction was also studied.

## 2. Experimental

### *2.1 Materials and methods*

In this research work, raw jute fiber was collected from local market of Bangladesh and then it was washed for removing any kind of dirt and then was dried in sun to drive away the moisture. To avoid any dissimilarity in properties of the composite, the middle portion of every jute fiber was selected so that they have the same diameter.

The polyester resin was used as the matrix material and Methyl Ethyl Ketone Peroxide was used as hardener.

The pure polyester sample was fabricated using resin and hardener. For composite fabrication required amount of continuous and aligned jute fibers (10 &15 percent by weight) were placed on the lower part of the mold and then the upper part was closed. Resin with 2% hardener was poured in the mold and after 7-10 minutes, when the resin starts to coagulate, various post stresses (0 kPa, 2.5 kPa, 5 kPa and 10 kPa) were applied. Then the whole setup was kept at room temperature for 24 hours.

**Figure 1:** Applying post stress during composite fabrication

**Figure 2:** Jute-polyester composite

### *2.2 Moisture Absorption Test*

Three specimens were made from each sample according to ASTM 570-98 standard (76.2mm x 16mm). For moisture absorption test samples were dried in a furnace at 100°C for 3 hours. It took almost 1hour for the furnace to reach 100°C. Counting of 3hours of time begins just after the temperature of the furnace reaches 100°C. After heating in the furnace for 3hours the samples were cooled for 10 minutes and then they were kept in a container full of tap water for 6 days. After 6 days their percentage of moisture absorption was measured using following formula.

$$\% \text{ Water Absorption} = \{(M_{wet} - M_{dry}) \div M_{dry}\} \times 100$$

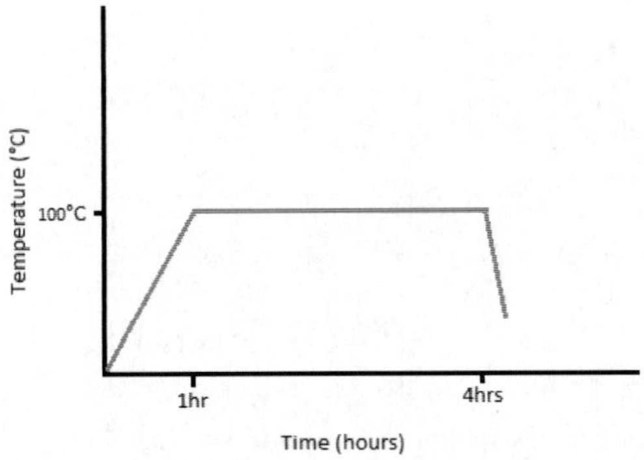


**Figure 3:** Heating cycle in moisture absorption test

*2.3 Visual Inspection*
Visual inspection of the samples was also done to identify any change on the surface of the samples after immersing into the water for 6 days.

## 3. Results and Discussion
*3.1 Moisture Absorption*
*3.1.1 Pure Polyester Sample*
Moisture absorption of pure polyester was conducted. Generally pure polyester absorbs very little amount of moisture. Average moisture absorption by pure polyester was found to be 0.7% at room temperature which matches with the result of other researchers [12].

*3.1.2 Jute Polyester Composite*
Moisture absorption by 5 wt% jute fiber reinforced polyester composite is found to be 1.8% which is much higher than pure polyester sample. Hydrophilic nature of jute fiber causes this excess moisture absorption. Moisture absorption of sisal and roselle fiber reinforced polyester increases linearly with increasing fiber content [13]. G Rajeshkumar and his co-workers reported the same effect for phoenix sp fiber reinforced epoxy composites [14]. This also happened in this research. Moisture absorption of 10 wt% jute fiber reinforced polyester was found to be 2.6%, 0.8% higher than 5 wt% jute fiber reinforced polyester and moisture absorption of 15 wt% jute fiber polyester was found to be 3.23%, 0.63% higher than 10 wt% jute fiber reinforced polyester. Moisture Absorption vs Fiber Content (figure 4) reveals that moisture absorption of jute fiber reinforced polyester increases almost linearly with increasing fiber content.

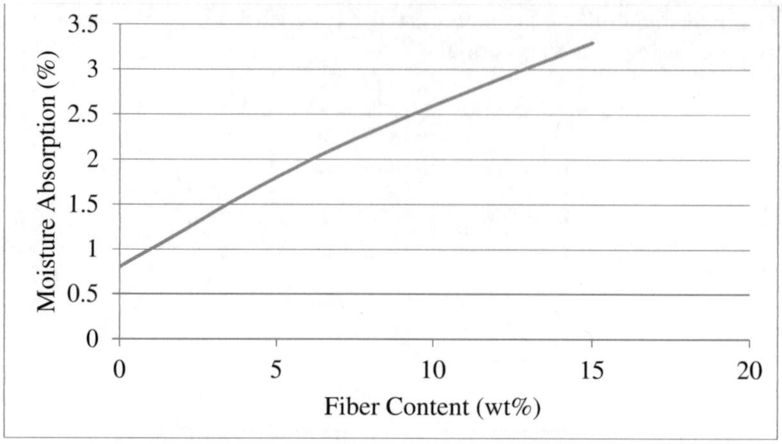


**Figure 4:** Moisture absorption vs Fiber Content

*3.2 Effect of Post Stress*

10 wt% jute fiber reinforced polyester fabricated without any post stress absorbs 2.6% moisture in room temperature. Moisture absorption was observed to decrease gradually with increasing post stress. A Avena and A R Bunsell also reported the same effect of pressure on moisture absorption by polymer composites [15]. Composite fabricated under 2.5 kPa post stress absorbs 2.33% moisture. For 5 kPa and 10 kPa post stress, it becomes 2.17% and 2.14% respectively. That means moisture absorption decreases almost 0.46% for 10 kPa post stress. Change in moisture absorption with increasing post stresses for 10 wt% jute fiber reinforced polyester can be easily visualized from figure 5.

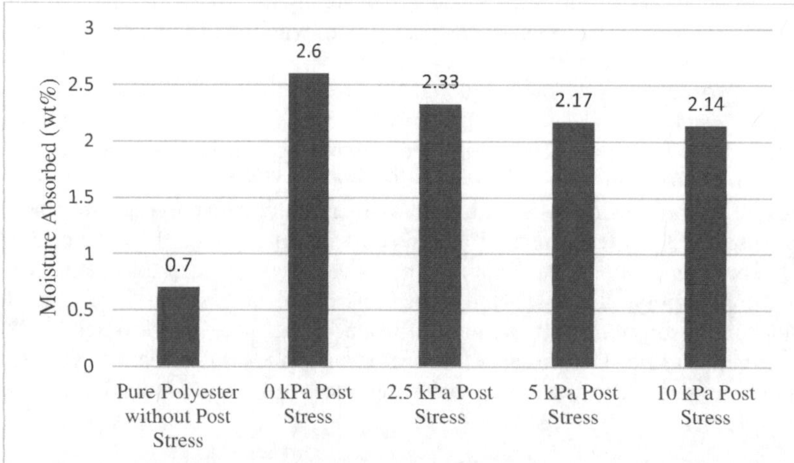


**Figure 5:** Change in moisture absorption with increasing post stress for 10 wt% jute-polyester composite

Moisture absorption by jute-polyester composites with varying post stresses relative to moisture absorption by pure polyester is presented in figure 6. Relative moisture absorbed by composite fabricated under 0 kPa and 10 kPa post stresses are 371% and 306% respectively.

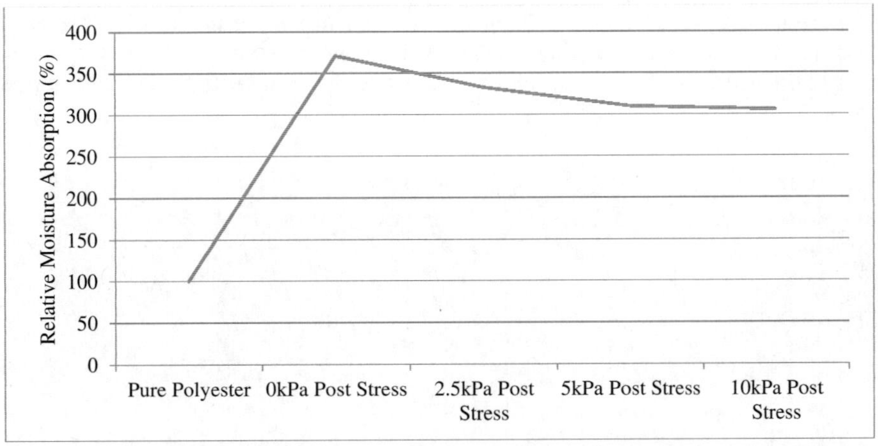


**Figure 6:** Percentage of moisture absorption by jute-polyester composites fabricated under different post stresses relative to pure polyester (for 10wt% fiber content)

Composite fabricated with 15 wt% jute fiber absorbs a higher amount of moisture as mentioned before. The composite that was fabricated without any post stress absorbs 3.23% (by weight) moisture at room temperature. With increasing post stress moisture absorption decreases like the previous case and for 10 kPa post stress, it becomes 1.98%. That means moisture absorption decreases almost 1.25% for 10 kPa post stress. From figure 7 it can be seen that the decrease in moisture absorption with increasing post stress is gradual. And it decreases rapidly for a larger amount of post-stress.

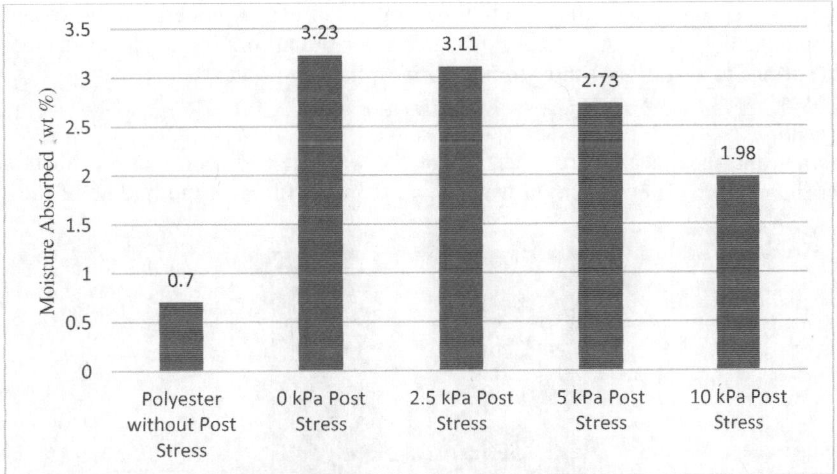


**Figure 7:** Change in moisture absorption with increasing post stress for 15 wt% jute-polyester composite

From the above data relative moisture absorption by jute-polyester composites fabricated under different post stresses to pure polyester is drawn and is presented in figure 7. The figure shows that relative moisture absorption decreases from 461% to 282% when post stress is increased from 0 kPa to 10 kPa. It is evident from figure 6 and 8, that the effect of post-stress on moisture absorption is more pronounced for higher jute fiber content. It happened because in case of higher fiber content it becomes difficult for the matrix to cover all the fibers. This causes more moisture uptake by the fibers. Applying pressure causes the matrix to cover all the tiny spaces inside the composite hence reducing the moisture absorption.

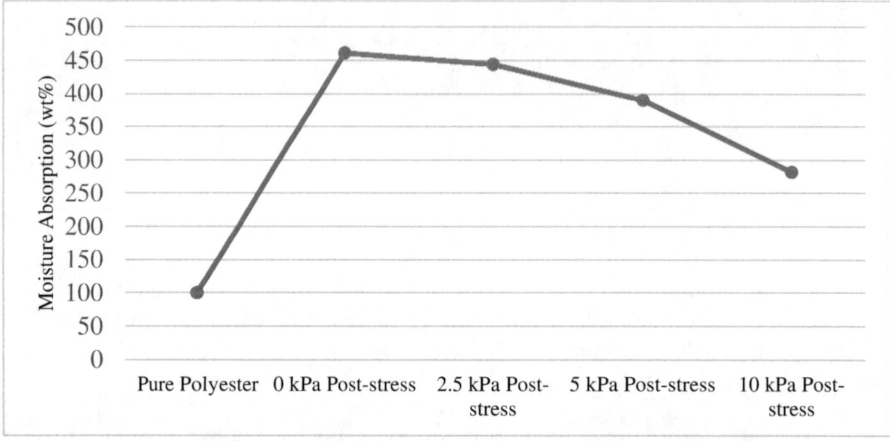


**Figure 8:** Percentage of moisture absorption by jute-polyester composites fabricated under different post stresses relative to pure polyester (for 15wt% fiber content)

So it is obvious that moisture absorption decreases with increasing post stress. This is due to the more effective impregnation of the jute fibers by the resin, which certainly occurred as a higher pressure is applied during the matrix setting. This effective impregnation causes better wetting of hydrophilic jute fibers by hydrophobic polyester resin. As a result, their moisture absorption tendency decreases as there is lesser exposed surface to absorb moisture from the atmosphere.

*3.3 Visual Inspection*
After 6 days when the specimens were taken out of the water it was seen that the fibers at the corners became darker than before. They absorbed water more than the fibers at the middle of the specimen and became darker.

**Figure 9:** Moisture absorption test sample;

(a) Before and (b) After immersing into water

## 4. Conclusion

This research project aimed to investigate the change in moisture absorption of jute fiber reinforced polymer matrix composite under different post stresses for 10 wt% and 15 wt% fiber content. The outcomes of this research are as follows:

- Generally pure polyester absorbs a little amount of moisture but jute polyester composites absorb moisture in a significant amount. Moisture absorption increases with increasing fiber content.
- For both 10 wt% and 15 wt% jute fiber reinforced polyester, moisture absorption decreases with increasing post stress.
- Composites containing a higher proportion of jute fiber show better results under higher post stresses than composites with lower jute fiber content.

## 5. Reference

[1] Nabi Saheb D and Jog J P 1999 *Adv.Polym. Technol.* **18** pp 351–63
[2] M A Masuelli 2013 *Introduction of Fiber-Reinforced Polymers – Polymers and Composites: Concepts, Properties and Processes* ed M Masuelli(INTECH) p 3.
[3] Arifur Rahman M, Parvin F, Hasan M, Hoque M E 2015 *Introduction to Manufacturing of Natural Fibre-Reinforced Polymer Composites* ed Salit M, Jawaid M, Yusoff N, HoqueM(Springer,Cham) p 17-43
[4] H M Khanlou et al. 2017 *J. Composite Materials* **0(0)** pp 1-16
[5] Karmaker A C, Hofeman A and Hinrichsen G. 1994 *J. Appl.Polym. Sci.* **54** p 1803
[6] Sapieha S, Caron M and Schreiber H P 1986 *J. Appl. Polym. Sci.* **32** p 5661
[7] Malkapuram R, Kumar V and Yuvraj S N 2008 *J. Reinf. Plast. Compos.* **28** pp 1169–89
[8] S M Wang and P A Cooper 2006 *Compos. Sci. Technol.* **66** pp 379–386
[9] P Laly and T Sabu 2004*J. Appl. Polym. Sci.* **91** pp 3856–65
[10] Mehdi Tajvidi, S K Najafi and N Moteei 2006 *J. Appl. Polym. Sci.* **99** pp 2199–2203
[11] P A Sreekumar, Selvin P Thomas, Jean marc Saiter, Kuruvilla Joseph, G Unnikrishnan and Sabu Thomas 2009 *Composites Part A: Appl. Sci. & Manufacturing* **40** pp 1777-1784
[12] H N Dhakal, Z Y Zhang and M O W Richardson 2007 *Compos. Sci. Technol.* **67** pp 1674–1683
[13] A Athijayamani, M Thiruchitrambalam, U Natarajan and B Pazhanivel 2009 *Materials Sci. Engineering A* **517** pp 344–353.
[14] G Rajeshkumar, V Hariharan, TP Sathishkumar, V Fiore and T Scalici 2016 *J. Industrial Textiles* **0(00)** pp 1-22.
[15] A Avena and A R Bunsell 1988 *Composites* **19** pp 355-57

**Acknowledgments**

The authors are grateful for a research grant in the course of this investigation to the authority of Department of Materials and Metallurgical Engineering, Bangladesh University of Engineering and Technology (BUET), Dhaka-1000.

# Mechanical Properties of Chemically Treated Coir and Betel Nut Fiber Reinforced Hybrid Polypropylene Composites

**M.M. Rahman, M. Mondol and M. Hasan**

Department of Materials and Metallurgical Engineering, Bangladesh University of Engineering and Technology, Dhaka 1000, Bangladesh

**Abstract.** Natural fiber reinforcement in polymer is one of the best ways to enhance mechanical properties of polymer. Hybrid composites provide the potential of achieving a balanced a pursuit of stiffness, strength, ductility and other mechanical properties. For this purpose coir and betel nut fiber were used as reinforcement in polypropylene matrix in present research. Raw coir and betel nut husk fiber were chemically treated with sodium hydroxide to increase adhesion of those fibers with polypropylene. Both raw and alkali treated fibers were utilized during composite preparation with the help of hot press technique. Fiber loading were varied at 5, 10 and 15 wt%, while coir and betel nut husk fiber ratios were varied at 1:1, 3:1 and 1:3. Tensile, flexural, impact, hardness and water absorption tests of prepared composites were subsequently conducted. All mechanical properties except tensile strength increased with increase in fiber loading. Properties of prepared composites were found better as compared to polypropylene matrix. Composites containing coir and betel nut at same ratio had better mechanical properties than composites containing those fibers at 1:3 and 3:1 ratio. On the other hand, sodium hydroxide treatment of raw coir and betel nut fiber increased mechanical properties of treated fiber composites as compared to raw fiber reinforced composite. Thus alkali treated 15 wt% coir and betel but fiber (at 1:1) reinforced polypropylene composite had the optimum set of mechanical properties among all prepared composites.

**Keywords:** Hybrid Composite; Coir and Betel Nut Fiber; Chemical Treatment; Mechanical Properties

## 1. Introduction

Hybrid composites using natural fibers have unique features that can be used to meet various design requirements in a more economical way than conventional composites. Balanced strength and stiffness, balanced thermal distortion, balanced bending, reduced weight and cost, reduced energy consumption, renewability, recyclability etc. are the advantages of natural fibers which gives the outstanding output of the composite properties. Eco-friendly nature of natural fiber composites have enhanced the use of natural fiber reinforced composites because of their minimized environmental pollution. By using a hybrid composite that contains two or more types of fiber, the advantages of one type of fiber could complement with what are lacking in the other. As a consequence, a balance in cost and performance can be achieved through proper material design. Natural fiber reinforced composites have some special features like low emission of toxic fumes when subjected to heat and during incineration at end of life, less abrasion damage to processing equipment, fibers have low density and high specific strength and stiffness etc [1].

Among various natural fibers, both coir and betel nut fibers are abundant in nature. Coir fiber is the natural fiber of the coir husk where it is thick and coarse but durable fiber. Coir fiber has a minimal effect on the environment due to their bio-degradable properties [2]. It possesses high failure strain, which provides a better strain compatibility between the fiber and the matrix in short fiber reinforced composites. It also has high weather resistance due to higher amount of lignin and absorbs water to a lesser extent compared to other natural fibers due to its less cellulose content. A special feature or coir fiber is that it can be stretched beyond its elastic limit without rupture due to helical arrangement of micro-fibrils [3]. On the other hand betel nut fibers are obtained from the husk of the betel nut fruit. Betel nut fibers are predominantly composed of hemicellulose. It is famous for its strength and lightness. Its main advantage is that it is renewable fiber. They have low density, high toughness, acceptable strength property and good thermal property [4]. The goal of present research is to determine mechanical properties of a proposed combined polymer composite which consist of a polypropylene matrix and a mixture of coir and betel-nut fibers.

## 2. Materials and Methods

### *2.1 Materials*

Commercial grade polypropylene (PP), coir and betel nut fibers were used in present research. All of them were collected from the local market. PP was white in colour and granular in form having a melting point of $160^0$C. Coir fiber was collected from coconut husk. Coconut husks were wetted for one week for retting. After retting the husks were crushed by hammer and fiber was collected manually from the husk. Fresh fiber was washed and dried under sun for 24 hours. Betel nut fibers were collected from ripe betel nut husk. The husks were wetted for two days for retting. Then fiber was collected manually from the husk. Finally they were cleaned and dried under sun.

### *2.2 Alkali treatment of natural fiber*

Alkali treatment (5%NaOH) is used to remove a certain amount of lignin, wax and oils covering of external surface of the natural fiber, depolymerizes cellulose and expose the short length crystallites and disruption of hydrogen bonding in the network structure, thereby increasing surface roughness [5].

$$\text{Fiber–OH+NaOH} \rightarrow \text{Fiber–O–Na+H}_2\text{O}$$

For this purpose, both fibers were soaked into 5% NaOH solution for 2 and half hours at room temperature. After that fiber was cleaned with distilled water. Finally 1% acetic acid solution treatment was done to remove excess NaOH. Then neutralized fiber was dried into the oven at $100^0$C for 2 and half hours to remove excess moisture.

### *2.3 Composite preparation*

Composites were prepared using hot press technique in a 150mm×150mm×3mm aluminum die. A hydraulic type hot press machine having maximum load of 35 kN and maximum temperature of $300^0$C was utilized. The fiber loading was varied at 0, 5, 10 and 15 wt% with ratio of betel nut to coir of 1:1. later composites were prepared with betel nut to coir ratio of 3:1 and 1:3 for 15 wt% fiber loading. Alkali treated fiber composite was also prepared using 15 wt% fiber loading and 1:1coir and betel nut fiber ratio. Firstly fibers were chopped to 3-5 mm length. Then required amount of fiber and PP were weighed in a balance. In order to allow the removal of moisture, fibers and polypropylene were dried in an oven at $80^0$C for 20 minutes before preparing each composite. Pre-mixed mixture was then placed inside the die. The fiber-matrix mixture was allowed to press at 30 kN pressure. The temperature was initially raised to $160^0$C and hold there for around 15-20 minutes, after that the temperature was raised to $(185\text{-}190)^0$C. The die was cooled to room temperature, pressure was released and composites were withdrawn from the die.

*2.4 Mechanical testing*
Tensile, flexural, Charpy impact, hardness and water absorption tests were conducted. For each test, three specimens were tested and the average values are reported. Tensile test were conducted according to ASTM-D 638-01 [6] using an universal tensile machine at a crosshead speed of 5 mm/min. The dimension of the specimen used was 100 mm x 19 mm x 3 mm. Static flexural test was carried out according to ASTM D 790-00 [7] using the same testing machine at same cross head speed as mentioned before. The dimension of the specimens used was 78 mm x 12.5 mm x 3 mm. Dynamic Charpy impact tests were conducted according to ASTM D 6110-97 [8] using an impact tester MT 3016. The dimension of the specimens used was 72 mm x 18 mm x 3mm. Hardness of the composite were measured using a shore hardness testing machine in shore D scale.

*2.5 Water absorption test.* In order to measure the water uptake characteristics of natural fiber reinforced the composites water absorption were carried out. For the water absorption test, the specimens were dried in an oven for a specified time and temperature and then placed in a desiccator to cool. Immediately upon cooling the specimens are weighed and immersed in distilled water according to ASTMD 570-99 [9] for 24 hours. The final weight of the specimens was then taken. The increase in the weight of the specimens was calculated using the following equation:

$$\text{Percent Water Absorption} = \frac{(\text{Wet weight} - \text{Dry weight})}{\text{Dry weight}} \times 100\% \quad \text{.........................................(4)}$$

## 3. Results and Discussion

*3.1 Tensile and flexural properties*
Tensile test and three point flexural test were performed to evaluate tensile properties and flexibility of prepared composites. Tensile and flexural properties against raw fiber loading are shown in Figure 1. Tensile strength decreased with increase in fiber loading [10-14]. As fiber loading increased, the interfacial area between the fiber and matrix increased that is weak because of weak interfacial bonding between cellulose based hydrophilic fiber and hydrophobic matrix. With increasing fiber loading, the bond become continuously weak. As a result tensile strength decreased (Figure 1 (a)). The same trend was also observed by other researchers [15-17]. According to Figure 1 (b), Young's modulus increased with an increase in fiber loading [10-13]. This is because with an increase in fiber content, brittleness of the composite increased and stress/strain curves becomes steeper. Poor interfacial bonding creates partially separated micro spaces which obstruct stress propagation between the fiber and the matrix [17]. As the fiber loading increases, the degree of obstruction increases, which in turn increased stiffness.

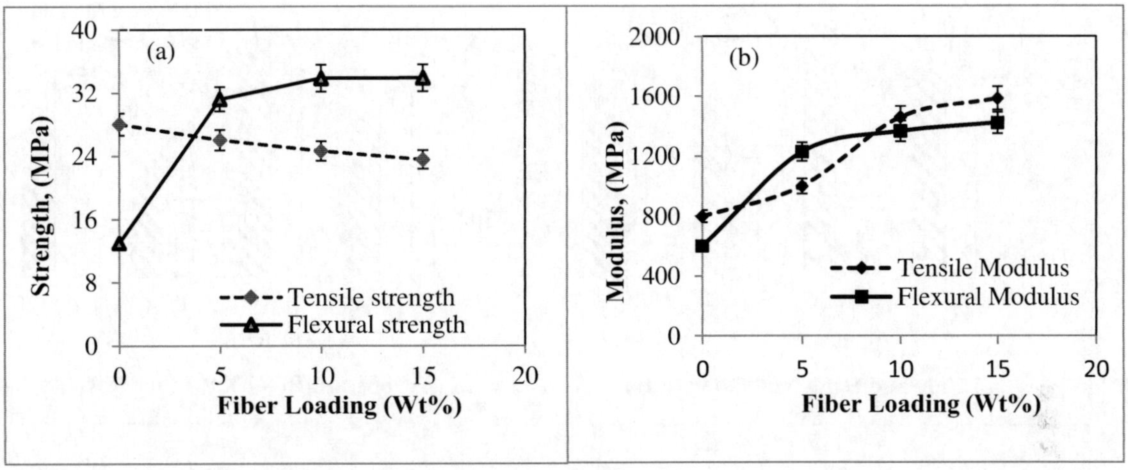


**Figure 1.** Variation of (a) tensile and flexural strength and (b) Young's modulus and flexural modulus against fiber loading.

On other hand, flexural properties (flexural strength and modulus) increased with increase in fiber content (Figure 1). This may be due to the favorable entanglement of polymer chain with filler which has overcome the weak filler matrix adhesion [16]. After 10% fiber loading, flexural properties did not change significantly. The reason may be weak filler matrix adhesion. With increase in fiber loading, they act as flaws and are not perfectly aligned with the matrix. Among different fiber ratio, better tensile sand flexural properties were found for coir and betel nut fiber ratio of 1:1 because of better compatibility of fibers with PP matrix and better fiber dispersion in the matrix (Figure 2). After alkali treatment tensile strength decreased and Young's modulus increased which conform better mechanical interlock of fiber and matrix [5]. Alkali treatment also ensures higher interfacial interaction between fiber and matrix, which increases effective surface area available of contact with the matrix and possibility of load transfer between the matrix and the reinforcing fibers. This resulted in increase of flexural strength and modulus (Figure 2).

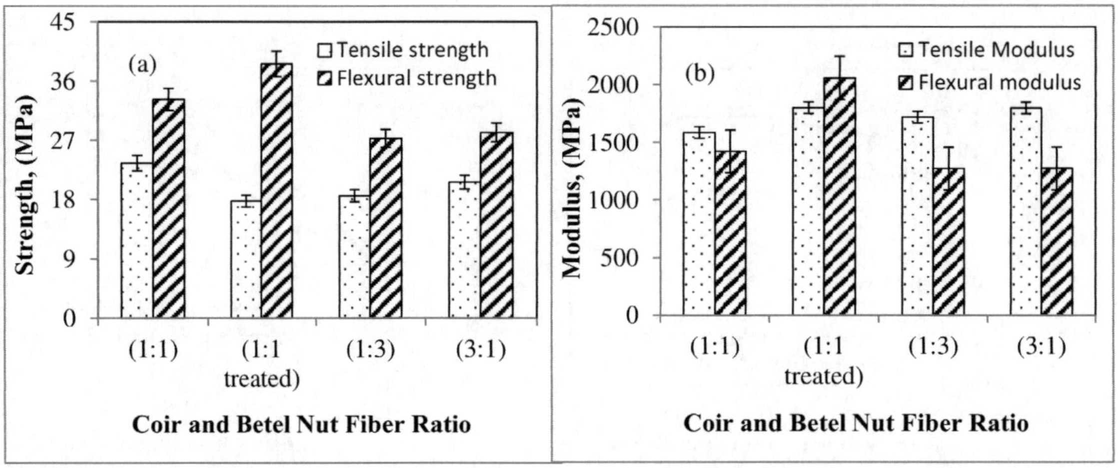


**Figure 2.** Variation of (a) tensile and flexural strength and (b) Young's modulus and flexural modulus against fiber ratio and alkali treatment.

*3.2 Hardness Test Results*

Incorporation of fiber into the PP matrix reduce the flexibility of the matrix which result in more rigid composite [15]. Hence with increase in fiber content, hardness values increased (Figure 3 (a)). Hardness of 25% coir-75% betel nut fiber reinforced composite was higher as compared to 75% coir-25% betel nut fiber reinforced composite (Figure 3 (b)). Due to combined effect of coir and betel fiber, highest hardness value was found in 50% coir-50% betel nut fiber reinforced composite. Alkali treatment of fiber resulted in collapse of the amorphous cellular structure into crystalline structure, which leads to better packing of cellulose chains and reduction of void, as well as better adhesion between the matrix and the filler [5]. Thus hardness increased after alkali treatment.

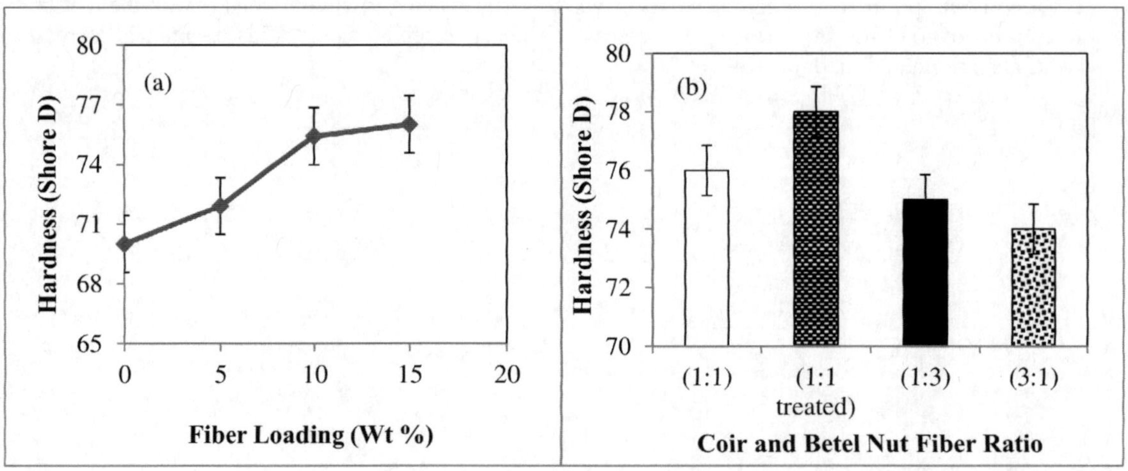


**Figure 3.** Variation of hardness against (a) fiber loading and (b) fiber ratio and alkali treatment.

*3.3 Charpy Impact Test Results*

Impact strength of fiber reinforced polymeric composites depends on the nature of the fiber, polymer and fiber-matrix interfacial bonding [5]. Impact strength of prepared composites increased with an increase in fiber loading (Figure 4 (a)), which is in agreement with other researcher [18, 19]. One of the factors of impact failure of a composite is fiber pull out. With the increase in fiber loading, stronger force is required to pull out the fibers. This in turn increases the impact strength. Alkaline treatment make the fiber surface rough, which help better fiber-matrix interfacial bonding. As a consequence impact strength of alkali treated fiber reinforced composites was higher than raw fiber reinforced composites (Figure 4 (b)).

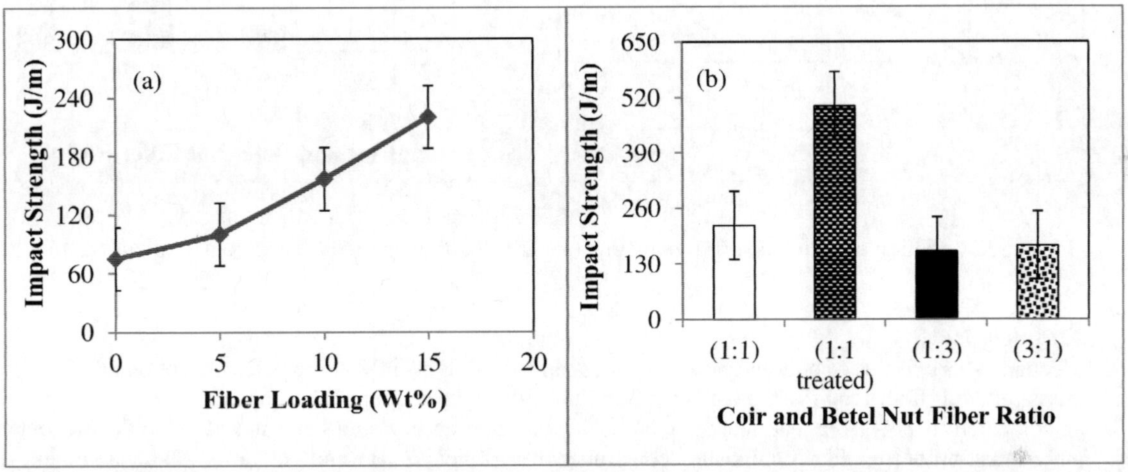


**Figure 4.** Variation of impact strength against (a) fiber loading and (b) fiber ratio and alkali treatment.

*3.4 Water Absorption Characteristics*

Polymer is hydrophilic in nature, however natural fiber is hydrophobic because of cellulose content. Water absorption property of composite depends on hydroxyl group in fiber. With increase in composite fiber content, number of hydroxyl groups in composite increases. As a result water absorption also increase [5] (Figure 5 (a)). Betel nut fiber contains high percentage of cellulose (53.3%) as compared to coir fiber (32-43%) [18]. Thus water absorption increased with increase in betel nut fiber ratio in composites (Figure 5 (b)). After alkali treatment, water absorption decreased due to decrease in number of hydroxyl groups in treated fiber reinforced composite [17].

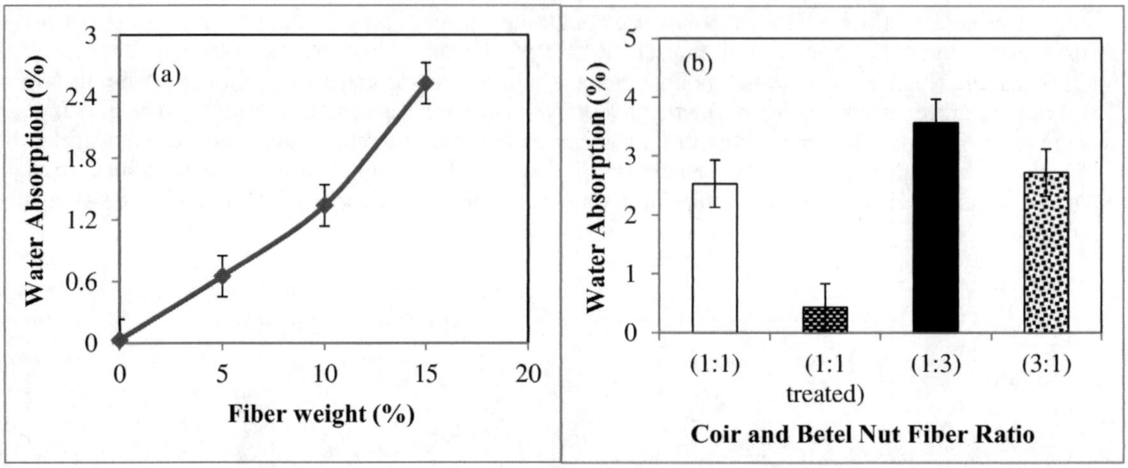


**Figure 5.** Variation of water absorption against (a) fiber loading and (b) fiber ratio and alkali treatment.

## 4. Conclusion

Mechanical properties of coir and betel nut fiber reinforced hybrid PP composites except tensile strength increased with fiber loading. On the other hand all properties of prepared composites were found better as compared to polypropylene matrix. Composites containing coir and betel nut at 1:1 ratio had better mechanical properties than composites containing those fibers at 1:3 and 3:1 ratio. Alkali treatment of raw coir and betel nut fiber increased mechanical properties of treated fiber composites as compared to raw fiber reinforced composite. Thus alkali treated 15 wt% coir and betel but fiber (at 1:1) reinforced polypropylene composite had the optimum set of mechanical properties among all prepared composites.

## 5. References

[1] Joshi S V, Drzal L T, Mohanty A K and Arora S 2004 *Are natural fiber composites environmentally superior to glass fiber reinforced composites?* Composites A, Vol. 35(3), pp. 371–376.

[2] Konduru S, Evans M R and Stamps R H 1999 *Coconut husk and processing effects on chemical and physical properties of coconut coir dust,* Hort Science, Vol. 34(1), pp. 88–90.

[3] Verma D and Gope P C 2015 *The use of coir/coconut fibers as reinforcements in composites,* Biofiber Reinforcements in Composite Materials, pp. 285-319.

[4] Yusriah L, Sapuan S M, Zainudin E S and Mariatti M 2012 *Exploring the potential of betel nut husk fiber as reinforcement in polymer composites: effect of fiber maturity*, Procedia Chemistry, Vol. 4, pp. 87-94.

[5] Li X, Tabil L G and Panigrahi S 2007 *Chemical treatments of natural fiber for use in natural fiber-reinforced composites: a review,* Journal of Polymers and the Environment, Vol. 15(1), pp. 25-33.

[6] ASTM Standard D 638-01, 2002 *Standard test methods for tensile properties of plastics*, Annual Book of ASTM Standard, Vol. 08.

[7] ASTM Standard D 790-00, 2002 *Standard test methods for flexural properties of unreinforced and reinforced plastics and electrical insulating materials,* Annual book of ASTM standard, Vol. 08.

[8] ASTM Standard D 6110-97, 2002 *Standard test methods for determining the Charpy impact resistance of notched specimens of plastics,* Annual Book of ASTM Standard, Vol. 08.
[9] ASTM Standard D 570-99, 2002 *Standard test methods for water absorption of plastics,* Annual book of ASTM standard, Vol. 08.
[10] Yang H S, Kim H J, Son J, Park H J, Lee B J and Hwang T S 2004 *Rice-husk flour filled polypropylene composites; mechanical and morphological study,* Composite Structures, Vol. 63, pp. 305-312.
[11] Lou C W, Lin C W, Lei C H, Su K H, Hsu C H, Liu Z H and Lin J H 2007 *PET/PP blends with bamboo charcoal to produce functional composites*, Journal of Materials Processing Technology, Vol. 192-193, pp. 428-433.
[12] Rana A K, Mandal A and Bandyopadhyay S 2003 *Short jute fiber reinforced polypropylene composites: effect of compatibilizer, impact modifier and fiber loading*, Composites Science and Technology, Vol. 63, pp. 801-806.
[13] Joseph S, Sreekala M, Oommen Z, Koshy P and Thomas S 2006 *A comparison of mechanical .properties of phenol formaldehyde composites reinforced with banana fibers and glass fibers,* Composites Science and Technology, Vol. 62, pp. 1857-1868.
[14] Ajay K, Chauhan S S, Modak J M and Chanda M 2007 *Mechanical properties of wood fiber reinforced polypropylene composites: effect of a novel compatibilizer with isocyanate functional group,* Composites A, Vol. 38, pp. 227-233.
[15] Jamil Md S, Ahmed I and Abdullah I 2006 *Effect of rice husk filler on the mechanical and thermal properties of liquid natural rubber compatibilized high-density polyethylene natural rubber blends,* Journal of Polymer Research, Vol. 13, pp. 315-321.
[16] Rahman M R, Haque M M, Islam M N and Hasan M 2008 *Improvement of physico-mechanical properties of jute fiber reinforced polypropylene composites by post treatment,* Composites A, Vol. 39, pp.1739-1747.
[17] Yang H S, Kim H J, Son J, Park H J, Lee B J and Hwang T S 2006 *Water absorption behavior and mechanical properties of lignocellulosic filler polyolefin biocomposites,* Composite Structures, Vol. 72, pp. 429-437.
[18] Sanadi A R, Caulfield D F, Jacobson R E and Rowell R M 1995 *Renewable agriculture fibers as reinforcing fillers in plastics: mechanical properties of kenaf fiber polypropylene composites,* Industrial Engineering and Chemistry Research, Vol. 34, pp. 1889-1896.
[19] Yang H, Yan R, Chen H, Lee D H and Zheng C 2007 *Chararteristics of hemicellulose, cellulose and lignin pyrolysis,* Fuel, Vol. 86(12-13), pp. 1781-1788.

**Acknowledgment**
The authors would like to thank department of Materials and Metallurgical Engineering, Bangladesh University of Engineering and Technology for their support during the work.

# Mechanical, Thermal and Morphological Properties of Pineapple and Betel Nut Husk Fiber Reinforced Hybrid Polypropylene Composites

**M. Akter, E. Jahan and M. Hasan**

Department of Materials and Metallurgical Engineering, Bangladesh University of Engineering and Technology, Dhaka 1000, Bangladesh

**Abstract:** Hybrid composites have extensive engineering application where strength to weight ratio, low cost and ease of fabrication are required. In recent times hybrid composites have been established as highly efficient, high performance structural materials. Structures made of composites have a long life and need little maintenance. Present research presents a review of the current status of hybrid composite materials technology in terms of mechanical, thermal and morphological properties. Hybrid composites were manufactured with pineapple and betel nut husk as hybrid fibers and polypropylene as matrix using a hot press machine. Fiber loading was varied at 5, 10 and 15 wt%, while pineapple and betel nut husk fiber ratio was set at 1:1. For mechanical characterization, tensile, flexural and hardness tests were conducted with the composites. Thermal properties were measured using thermo gravimetric analysis. To observe surface morphology of the samples, scanning electron microscope was used. Tensile test of the composites showed a decreasing trend of tensile strength with increasing fiber content. Tensile modulus initially increased with fibers loading, then decreased at 10% fiber content. Flexural strength and flexural modulus were found higher for 10% fiber loaded composite as compared composites containing 5% and 15% fiber. From the TGA analysis, it is observed that 10% fiber loaded composites had highest thermal stability among all manufactured composites. According to scanning electron microscopy, 10% fiber loaded composite showed more adhesion as compared to 5% and 15% fiber loaded composites. Fiber agglomeration was also observed in 15% fiber loaded composite.

**Keywords**: Hybrid Composite; Mechanical Properties; Thermal Properties; Morphology

## 1. Introduction

Hybrid composites contain more than one type of fiber in a single matrix material. In principle, several different fiber types may be incorporated into a hybrid, but it is more likely that a combination of only two types of fibers would be most beneficial. They have been developed as a logical sequel to conventional composites containing one fiber. Hybrid composites have unique features that can be used to meet various design requirements in a more economical way than conventional composites. This is because expensive fibers like graphite and boron can be partially replaced by less expensive fibers such as glass and Kevlar. Some of the specific advantages of hybrid composites over conventional composites include balanced strength and stiffness, balanced bending and membrane mechanical properties, balanced thermal distortion stability, reduced weight and/or cost, improved fatigue resistance, reduced notch sensitivity, improved fracture toughness and/or crack arresting properties, and improved impact resistance [1].

Natural fibers are now dominate the automotive, construction and sporting industries by its superior mechanical properties. These natural fibers include flax, hemp, jute, sisal, kenaf, coir and many others. Various advantages of natural fibers are low density, low cost, low energy inputs and comparable mechanical properties and also better elasticity of polymer composites reinforced with natural fibers, especially when modified with crushed fibers, embroidered and 3-D weaved fibers. Experimental techniques can be employed to understand the effects of various fibers, their volume fractions and matrix properties in hybrid composites. These experiments require fabrication of various composites with the above mentioned parameters, which are time consuming and cost prohibitive [2].

One of the major scientific challenges for the composite engineers is the development of new stronger and tougher lightweight structural materials supporting latest technologies and design concepts for the complex shaped structures like aircraft, automotive structures, and large wind turbine blade structures. The development of composite materials improving their performance limits based on the reinforcement of two or more fibers (synthetic fiber with another synthetic fiber or synthetic fiber with natural fiber or synthetic fiber with metallic fibers) in a single polymeric matrix, which leads to the advanced material system called hybrid composites with a great diversity of material properties, is still in its infancy. This is a major challenge that can only be met through an understanding of the relationships between materials architecture and mechanical response, as well observing microstructure formation [3]. Pineapple fiber and betel nut fiber were used in present research. These fibers are available and very cheap n context to the economic condition of Bangladesh. The objective of the research was to develop hybrid composite by using two natural fibers, which are abundant, inexpensive and green in nature. The purpose was to characterize pineapple and betel nut fibers reinforced PP hybrid composite by finding out their mechanical, thermal and morphological characteristics.

## 2. Materials and Methods

### *2.1 Materials*

Pineapple fibers were collected from the pineapple leaf. The leaves were smashed and then the fibers were collected. Betel nut fibers were collected from the upper portion of the betel nut. Then they were cleaned and dried. Polypropylene is a very common thermoplastic, which is very available in the local market. Commercial grade Polypropylene was thus collected from local market.

### *2.2 Composite Preparation*

Discontinuous fiber composites were prepared in present research. Hybrid composites of polypropylene matrix were prepared by varying fiber content at 5%, 10% and 15 wt%, while of betel nut and pineapple fibers were used at 1:1. Hot press technique was utilized for composite preparation. At first fibers were weighted according to required fraction needed. Then both fibers were cut into 3-5 mm size in length. Then to allow the removal of moistures, fibers were dried in an oven at 80°C for 20 minutes before preparing each composite. The required amount of polyproylene was weighed. To prevent voids, water bubbles, poor fiber matrix adhesion the polypropylene was dried in an oven at 80°C for 20 minutes before preparing each composite. Both fibers were mixed in a container [4]. Mold surface was rubbed with emery paper and cleaned very carefully. Mold releasing agent was sprayed on the mold surface. Then polypropylene was distributed on the female mold, then fibers were given as layer on the polypropylene. Remaining part of polypropylene was distributed as the upper layer. Then the mold was closed by male mold. The mold was placed in the hot pressing machine. The capacity of the machine is maximum load 30kN maximum operating temperature 300°C. The mixtures of fiber and matrix were allowed to press at 30kN pressure and the temperature was initially raised to 160°C. When the temperature was reached at 160°C then sample was put in that temperature at 15-20 minutes, after that the temperature was raised to 190°C. When the temperature was reached at 190°C, then water was supplied to cool the mold to room temperature. After cooling the mold or die the pressure was released and the specimen was carefully withdrawn from the die [5].

*2.3 Mechanical Tests*

Tensile tests were conducted according to ASTM D 638-01using an Instron machine. Each test was continued until the samples become failed. By this test, tensile properties of the composites were determined. The three point bend test is generally performed on composite material. Specimens were prepared according to ASTM D 790-00 using the same Instron machine. Size was specified by using a grinding machine. The hardness of polymer composite is most commonly measured by the Shore (Durometer) test. This method measures the resistance of composites toward indentation and provides an empirical hardness value that does not necessarily correlate well to other properties or fundamental characteristics. In present research shore D hardness scale was used.

*2.4 Scanning Electron Microscopy (SEM)*

Surface morphology of the composites of varying fiber percentage was observed by using a scanning electron microscope. The surface of the fiber was made conductive by giving platinum coating using a sputtering machine. The fiber was then observed in vacuum condition into the SEM machine.

*2.5 Fourier Transform Infrared Spectroscopy (FTIR)*

Powder sample was made to perform FTIR, Potassium bromide (KBr) was taken as a reagent which was mixed with the sample (KBr: sample = 100:1). The mixture was taken into a dice. Then it was pressed by a hand pressing machine. The infrared spectra of composites were recorded on a Nicolet 380 spectrophotometer with co-addition of 32 scans.

*2.6 Thermogravimetric Analysis (TGA)*

TGA measures the amount of weight change of a material, either as a function of increasing temperature, or isothermally as a function of time, in an atmosphere of nitrogen, helium, air, other gas, or in vacuum. Inorganic materials, metals, polymers and plastics, ceramics, glasses and composite materials can be analyzed. Temperature ranges from 25°C to 900°C routinely. The maximum temperature was set at 500°C during present research.

## 3. Result and Discussion

*3.1 Mechanical Properties*

*3.1.1 Tensile Test Results.* Figure 1 shows the effect of fiber loading on tensile strength of the composites. The tensile strength decreased with an increase in fiber loading (Figure 1 (a)). As the fiber loading increased, the interfacial area between the fiber and the matrix increased, which was weak because of worsening interfacial bonding between the cellulose based hydrophilic filler (PAL and BNH) and hydrophobic matrix PP [6, 7]. This consequently decreased the tensile strength. It is observed that the Young's modulus increased with an increase in fiber loading then again decreased with an increase in fiber loading (Figure 1 (b)). This is because with an increase in fiber content, the brittleness of the composite increased and stress-strain curves became steeper. Poor interfacial bonding creates partially separated micro spaces which obstruct stress propagation between the fiber and the matrix. As the fiber loading increases, the degree of obstruction increases, which in turn increased the stiffness [8]. The elongation at break reduced with an increase of fiber weight percentage due to the low percentage of elongation at break of the fibers compared to PP (Figure 1 (c)).

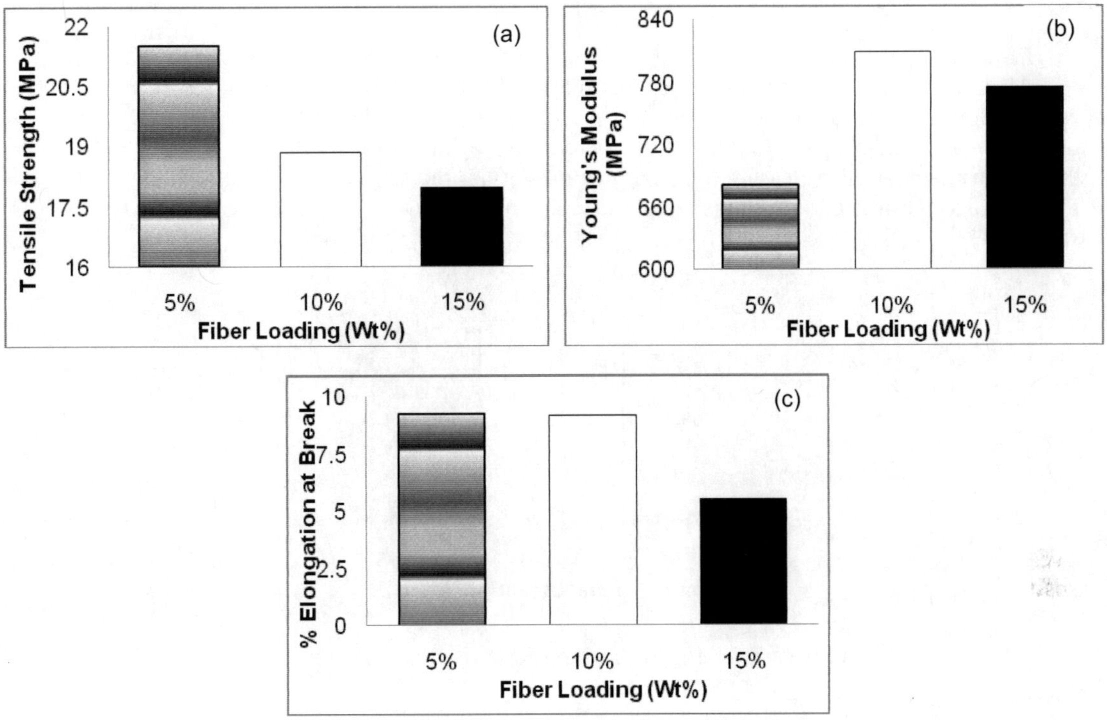


**Figure 1.** Variation of (a) tensile strength, (b) Young's modulus and (c) % elongation at break against fiber loading.

*3.1.2 Flexural Test Results.* Flexural properties of the composite samples were measured for each fiber content (5, 10, 15 weight %) with the help of stress-strain curves and respective equation. The flexural strength and flexural modulus of raw pineapple leaf fiber and betel nut husk fiber (PAL: BNH=1:1) reinforced hybrid polypropylene composites at different fiber loadings are shown in Figures 2 (a) and 2 (b) respectively. The flexural strength increased with an increase in fiber loading. This may be due to the favorable entanglement of the polymer chain with the filler, which has overcome the weak filler matrix adhesion with increasing filler content. The flexural modulus values of the composites for different fiber loadings are shown in Figure 2 (b). It is observed that the flexural modulus increased with an increase in fiber loading, then again decreased with an increase in fiber loading.

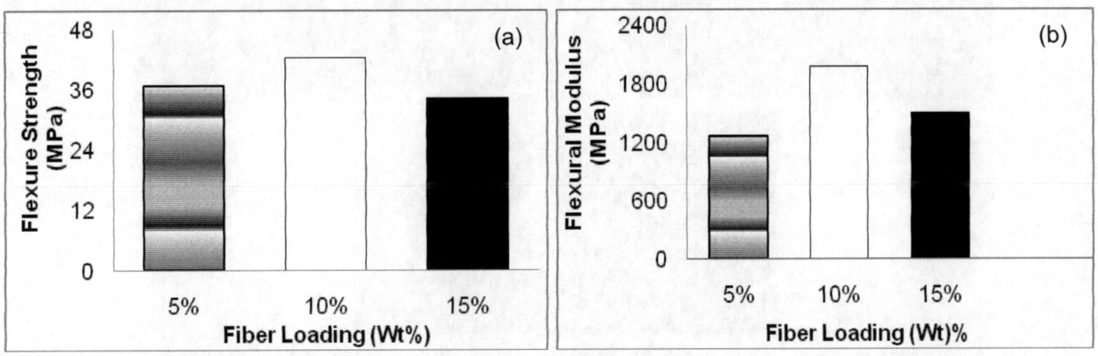


**Figure 2.** Variation of (a) flexural strength and (b) flexural modulus against fiber loading.

*3.1.3 Hardness Test Results.* Hardness of a composite depends on the distribution of the filler into the matrix. Usually the more flexible matrix causes the resultant composites to exhibit lower hardness. As shown in Figure 3, incorporation of fiber into the PP matrix has reduced the flexibility of the matrix resulting in more rigid composites. Due to the increase of stiffness of respective composite the hardness of the composites showed increasing trend with an increase in the fiber content [9, 10]. Better dispersion of the filler into the matrix with minimization of voids between the matrix and the filler also enhanced hardness [11].

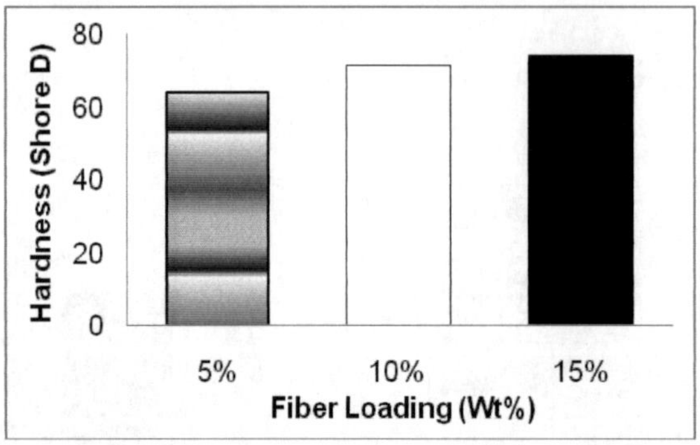


**Figure 3.** Variation of hardness against fiber loading.

*3.2 Morphological Properties*

In order to study surface morphology and interfacial adhesion between the hybrid fiber and PP of manufactured composites, SEM images were taken for different fiber loaded composites. SEM micrographs of composites with 5, 10 and 15 wt% fiber reinforced composites are shown in Figure 4. Fiber pull-out was observed in case of 5% and 15% fiber loaded composites (Figures 4 (a) and 4 (c) respectively). 15% fiber loaded composite also had fiber agglomeration. On the other hand, very good wetting between fiber and matrix is also observed in the 10% fiber loaded composite (Figure 4 (b)) [12]. Thus 10% fiber loaded composite showed more adhesion as compared to 5% and 15% fiber loaded composites.

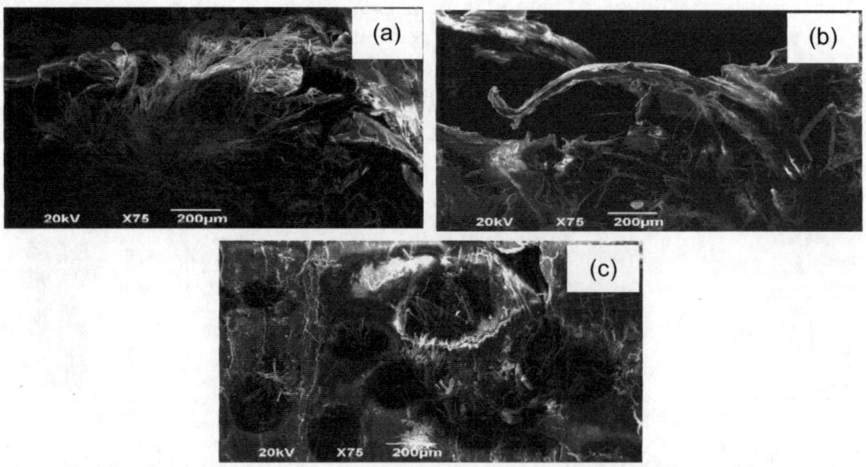


**Figure 4**. SEM micrographs of (a) 5, (b) 10 and (c) 15 wt% fiber loaded PP composites.

### *3.3 FTIR Analysis Results*

FTIR curves of 5, 10 and 15 wt% betel nut husk fiber and pineapple fiber reinforced composites are shown in Figure 5. According to the results there are no significant differences between the peaks in different loading fiber composites of FTIR spectra. The only noticeable thing is that the FTIR peaks of 15% loaded fibers are wider than others. It is observed that the absorption peak around 1740 cm-1, corresponding to the C=O stretching of hemi-cellulose in the raw fibers composites. Moreover, absorption peak of lignin is found corresponding to Syringyl (S) ring breathing with C-O stretching around 1372 cm-1, C=O stretching in 2838.7 cm-1. C-H stretching in aromatic methoxyl groups and aliphatic methyl and methylene groups of the side chains and the band at 2930 cm-1 is due to the C-H stretching of aromatic methoxyl groups and methyl and methylene groups of the side chains. The absorption peak at 3420 cm-1 is attributed to the hydroxyl stretching vibrations, aromatic C-H in-plane deformation (typical of S units) plus secondary alcohols plus C=O stretching around 1261 cm-1, C=C stretching of the aromatic ring of the guaiacyl unit at 1630 cm-1. The absorption peak at the range of 2900-2880 cm-1 is due to C-H vibration of $CH_2$ group of PP [13].

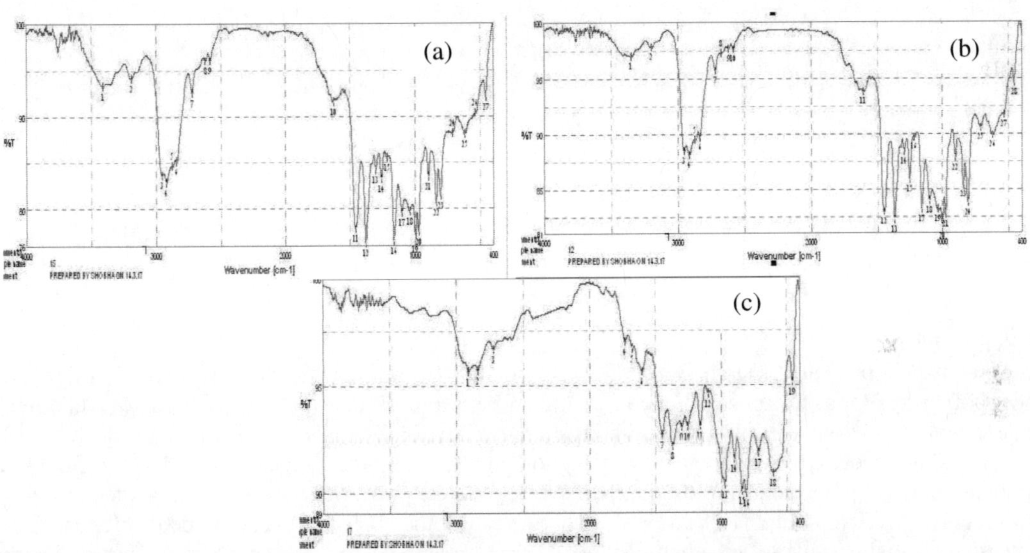


**Figure 5.** FTIR curves of (a) 5, (b) 10 and (c) 15 wt% fiber loaded PP composites.

### *3.4 TGA Results*

For the manufacturing of composites, the thermal stability of the fiber is very important. TGA curves of composites with 5, 10 and 15 wt% fiber loading are shown in Figure 6. In TGA curve of 5% fiber (PAL: BNH=1: 1) reinforced composite, moisture loss occurred at <100 °C. Thermal degradation occurred at around 225 to 358 °C. In 10% fiber (PAL: BNH=1: 1) reinforced composite, moisture loss occurred at <100 °C. Thermal degradation occurs at around 250 to 410 °C. In 15% fiber (PAL: BNH=1: 1) reinforced composite moisture loss occurred at <100 °C. Thermal degradation occurred at around 223 to 374 °C. Thus 10% fiber loaded composite had highest thermal stability among all prepared composites. This was due to better adhesion between fibers and matrix in case of 10% fiber loaded composite [14, 15].

IOP Conf. Series: Materials Science and Engineering **438** (2018) 012026 doi:10.1088/1757-899X/438/1/012026

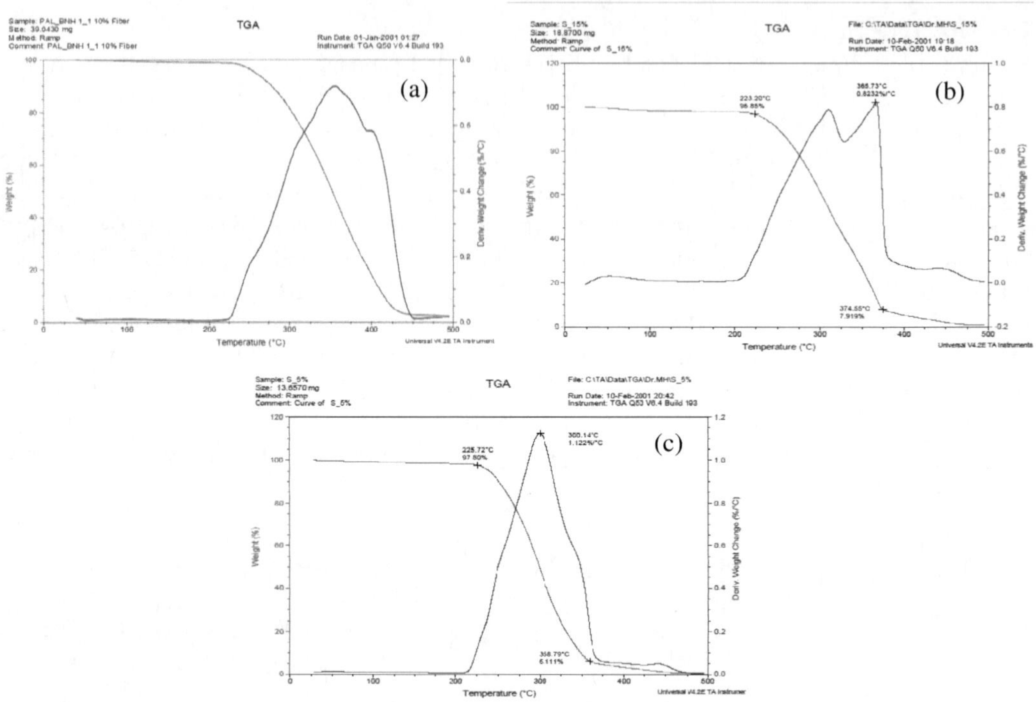


**Figure 6.** TGA curves of (a) 5, (b) 10 and (c) 15 wt% fiber loaded PP composites.

## 4. Conclusion

The objective of present research is to show improved physico-mechanical and thermal properties in hybrid polymer composite. Pineapple and betel nut fiber were used, which are cheap and available. The tensile strength of prepared composites decreased, while the Young's modulus, flexural strength and flexural modulus increased with increase in fiber loading. SEM images indicated that 10 wt% fiber loaded composite showed more adhesion as compared to 5 and 15 wt% fiber loaded composites. Very good wetting between fiber and matrix was also observed in the 10 wt% fiber loaded composite. FTIR spectroscopic analysis identified chemical bonds in a molecule by producing an infrared absorption spectrum for 5, 10 and 15 wt% fiber loaded composites. The spectra produce a profile of the sample, a distinctive molecular fingerprint that can be used to screen and scan samples for different components. According to TGA results, 10% fiber loaded composites had highest thermal stability among all manufactured composites.

## 5. References

[1] Jyoti P D and Mishra S C 2013, *Processing and properties of natural fiber-reinforced polymer composite*, Journal of Materials, Vol. 2013, pp. 1-6.

[2] Lai W L and Mariattim M 2008 *The Properties of woven betel palm (Areca catechu) reinforced polyester composites*, Journal of Reinforced Plastic and Composites, Vol. 27(9), pp. 925-935.

[3] Kuruvilla J 2007 *Natural fiber reinforced polymer composites: status and future*, Proc. *ICPP-*2007, pp. 316-327.

[4] Srinivasulu N V and Tejaswi V 2012 *Mechanical properties of polymer composite materials*, International Journal of Research in Engineering and Technology, Vol. 1, pp. 78-81.

[5] Asim M, Abdan K, Jawaid M, Nasir M, Dashtizadeh Z, Ishak M R and Hoque M E 2015 *A review on pineapple leaves fibre and its composites*, International Journal of Polymer Science, Vol. 2015, Article ID 950567.

[6] Yusriah L, Sapuan S M, Zainudin E S and Mariatti M 2012 *Exploring the potential of betel nut husk fiber as reinforcement in polymer composites: Effect of fiber maturity*, Procedia Chemistry, Vol. 4, pp. 87-94.

[7] Rahman M R, Haque M M, Islam M N and Hasan M 2009 *Mechanical properties of polypropylene composites reinforced with chemically treated abaca*, Composites A, Vol. 40, pp. 511-517.

[8] Prabhakaran R D, Andersen T L, Markussen C M, Madsen B and Lilholt H 2013 *Tensile and compression properties of hybrid composites–a comparative study,* Proc. of the International Conference on Composite Materials, Denmark, pp. 1029-1035.

[9] Haque M M, Islam M S, Islam M S, Islam M N, Huque M M and Hasan H 2010 *Physico-mechanical properties of chemically treated palm fiber reinforced polypropylene composites*, Journal of Reinforced Plastics and Composites Vol. 29, pp. 1734-1742.

[10] Haque M M, Rahman M R, Islam M N, Huque M M and Hasan H 2010 *Physico-mechanical properties of polypropylene composites reinforced with chemically treated coir and abaca fiber*, Journal of Reinforced Plastics and Composites Vol. 29, pp. 2253-2261.

[11] Ashik K P and Sharma R S 2015 *A review on mechanical properties of natural fiber reinforced hybrid polymer composites,* Journal of Minerals and Materials Characterization and Engineering, Vol. 3(5), pp. 420-426.

[12] Daud Z, Hatta M M Z, Kassi M A S and Aripi M A 2014. *Analysis of the chemical compositions and fiber morphology of pineapple (Ananascomosus) leaves in Malaysia*, Journal of Applied Sciences, Vol. 14, pp. 1355-1358.

[13] Siddika S, Monsura F, Hasan M and Hassan A *2014 Effect of reinforcement and chemical treatment of fiber on the properties of jute-coir fiber reinforced hybrid polypropylene composites,* Fibers and Polymers, Vol. 15(5), pp. 1023-1028.

[14] Bucci D Z, Alberti L D, Souza D F and Barcellos I O 2014 *Studies on mechanical and thermal properties of PHB biocomposites with rice hull ash*, Materials Science Forum, Vol. 775-776, pp. 557-561.

[15] Jamil Md S, Ahmed I and Abdullah I 2006 *Effect of rice husk filler on the mechanical and thermal properties of liquid natural rubber compatibilized high-density polyethylene natural rubber blends,* Journal of Polymer Research, Vol.13, pp.315-321.

**Acknowledgment**
The authors are thankful to the department of Materials and Metallurgical Engineering, Bangladesh University of Engineering and Technology for their technical and financial support during the work.

# Physico-Mechanical Properties of Pineapple Leaf and Banana Fiber Reinforced Hybrid Polypropylene Composites: Effect of Fiber Ratio and Sodium Hydroxide Treatment

**S. Das, M. Rahman and M. Hasan**

Department of Materials and Metallurgical Engineering, Bangladesh University of Engineering and Technology, Dhaka 1000, Bangladesh

**Abstract.** Natural fibers are becoming abundant and viable substitute for the synthetic fibers day by day, as synthetic fibers are expensive and nonrenewable. The main aim of present research is to evaluate the effect of sodium hydroxide treatment and banana and pineapple leaf fiber ratio on mechanical and physical properties of pineapple leaf-banana fiber reinforced hybrid polypropylene composites. Hybrid composites were prepared using a hot press machine. Raw pineapple leaf and banana fiber were chemically treated with 5% sodium hydroxide to increase adhesion between the fibers and polypropylene. Both raw and alkali treated fibers at 5 wt% were utilized during composite preparation. Pineapple leaf and banana fiber ratios were varied in 1:1, 3:1 and 1:3. Morphological analysis (scanning electron microscopy) and mechanical (tensile, flexural and hardness) tests of prepared composites were subsequently conducted. Better adhesion between the fibers and polypropylene was observed in case of treated fiber composites as compared to raw fiber composites. Scanning electron microscopic analysis also showed better bonding between the fibers and polypropylene when pineapple leaf and banana fiber ratio was 3:1. Subsequently thebest set of mechanical properties was attained for that particular fiber ratio.

**Keywords:** Banana and pineapple leaf fiber; Polypropylene; Hybrid composite; Sodium hydroxide treatment.

## 1. Introduction

Natural fibers are ecofriendly, lightweight, strong, renewable, cheap and biodegradable. Natural fiber polymer consists a polymer matrix reinforced with high strength natural fiber. Natural fiber composites are attractive to industry because of their low density, low cost and ecological advantages over synthetic composites.Natural fiber containing composites are using in transportation (automobiles, railway coaches, aerospace), military applications, building and construction industries (ceiling paneling, partition boards), packaging, consumer products, etc. Although synthetic fiber shows higher properties than natural fibers, all of these properties of natural fiber make natural fiber more attractive [1]. Hybrid composite can have more than one reinforcing phase and a single matrix phase or single reinforcing phase with multiple matrix phases or multiple reinforcing and multiple matrix phases. Hybrid composites have received considerableattention in the scientific community recently. By using hybrid composites better properties can be achieved than binary systems containing one type of fiber dispersed in a matrix. So, hybrid composites can give a balanced pursuit of stiffness, strength and ductility, as well as bending and membrane related mechanical properties with weight savings, reduced notch sensitivity, improved fracture toughness,longer fatigue life and excellent impact resistance.The physical properties of natural fiberreinforced composites largely depend on the type of matrix, fiber weight, fiber ratio, bonding between fiber and matrix [2]. However, the main disadvantages of natural fibers in composites are the poor compatibility between fiber and matrix and the relatively high moisture sorption. Therefore, chemical treatments are considered in modifying the fiber surface properties [3]. Banana and pineapple leaf fibers, both are available and low cost natural fiber. Both fibers are bio degradable and light weight. Pineapple fruits are very important, but leavewaste materials of fruit which is producing natural fibers. Banana and pineapple leaf fiber both are lignocellulosic material. Cellulose and lignin content of banana fiber are 44-54% and 11-22% respectively [4]. On the other hand, pineapple leaf fiber has a cellulose content of 67-85% and lignin content of 4.4-15% [5]. Thus the pineapple leaf fiber is more strong fiber than banana fiber [6]. The aim of this research is to develop a new hybrid using inexpensive and eco-friendly banana and pineapple leaf fiber. Effects of fiber ratio and chemical treatment on physico-mechanical properties of pineapple leaf and banana fiber reinforced hybrid polypropylene composite were observed.

## 2. Materials and Methods

### *2.1 Materials*

The thermoplastic polymer polypropylene (PP), supplied by the local marketin the form of homopolymer pellets, was used as matrix material. The PP was white in color.The melting point of the commercial PP lies in the range of 160-170°C, while the specific gravity is 0.90-0.91 [2]. Banana fiber was extracted from the pseudo stem sheath of the plant. Pineapple leaf fiber was collected from pineapple leaf, which is actually a left out portion of pineapple. When extracted, the fibers contained moisture. So, they were dried after extraction. The chemical used to both pineapple leaf and banana fiber is 5% NaOH.

### *2.2 Chemical treatment of natural fiber*

Firstly 5% solution was prepared into a beaker. Then the dried extracted fibers were drowned into the solution and stirred properly, so that all fibers were wet with the solution. Then the beaker containing the solution with fiber was taken to the oven for heating. It was heated at 70°C for about two and half hours. After two and half hours the beaker was taken out from the oven. Then the fibers were washed at first with tap water. Then they were washed with distilled water. The fibers contained water and evaporation was needed. That is why they were taken to the oven for the complete evaporation of moisture. They were heated for two hours in an oven at 180°C. After complete evaporation of moisture they were taken out from the oven.

### *2.3 Manufacturing of composites*

Hybrid composite of polypropylene and banana-pineapple leaf fiber was prepared using hot press machine. Hybrid composite was prepared for pineapple leaf and banana fiber in 3:1, 1:1 and 1:3 rations and 5wt% fibers. Composites were prepared by taking 5% NaOH treated pineapple and banana fiber at a ratio of 3:1 using the same hot press technique. Fibers were weighted according to the required volume fraction needed. Then both fibers were cut into 3 mm size in length. The required amount of polypropylene was weighted. To prevent voids, water bubbles and poor fiber matrix adhesion of the polypropylene was dried in an oven at about 80°C for 20 minutes. Mold surface was cleaned very carefully and the mold releasing agent was sprayed over the mold surface properly for the easy removal of the product. Then at first a layer of polypropylene was poured into the female mould. Chopped fiber was given to the layer. Then again a layer of polypropylene was given on the chopped fiber. The female mould with randomly oriented fiber and polypropylene were covered with the male mould according to the indication provided in both mould. Then the die was placed in a hot pressing machine. In the present study, a hydraulic type machine having a capacity of maximum load of 50 kN and maximum temperature of 300°C was used. The fiber matrix mixture was allowed to press at 30 kN pressures. The temperature was initially raised to 160°C and hold there for around 20-25 minutes. After that the temperature was raised to 190°C . The die was cooled to room temperature, pressure was released and the specimen was carefully withdrawn from the die. The same procedure was applied to other percentage of fiber. For making composite with treated fiber, the above procedure was followed.

*2.4 Mechanical testing*
Tensile, flexure and hardness tests of prepared composites were conducted. Tensile test was conducted according to ASTM D 638-01 using an Instron machine at a cross head speed of 5mm/min. Each test was continued until tensile failure. Static flexural tests were carried out according to ASTM D 790-00 using the same testing machine mentioned above at same crosshead speed. The hardness of composites was measured using a shore hardness testing machine on scale D.

*2.5 Scanning Electron Microscopy Analysis*
Surface morphology of the raw and chemically treated fiber composites were observed under a scanning electron microscope (Philips XL 30). The composites surface was initially made conductive by applying gold coating using a sputtering machine. The composites were then taken inside SEM, vacuum was created and micrographs were taken.

## 3. Results and Discussion

*3.1 Effect of fiber ratio*
*3.1.1 Tensile properties.* Tensile properties of composites were measured for pineapple leaf and banana fiber ratio of 3:1, 1:1 and 1:3 using stress strain curve. The variation of tensile strength, Young's modulus and % elongation at break for the different pineapple leaf and banana fiber ratio are shown in Figure 1. Tensile strength of composites containing pineapple leaf and banana at 3:1 was higher as compared to composites containing those two fibers at 1:1 and 1:3 ratios (Figure 1 (a)). Banana fiber has low tensile strength as compared pineapple leaf fiber due to its low cellulose content. The cellulose content of pineapple leaf fiber is around 70-82% and the cellulose content of banana fiber is 48-50%The tensile strength of pineapple leaf fiber is 413-1627 MPa, while that ofbanana fiber is 198-780 MPa [4, 7]. Again Young's modulus of composites containing pineapple leaf and banana at 3:1 was higher as compared to composites containing those two fibers at 1:1 and 1:3 ratio (Figure 1 (b)). Young's modulus of pineapple leaf fiber is 34.5-82.5 GPa and Young's modulus of banana fiber is 6.6-25.6 GPa. So, higher concentration of pineapple fiber incorporation demands higher stress for the same deformation and results in the enhancement of Young's modulus [1]. Finally composite containing 75% banana fiber had highest % elongation at break as comparedto other two composites pineapple fiber due to its lowest stiffness (Figure 1 (c)).

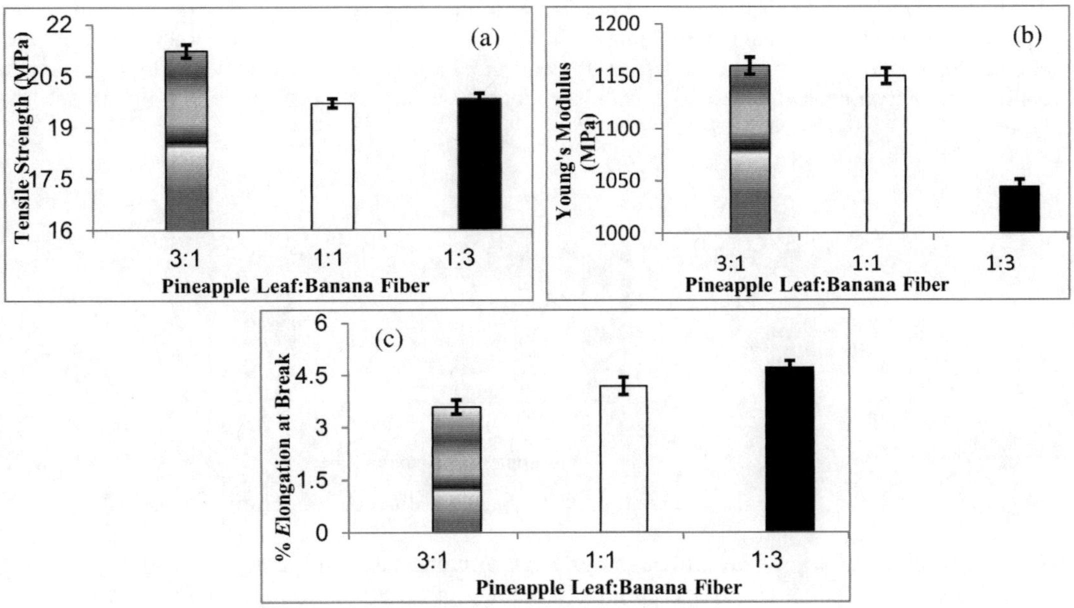


**Figure 1.** Variation of (a) tensile strength, (b) Young's modulus and (c) % elongation at break for different fiber ratio.

*3.1.2 Flexural properties.* Flexural properties of prepared composites were measured for pineapple leaf and banana fiber ratio of 3:1, 1:1, and 1:3 with the help of flexural stress/strain curves and respective equations. Flexural strength value of composites containing pineapple leaf and banana fiber at 3:1 ratio was higher as compared to compositecontaining pineapple leaf and banana fiber at 1:3 ratio (Figure 2 (a)). Thus with increase in pineapple fiber content,flexural strength increased. The same trend was also obtained for flexural modulus (Figure 2 (b)). Pineapple leaf fiber contains higher cellulose (70-82%) as compared to banana fiber (48-50%). It also contains lower amount of lignin (4.6-12%) as compared to banana fiber (19.17%). These may bethe reason of higher bending properties of composites containing higher pineapple leaf fiber [7, 8].

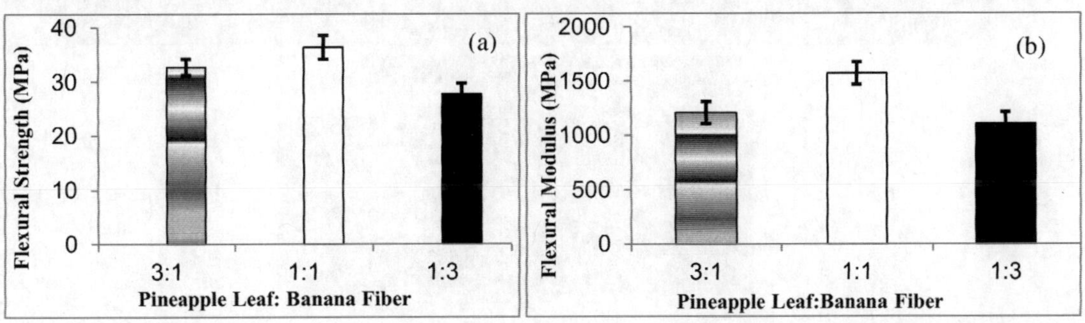


**Figure 2.** Variation of (a) flexural strength and (b) flexural modulus for different fiber ratio.

IOP Conf. Series: Materials Science and Engineering **438** (2018) 012027 doi:10.1088/1757-899X/438/1/012027

*3.1.3 Hardness results.* Variation of hardness of various prepared composites is shown in Figure 3. Hardness of composites containing pineapple leaf and banana at 3:1 was higher as compared to composites containing those two fibers at 1:1 and 1:3 ratios. The reason may be the higher cellulose content of pineapple leaf fiber. The cellulose content is mainly responsible for higher mechanical properties, such as hardness [7, 8].

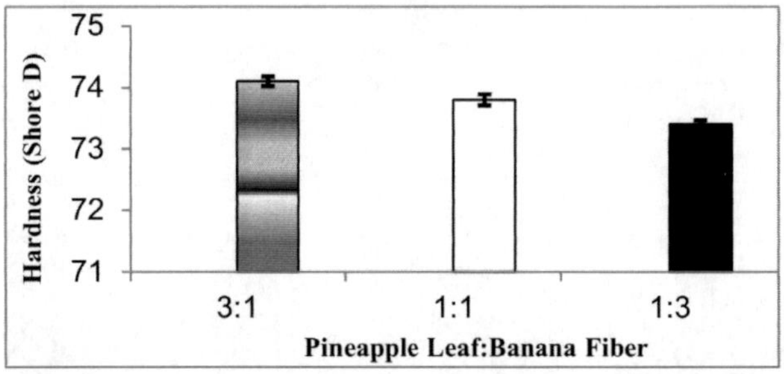


**Figure 3.** Variation of hardness for different fiber ratio.

*3.1.4 Surface morphology.* The surface morphology of composites with pineapple leaf and banana fiber at 3:1, 1:1 and 1:3 ratios was observed under scanning electron microscope and is shown in Figure 4. From the SEM images, better bonding is observed for composites containing pineapple leaf and banana fiber at 3:1 ratio as compared to other two composites. The reason is that the pineapple leaf fiber contains higher cellulose and lower lignin as compared to banana fiber. Thus pineapple leaf fiber has higher amount of cellulose exposed on the fiber surface, so it has higher number of possible reaction sites. Pineapple leaf fiber also has lower amount of lignin, wax and oil at the surface, which increases surface roughness. As a result, wetting of the fiber as well as bonding of fiber with matrix improves [9, 10].

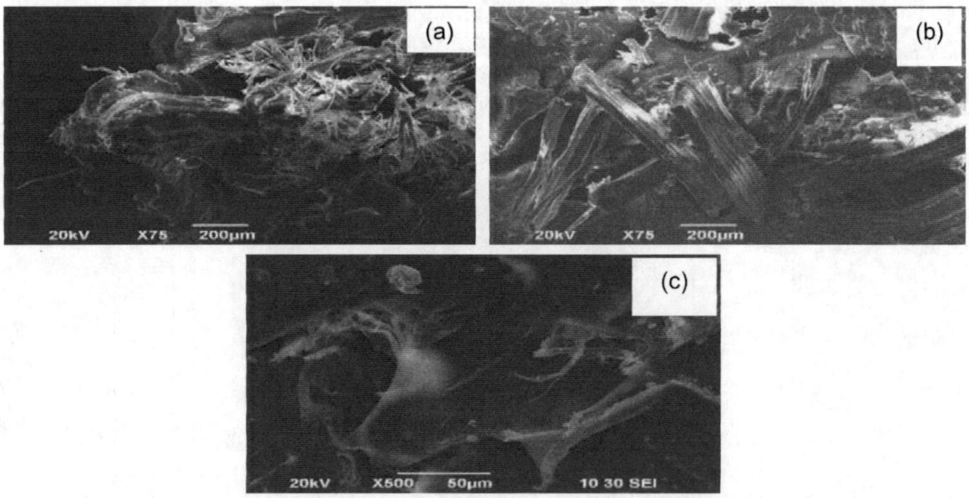


**Figure 4.** SEM micrographs of composites containing pineapple leaf and banana fiber at (a) 1:3, (b)1:1 and (c) 3:1 ratio.

### 3.2 Effect of chemical treatment

*3.2.1 Tensile properties.* Tensile properties of sodium hydroxide (alkali) treated composite was also measured using stress strain curve. Main purpose ofalkali treatment is to disrupt hydrogen bonding in the network structure and remove some hemicellulose, lignin, wax and oils, thereby increasing surface roughness and reducing its hydrophilic nature. During alkali treatment alkali reacts with a cementing material specially hemicellulose, which leads to the splitting of the fibers into finer filaments. This reaction with hemicellulose also increases the amount of cellulose. As a result, wetting of the fiber as well as bonding of fiber with matrix improves [3, 10]. Thus alkali treatment of fibers increased both strength and stiffness of the prepared composites (Figure 5). On the other hand, alkali treatment decreased % elongation at break due to increase in stiffness of respective composite.

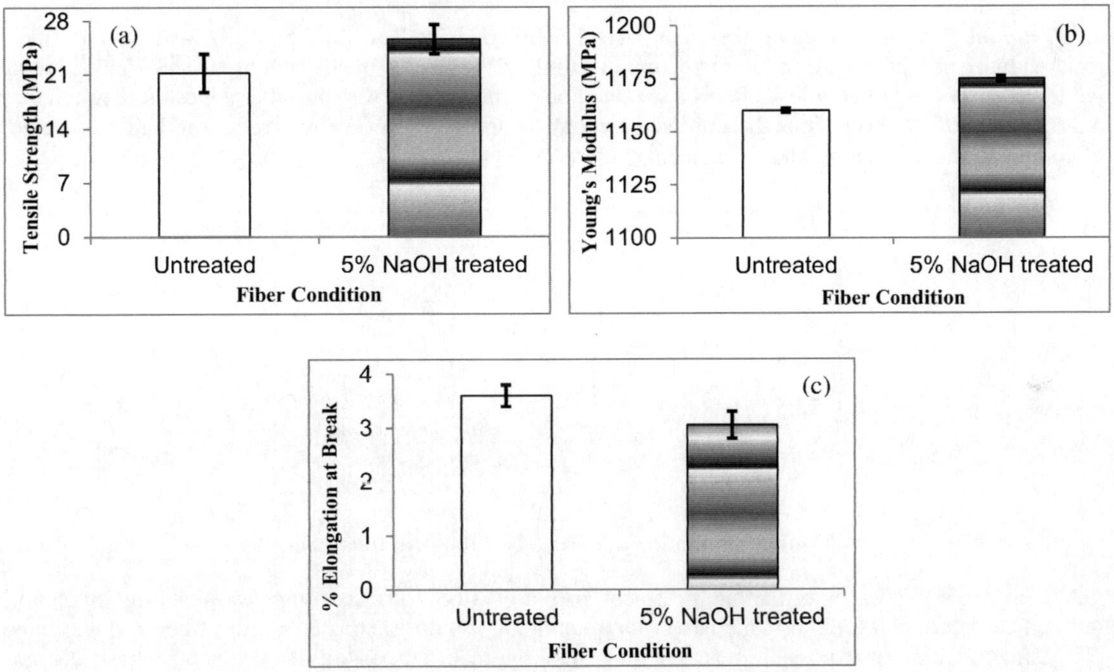


**Figure 5.** Variation of (a) tensile strength, (b) Young's modulus and (c) % elongation at break with alkali treatment.

*3.2.2 Flexural properties.* Alkali treatment increases the interfacial interaction between the fiber and matrix.This in turn increases the effective surface area available to contact with the matrix and possibility of load transfer between the matrix and the reinforcing fibers. As a result of flexural strength and modulus increase [10, 11]. Flexural strength and flexural modulus of treated pineapple leaf and banana fiber composites was higher as compared to those of untreated composites (Figure 6).

IOP Conf. Series: Materials Science and Engineering **438** (2018) 012027 doi:10.1088/1757-899X/438/1/012027

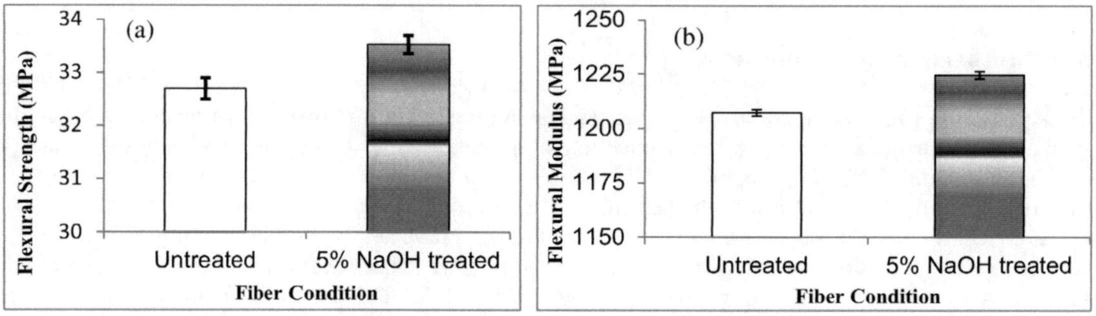


**Figure 6.** Variation of (a) flexural strength and (b) flexural modulus with alkali treatment.

*3.2.3 Hardness results.* Alkaline treatment, which was conducted on pineapple leaf and banana fiber resulted inthe collapse of the cellular structure due to the demotion of the cementing material.This in turn led to better packing of cellulose chains and reduction of the void, as well as better adhesion between the matrix and the filler [11]. Thus the interface can transfer pressure more effectively. The hardness value of composite increased with alkali treatment (Figure 7).

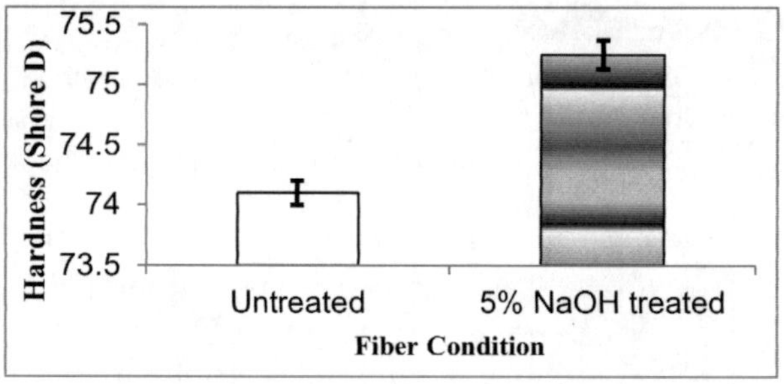


**Figure 7.** Variation of hardness with alkali treatment.

*3.2.4 Surface morphology.* Alkaline treatment increases fiber surface roughness resulting in better mechanical interlocking. It also increases the amount of cellulose exposed on the fiber surface, thus increasing the number of possible reaction sites. Also the main purpose of alkali treatment is to disrupt hydrogen bonding in the network structure and remove some hemicellulose, lignin, wax and oils, thereby increasing surface roughness and reducing its hydrophilic nature [12, 13]. SEM micrographs in Figure 8 show better bonding of fiber with matrix in case of treated fiber composite.

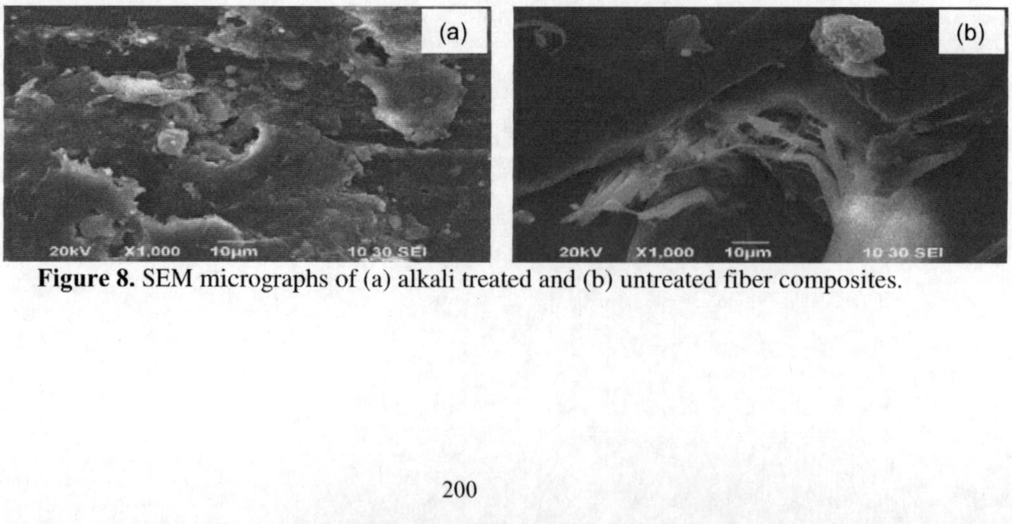


**Figure 8.** SEM micrographs of (a) alkali treated and (b) untreated fiber composites.

## 4. Conclusion

In present research, effect of pineapple leaf and the banana fiber ratio and alkali treatment of those fibers on physical and mechanical properties of pineapple leaf and banana fiber reinforced hybrid polypropylene composites are observed. Composites were prepared with pineapple leaf and banana fiber in 3:1, 1:1 and 1:3 ratio and treated pineapple leaf and banana fiber at 3:1 ratio for 5 fiber wt% using hot press machine. Mechanical properties (tensile, flexural, hardness) were higher for 3:1 fiber ratio compositeas compared to the other fiber ratio. SEM micrographs showed better bonding between fiber and matrix in case of 3:1 fiber ratio. Treated fiber composites had better bonding between fiber and matrix and better mechanical properties as compared to untreated fiber composites.

## 5. References

[1] Premalal H G B, Ismail H and Baharin A 2002 *Comparison of the mechanical properties of rice husk powder filled polypropylene composites with talc filled polypropylene composites*, Polymer Testing, Vol. 21(7), pp. 833-839.

[2] Netravali A N 2004 *Natural Fibers, Plastics and Composites*, Kluwer Academic Publisher, USA.

[3] Xue L, Lope G. Tabil, Panigrah S 2007 *Chemical treatments of natural fiber for use in natural fiber-reinforced composites: a Review*, Journal of Polymers and the Environment, Vol. 15(1), pp. 25-33.

[4] Manilal V B and Sony J 2011 *Banana pseudostem characterization and its fiber property evaluation on physical and bioextraction*, Journal of Natural Fibers, Vol. 8, pp. 1-12.

[5] Asim M, Adnan K, Jawaid M, Nasir M, Deshtizadeh Z, Iskak M R and Hoque M E 2015 A review on pineapple leaves fibre and its composites, International Journal of Polymer Science, Vol. 2015, Article ID 950567

[6] Haque M M, Islam M S, Islam M S, Islam M N, Huque M M and Hasan H 2010 *Physico-mechanical properties of chemically treated palm fiber reinforced polypropylene composites*, Journal of Reinforced Plastics and Composites Vol. 29, pp. 1734-1742.

[7] Haque M M, Rahman M R, Islam M N, Huque M M and Hasan H 2010 *Physico-mechanical properties of polypropylene composites reinforced with chemically treated coir and abaca fiber*, Journal of Reinforced Plastics and Composites Vol. 29, pp. 2253-2261.

[9] Jamil Md S, Ahmed I and Abdullah I 2006 *Effect of rice husk filler on the mechanical and thermal properties of liquid natural rubber compatibilized high-density polyethylene natural rubber blends,* Journal of Polymer Research, Vol.13, pp.315-321.

[10] Yang H S, Kim H J, Son J, Park H J, Lee B J and Hwang T S 2006 *Water absorption behavior and mechanical properties of lignocellulosic filler polyolefin biocomposites,* Composite Structures, Vol. 72, pp. 429-437.

[11] Rahman M R, Haque M M, Islam M N and Hasan M 2009 *Mechanical properties of polypropylene composites reinforced with chemically treated abaca*, Composites A, Vol. 40, pp. 511-517.

[12] Khan J A, Khan M A and Islam R 2012 *Effect of mercerization on mechanical, thermal and degradation characteristics of jute fabric-reinforced polypropylene composites*, Fibers and Polymers, Vol. 13, pp. 1300-1309.

[13] Yang H S, Kim H J, Son J, Park H J, Lee B J and Hwang T S 2006 *Water absorption behavior and mechanical properties of lignocellulosic filler–polyolefin bio-composites*, Composite Structures, Vol. 72, pp. 429-437.

**Acknowledgment**

The authors would like to thank department of Materials and Metallurgical Engineering, Bangladesh University of Engineering and Technology for their support during the work.

IOP Conf. Series: Materials Science and Engineering **438** (2018) 012028 doi:10.1088/1757-899X/438/1/012028

# Mechanical, Morphological and Thermal Properties of Alkali Treated Ladies Finger Fiber Reinforced Polypropylene Composites

**M.I. Hossain and M. Hasan**

Department of Materials and Metallurgical, Bangladesh University of Engineering and Technology, Dhaka 1000, Bangladesh

**Abstract:** Fiber reinforced polymer composites have played a dominant role for a long time in a variety of applications for their high specific strength and modulus. The fiber which serves as a reinforcement in reinforced plastics may be synthetic or natural. Although glass and other synthetic fiber-reinforced plastics possess high specific strength, their fields of application are very limited because of their inherent higher cost of production. Present research investigated the effect of alkali treatment and loading of ladies finger fiber on mechanical and thermal properties of ladies finger fiber reinforced polypropylene composites. Composites were prepared using compression molding technique. Ladies finger fiber loading were varied at 5, 10 and 15 wt%. Both raw and alkali treated ladies finger fiber were utilized to reinforce polypropylene. Mechanical (tensile, flexural and hardness) and water absorption tests and thermogrvimetric and scanning electron microscopic analysis of the prepared composites were subsequently carried out. Tensile and flexural strength decreased with fiber loading, while opposite trend was observed in case of Young's modulus, flexural modulus, hardness and water absorption. Alkali treatment of ladies finger enhanced mechanical properties of prepared composites due to better adhesion between the fiber and matrix in treated fiber composites. According to thermogrvimetric analysis, thermal stability of composites increased with fiber loading.

**Keywords:** Alkali Treatment; Ladies Finger Fiber-Polypropylene Composite; Mechanical Properties; TGA; SEM

## 1. *Introduction*

Replacement of the conventional synthetic fibers with natural fibers as reinforcement in polymer matrix composites could lead to a green, renewable path of applications. Natural fibers have received great attention from researchers and industrialists due to their biodegradability, better mechanical properties, easy manufacturing, and overall cost effective quality [1–3]. While most of the synthetic fibers are costly and might cause skin irritation, natural fibers are completely free from these problems. Natural fiber reinforced polymer composites has increased noticeably since last few decades. Fiber reinforced composites are presently used in automotive application, aerospace parts, naval architecture and membrane as they are less corrosive, less abrasive, and have good mechanical properties. Hot pressing, fiber extrusion and hand lay-up methods are the most common composite manufacturing techniques used in preparing natural fiber reinforced polymer composites [4–6]. However, with all these advantages and promise ahead, the biggest obstacle that is yet to overcome is assuring uniformity. As nearly all natural fibers are collected as husk or skin, assuring uniformity in surface condition, strength and other properties are in question. To ensure homogeneity in properties, natural fibers are required to undergo some treatments varying from chemical to physical. In many cases, natural fibers are chemically treated before composites manufacturing in order to increase the interfacial adhesion between the fiber and polymer, which in turn increases the mechanical properties [7-9]. Current day's polymers of two kinds are widely used: thermoplastic and thermosetting. Polypropylene (PP), polyethylene (PE) and polyvinyl chloride (PVC) are three of the most popular types of thermoplastic materials, while epoxy, polyester, phenolic dominates the thermosetting domain. To be used as matrices for bio-fibers, the thermoplastic materials are commonly used. In present research raw and alkali treated ladies finger fiber reinforced polypropylene composites were prepared by varying fiber loading. Prepared composites were subsequently characterized using mechanical tests and thermegravimetric analysis.

## 2. Materials and Methods

### 2.1 Materials

The matrix materials PP were collected in pellet form from Polyolefin Company Private Limited, Singapore. The reinforcing agent (ladies finger fiber) was collected from a village of Bangladesh. Sodium hydroxide (NaOH) was used during chemical treatment ladies finger of fiber. It was purchased from Merck, Germany.

### 2.2 *Alkali Treatment of Ladies Finger Fiber*

Ladies finger fiber solution treated in 5% NaOH solution. 5% NaOH solution was prepared by adding 5 gm NaOH to 100 ml water in a beaker. Fiber was then treated in 5% NaOH solution at 70°C for two and half hours. The fiber was washed and dried in an oven at 80°C for 20 minutes.

### 2.3 Composite Preparation

Firstly reinforced composites of polypropylene matrix and varying amount of (5%, 10% and 15%) were manufactured using hot press technique. Initially fibers were weighted according to the required weight fraction. Then both fibers were cut into 3-5 mm size in length. In order to allow the removal of moisture, fibers were dried in an oven at 100°C for 20 minutes before preparing each composite. The required amount of polypropylene was weighted. Mold surface was cleaned very carefully and mold releasing agent (silicon spray) was sprayed over the mould surface properly for the easy removal of the product. Then the fibers and polypropylene were mixed thoroughly on to the lower die. The lower die was covered by the upper die according to the indication provided in the both mold. The die was placed in a hot press machine. In present research a hydraulic type machine having capacity of maximum load of 50 kN and maximum temperature of 300°C was used. The fiber-matrix mixture was allowed to press at 30 kN pressure. The temperature was initially raised to 160°C and hold there for around 20-25 minutes, after that the temperature was raised to (190-195)°C depending on the thickness required. The die was cooled to room temperature, pressure was released and the specimen was carefully withdrawn from the die.

### 2.4 Characterization Techniques

To characterize the properties of ladies finger fiber reinforced PP composites mechanical, water absorption and thermal analysis were performed. The mechanical analysis involved tensile, flexural and hardness tests. During tensile test, a universal testing machine (Model MSC-5/500, Agawn Seiki Co., Tokyo, Japan) was used along with ASTM D 638-10 method maintaining 10 mm/min crosshead speed. The same universal testing machine was used for the flexural test with same crosshead speed where the method used was ASTM D 790-10. The experimental work involved a Shore hardness tester for measurement of hardness of various composites. Following ASTM D 570- 98, water absorption tests was performed. Rectangular specimens were immersed in water for 24 hours. The weight of the specimen was measured before and after the test. Finally the difference between final and initial weight of the specimen was calculated to observe the water uptake capacity of composites. Surface morphology of prepared composites was evaluated using scanning electron microscopic analysis. Surface of the composites were made conductive by giving platinum coating using a sputtering machine. Composite surfaces were then observed in vacuum condition into the SEM machine. TGA method used was based on continuous measurement of weight during increase of sample temperature in an inert atmosphere. Data were recorded as a thermogram of weight versus temperature.

## *3. Results and Discussion*

### *3.1 Tensile Properties*

Figure 1 (a) shows the variation of tensile strength with fiber loading. The figure shows that tensile strength gradually decreased with increase in fiber loading. As fiber content increases, the weak interfacial area between the fiber and matrix increases, which decreases tensile strength [10]. It can be noticed that the tensile strength of chemically treated fiber reinforced composite was higher than those of raw fiber reinforced composite. The increased tensile strength of chemically treated composite was due to stronger adhesion between treated fiber and matrix [10]. The range of tensile strength of prepared composites was 22-35 MPa, while the same of ladies finger fiber reinforced polymer composite found elsewhere was 23-69 MPa [11].

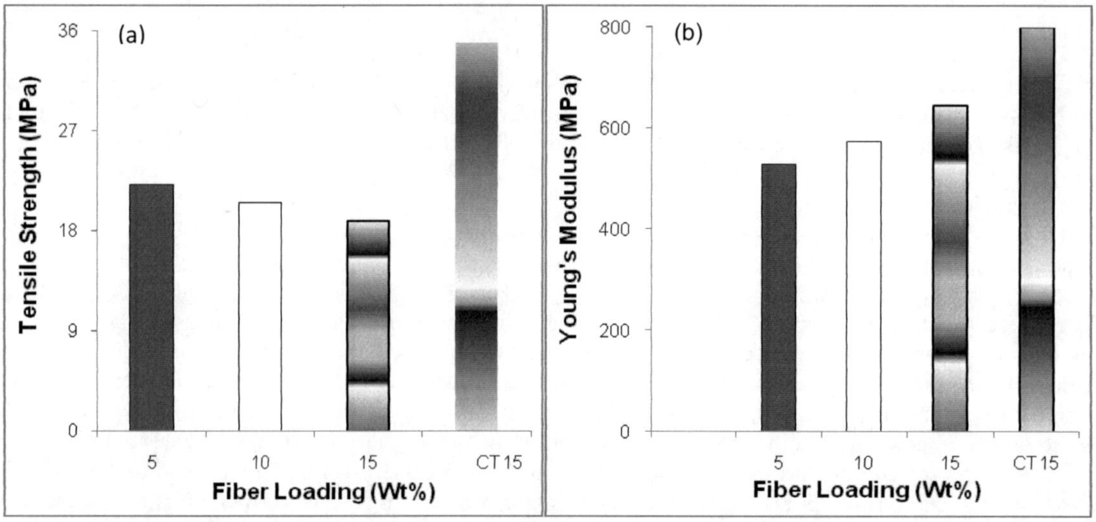


**Figure 1.** Variation of (a) tensile strength and (b) Young's modulus against fiber loading.

Figure 1 (b) shows the variation of Young's modulus against fiber loading. Young's modulus increased with increase fiber loading. During tensile loading, partially separated microspaces are created, which obstructs stress propagation between the fiber and the matrix. As the fiber loading increases, the degree of obstruction increases, which in turn increases the stiffness [12]. The Young's modulus of composites made of treated ladies finger fiber reinforced composites was higher than those of untreated ladies finger composite due to the change in the structure of the cellulose unit of ladies finger. The chemical treatment of ladies finger fiber reduced the hydroxyl group of the cellulose unit by coupling with basic sodium sulfate salt [10]. The range of Young's modulus of prepared composites was 520-780 MPa, while the same of ladies finger fiber reinforced polymer composite found elsewhere was 507-1025 MPa [11].

*3.2 Flexural Properties*

Variation of flexural strength and flexural modulus of ladies finger fiber reinforced polypropylene composite against fiber loading are shown in Figure 2. The flexural strength decreased with increase fiber loading (Figure 2 (a)). As fiber loading increased, the interfacial area between the fiber and matrix increased, which was weak because of worsening interfacial bonding between cellulose based hydrophilic filler (ladies finger) and hydrophobic PP matrix. This consequently decreased the flexural strength [12]. The range of flexural strength of prepared composites was 20-26 MPa, while the same of ladies finger fiber reinforced polymer composite found elsewhere was 39-118 MPa [11]. According to Figure 2 (b), flexural modulus increased with fiber loading. Since ladies finger is high modulus material, higher fiber concentration demands higher stress for the same deformation. So the incorporation of the filler (ladies finger) into the soft polypropylene matrix results into the increase in modulus [10]. The range of flexural modulus of prepared composites was 1200-1700 MPa, while the same of ladies finger fiber reinforced polymer composite found elsewhere was 3370-10135 MPa [11].

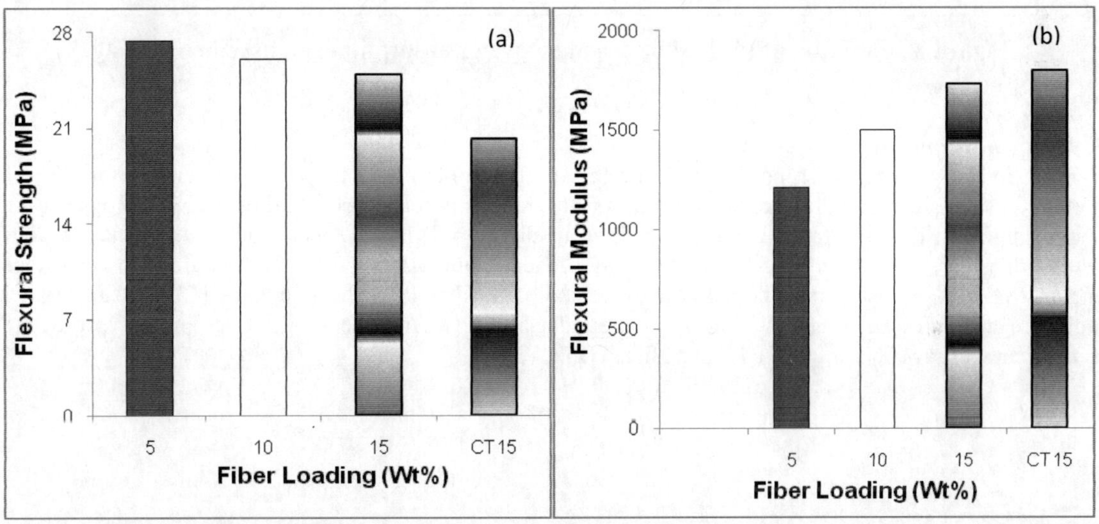


**Figure 2.** Variation of (a) flexural strength and (b) flexural modulus against fiber loading.

*3.3 Hardness Properties*

Variation of hardness of prepared composites is shown in Figure 3 (a). Hardness of ladies finger fiber reinforced PP composites increased with increase in fiber loading. Due to the increase of stiffness of respective composite, hardness of PP composites showed a slight increasing trend with an increase in the fiber content [10].

*3.4 Water Absorption Characteristics*

Natural fiber has a tendency to absorb water due to the presence of hydrophilic hydroxyl groups of cellulose, hemicelluloses and lignin. The experimented data shown in Figure 3 (b). It can be seen that with increase in fiber loading, the water absorption increased. It is also observed that the water absorption tendency decreased after chemical treatment due decrease of hydroxyl groups in ladies finger fiber with alkali treatment [10].

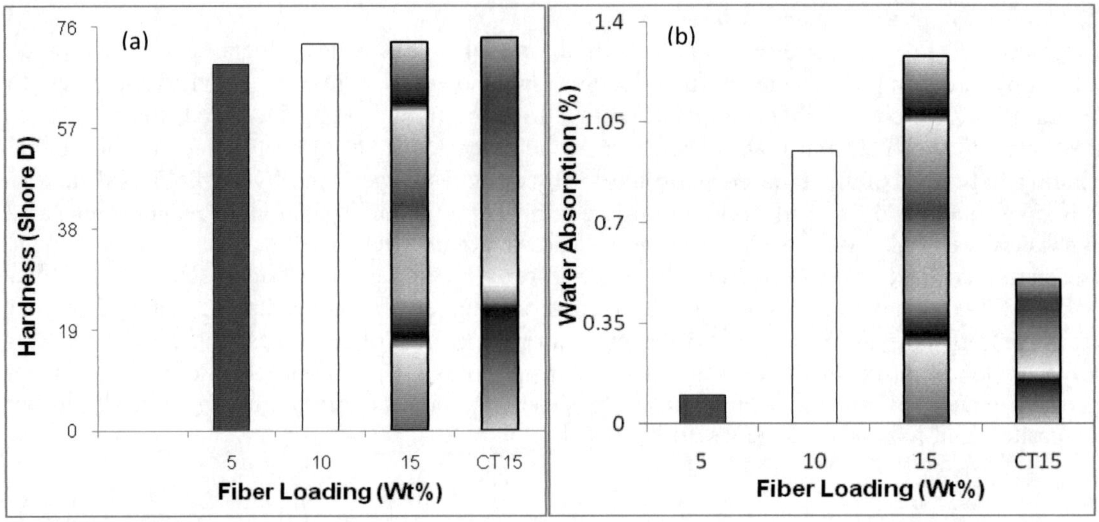


**Figure 3.** Variation of (a) hardness and (b) water absorption against fiber loading.

*3.5 Surface Morphology*

In order to study surface morphology and interfacial adhesion between the ladies finger fiber and PP of prepared composites, SEM images were taken for different fiber loaded composites. SEM micrographs of composites with raw and treated fiber are shown in Figure 4. 15 wt% fiber loaded composite showed better adhesion as compared to 5% and 10% fiber loaded composites. Very good wetting between fiber and matrix is also observed in both raw and treated 15% fiber loaded composites [13]. Alkali treated ladies finger reinforced composites showed better adhesion between ladies finger fiber and polypropylene as compared to raw fiber reinforced composites [12].

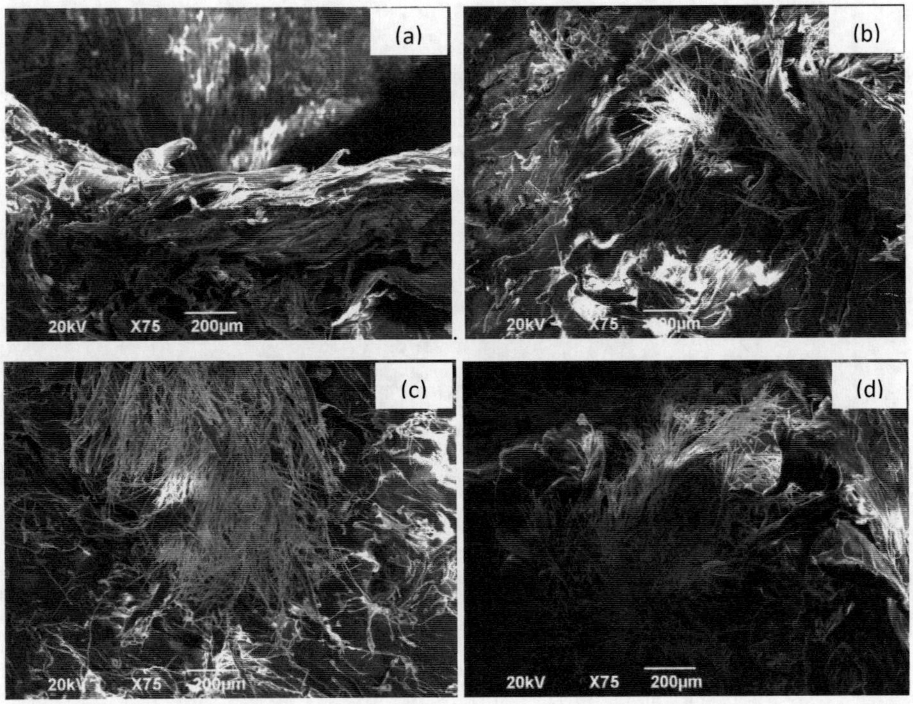


**Figure 4.** SEM micrographs of (a) 5% raw, (b) 10% raw, (c) 15% raw and (d) 15% treated ladies finger fiber reinforced PP composites.

### *3.6 TGA Analysis Results*

TGA analysis was performed for both raw and treated ladies finger fiber PP composites to observe the thermal stability of those composites (Figure 5). Composite specimens showed thermal stability at the range of 250–375°C. Stability of composites increased with increase in fiber content. No significant difference was observed between the raw and treated ladies finger fiber composites. from the DTG curves of 5% fiber composite, the sharp peak is found at about 335$^0$C. For 10% fiber composite, two sharp peaks are found at about 325$^0$C and 350$^0$C. For 15% fiber composite, two sharp peaks are found at about 350$^0$C and 375$^0$C. On the other hand, for 15% treated fiber composite a sharp peak is found at about 350$^0$C.

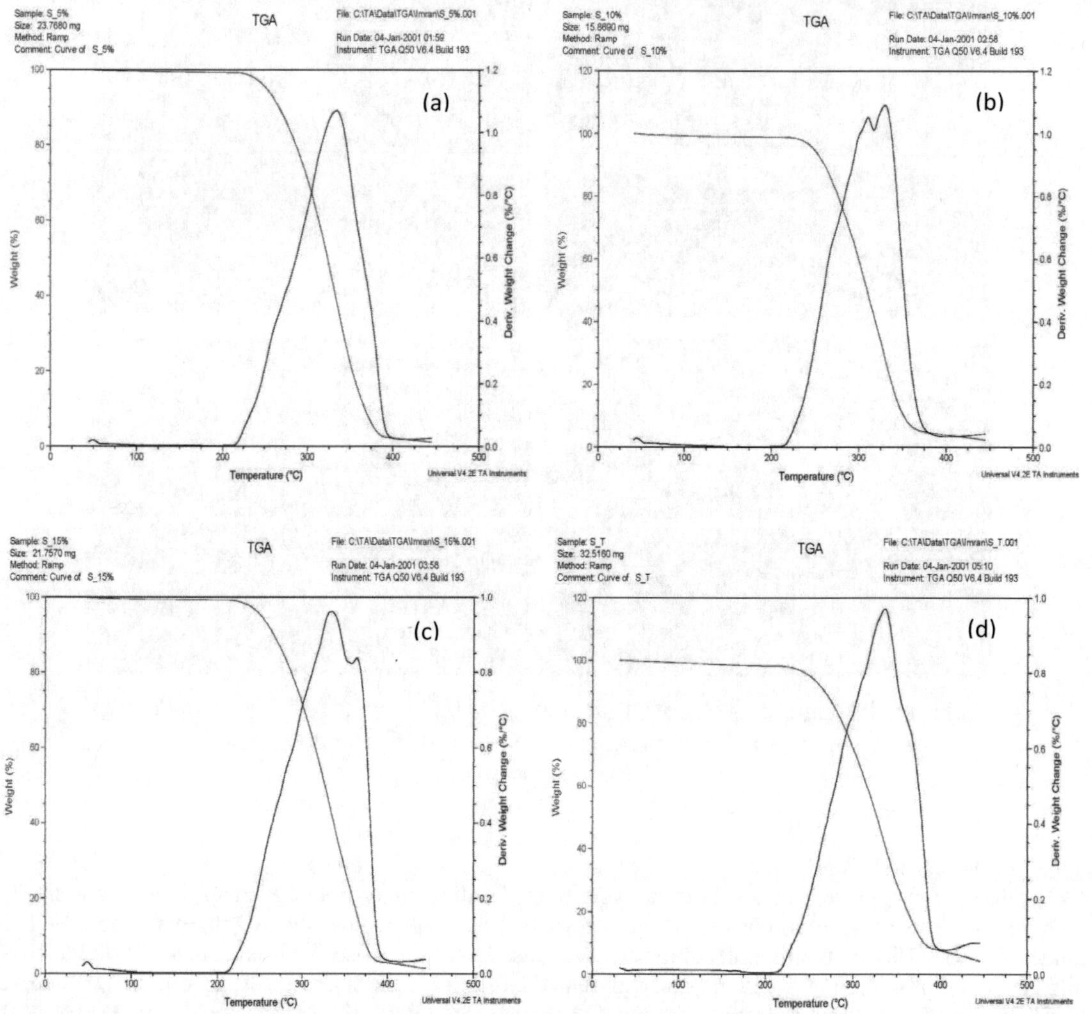


**Figure 5.** TGA curves of (a) 5% raw, (b) 10% raw, (c) 15% raw and (d) 15% treated ladies finger fiber reinforced PP composites.

## *4. Conclusion*

Present research investigated mechanical and thermal properties of ladies finger reinforced polypropylene composite. The fiber loading was varied at 5, 10 and 15 wt% during composite preparation. Tensile and flexural strength decreased with fiber loading, while opposite trend was observed in case of Young's modulus, flexural modulus, hardness and water absorption. Chemical treatment increased the adhesion between ladies finger fiber and polypropylene matrix. Thus, chemically treated ladies fiber reinforced polymer composites had better properties as compared to raw ladies finger fiber reinforced polymer composites. Thermal stability of composites increased with fiber loading, while no significant difference was observed between raw and treated fiber reinforced composites.

***References***

1) Chou T W 1992 Micro Structural Design of Fiber Composite (USA: University of Delaware) pp. 231-284.
2) Premalal H G B, Ismail H and Baharin A 2002 *Comparison of the mechanical properties of rice husk powder filled polypropylene composites with talc filled polypropylene composites*, Polymer Testing, Vol. 21(7), pp. 833-839.
3) Wolf C J 1998 Encyclopedia of Chemical Technology (UK: John Wiley and Sons).
4) Gururaja M N and Rao A H N 2012 *A review on recent applications and future prospectus of hybrid composites*, International Journal of Soft Computing and Engineering, Vol. 1(6), pp. 2231-2307.
5) Ajay K, Chauhan S S, Modak J M and Chanda M 2007 *Mechanical properties of wood-fiber reinforced polyethylene composites: effect of a novel compatibilizer with isocyanate functional group,* Composite A Vol. 38, pp. 227-233.
6) Malkapuram R, Kumar V and Negi Y S 2009 *Recent development in natural fiber reinforced polyethylene composites*, Journal of Reinforced Plastics and Composites, Vol. 28(10), pp. 1169-1189.
7) Shahinur S, Hasan M and Ahsan Q 2011 *Proc. Int. Symp. on Green Manufacturing and Applications (South Korea).*
8) De Velde K V and Kiekens P 2002 *Development of a flax/polyethylene composite with optimal mechanical characteristics by fiber and matrix modification*, Journal of Thermoplastic Composite Materials, Vol. 15, pp. 281.
9) Enriuqez J K E D V, Santiago P J M, Ong T F and Chakraborty S 2009 *Fabrication and characterization of high-density polyethylene*, Journal of Thermoplastic Composite Materials, Vol. 23(3), pp. 361-373.
10) Haque M M, Islam M N, Hasan Islam M S, Islam M S and Huque M M 2010 *Coir fiber reinforced polypropylene composites: physical and mechanical properties*, Advanced Composite Materials, Vol. 19, pp. 91-106.
11) Srinivasababu N 2015 *An Overview of okra fibre reinforced polymer composites,* IOP Conference Series: Materials Science and Engineering, Vol. 83, DOI 10.1088/1757-899X/83/1/012003.
12) Haque M M, Islam M S, Islam M S, Islam M N, Huque M M and Hasan M 20100 *Physico-mechanical properties of chemically treated palm fiber reinforced polypropylene composites*, Journal of Reinforced Plastics and Composites, Vol. 29(11), pp. 1734-1742.
13) Daud Z, Hatta M M Z, Kassi M A.S and Aripi M A 2014. *Analysis of the chemical compositions and fiber morphology of pineapple (Ananascomosus) leaves in Malaysia*, Journal of Applied Sciences, Vol. 14, pp. 1355-1358.

***Acknowledgment***

The authors are thankful to the department of Materials and Metallurgical Engineering, Bangladesh University of Engineering and Technology for their support during the work.

# Properties of Jute and Sheep Wool Fiber Reinforced Hybrid Polypropylene Composites

**J. Tusnim, N.S. Jenifar and M. Hasan**

Department of Materials and Metallurgical Engineering, Bangladesh University of Engineering and Technology, Dhaka 1000, Bangladesh

**Abstract.** Use of plant and animal-based biodegradable resources as reinforcement in polymeric composites has incontestable advantages compared to synthetic fiber composites- including low density, low cost, continuous supply, easy and safe handling and supports their potential across a wide range of applications. Hybrid fiber reinforced composites are generally prepared to enhance different properties as compared to single-fiber reinforced composites. In present research, sheep wool fiber and jute fiber reinforced hybrid polypropylene composites were prepared. Their mechanical properties were examined first by varying the fiber and polypropylene ratio and then by varying wool and jute fiber ratio keeping the polypropylene quantity unchanged. Here, 5%, 10% and 15% of the fibers in a ratio of 1:1 were copulated with polypropylene in a hot press machine. For mechanical characterization, tensile, flexural, impact and hardness tests were conducted. From these tests, it was observed that the composite with 15% fiber content had the best properties. For further enhancement, two separate samples with 1:3 and 3:1 fiber ratios and 85% polypropylene were prepared. Running the aforementioned tests, it was revealed that the sample with wool and jute fiber at 1:3 ratio had the best set of properties.

**Keywords.** Hybrid composite; Sheep wool fiber; Jute fiber; Polypropylene; Mechanical properties

## 1. Introduction

Recently, people have realized that the whole world will be threatened by the over consumption of natural resources if the environment is not conserved. For that reason there has been a rapid growth in research and innovation in the natural fiber composite (NFC) area. It has also become attractive as an alternative reinforcement for fiber reinforced polymer (FRP) composites due to their low cost, fairly good mechanical properties, high specific strength, non-abrasive, eco-friendly and bio-degradability characteristics. They are exploited as a replacement for the conventional fiber, such as glass, aramid and carbon [1]. Among all the thermoplastics, polypropylene (PP) has great properties like good fatigue resistance, very good abrasion resistance, good surface hardness, lower density, higher softening point and higher rigidity. PP does not create stress-cracking problems and it has also excellent electrical and chemical resistance at higher temperatures [2]. As natural fiber, jute and sheep wool fiber both are easily available and also cheap in Bangladesh. Wool is the most commonly used animal fiber which is obtained from the soft, hairy covering of sheep. They have good elasticity and resilience. Jute is annually regenerative, ligno-cellulosic self-composite biopolymer bast fiber. It is nonabrasive, has low density and high strength. Because of these reasons, it was decided to use jute fiber as a reinforcing fiber in the thermoplastic composite [2, 3].

Polymer matrix composites (PMCs) are used widely because they are strong and lightweight materials. The fibers provide strength and stiffness to the material, while the matrix holds the fibers in place and transfers internal load between them. Only simple equipments are required to produce PMCs. PMCs do not require high pressures or high temperatures. Therefore the possibilities of fiber damage and degradation are minimum [4]. The behaviour of hybrid composites is a weighed sum of the individual components in which there is a more favourable balance between the inherent advantages and disadvantages. Also, using a hybrid composite that contains two or more types of fiber, the advantages of one type of fiber could compliment with what are lacking in the other. As a consequence, a balance in cost and performance can be achieved through proper material design [5]. Thus, combination of these two fibers in making composite is a new work. The main goal of present research is to prepare hybrid composites from jute and sheep wool fiber and determine their mechanical properties. The effects of fiber loading and fiber ratio on mechanical properties of prepared composites are also evaluated.

## 2. Methodology

### *2.1 Collection of Fibers and Polypropylene Resin*

At first, jute and sheep wool both were collected from the local market. Jute single fibers were separated from the bundle. Then the fibers were dried in normal atmosphere to remove moisture. Commercial grade polypropylene (PP) was collected from the local market. It was white in color and granular in form, having a melting point of 160$^0$C.

### *2.2 Composite Preparation*

Hybrid composite was prepared in five different steps. At first fibers were weighted according to the required weight fraction needed. Then both fibers were cut into 3mm size in length. The required amount of polypropylene was weighted. To prevent voids, water bubbles, poor fiber matrix adhesion of the polypropylene, the fibers were dried in an oven at about 80-110$^0$C for around 40 minutes. Then the mould surface was cleaned very carefully and a mould releasing agent (silicon spray) was sprayed all over the mould surface properly for the easy removal and a good surface finishing of the product. After preparing the mould, a layer of polypropylene was poured onto one part of it. Chopped fibers were evenly spread out on this layer. Then finally a second layer of polypropylene was given on the chopped fiber. Care was taken so as to ensure uniform distribution of the fibers and the matrix throughout the entire sample. Then the mould with randomly oriented fiber and polypropylene were covered with another part of the mould according to the indication provided in both the mould sections. Finally the die was placed in a hot press machine. A hydraulic type machine having capacity of maximum load of 50KN and maximum temperature of 300°C was used. The fiber matrix mixture was allowed to press at 30KN pressure. The temperature was initially raised to 160°C and held for 20 minutes. After that the temperature was raised to 195°C. As soon as this temperature was reached, the dial was brought down to 0$^0$C and water cooling system was turned on. The die was then cooled to room temperature, pressure was released and the specimen was carefully withdrawn from the die. This same procedure was applied for other percentage and ratios of fibers.

### *2.3 Mechanical Tests*

*2.3.1 Tensile Test.*Tensile tests were carried out according to ASTM D 638-01 using an Instron machine, having a maximum capacity 50 kN. Each test was continued until tensile failure. The tensile properties as obtained from this test method indicates the stress of the composite specimen, their fracture characteristics under tensile load, load strain behaviour bearing properties of fiber and the matrix materials. To perform the tensile test, at first the dimensions of the specimen were taken. Then the test machine was calibrated following the standard chart. After that the ultimate tensile strength (UTS) was calculated from the following formula:

$$\text{UTS} = \frac{maximum\ load}{initial\ area}(\text{MPa}) \quad \text{or} \qquad \sigma = {Pmax}/{A}\ (\text{MPa}) \tag{1}$$

The modulus was calculated by taking the slope of the stress-strain curve.

*2.3.2 Flexural Test.* Flexural test of the composite specimens was conducted according to ASTM D 790-00 using the same Instron machine. The loading nose and supports were aligned in such a way that the axis of the cylindrical surfaces was parallel and the loading nose was midway between the supports. The load was applied to the specimen at the specified cross head motion and simultaneously load deflection data was taken. The test was terminated when rupture occurred at the outer surface of the test specimen. The flexural stress was calculated by means of the following equation:

$$\sigma_f = (3PL)/(2bd^2) \tag{2}$$

Where, $\sigma_f$ is the stress in the outer fibers at midpoint, P is the load, L is the length of support span, b is the width of beam, and d is the depth of beam. Again, flexural modulus was calculated by means of the following equation:

$$E_B = (L^3m)/(4bd^3) \tag{3}$$

Where, m is the gradient (i.e. slope) of the initial straight-line portion of the load deflection curve.

*2.3.3 Impact Test.* Impact strength is generally defined as the resistance of a material to fracture by a blow or impact loading, expressed in terms of the amount of energy absorbed before fracture. It is a standardized high strain-rate test which determines the amount of energy absorbed by a material during fracture which is a measure of a given material's toughness. The dynamic Charpy impact test of the composite was conducted using an impact tester MT 3016 and specimen were prepared according to ASTM D 6110-97.

*2.3.4 Hardness Test.* Hardness is defined as the ability of a material to resist plastic deformation. The hardness of the composite was measured using a shore hardness tester. Shore Durometer is typically used as a measure of hardness in polymer, elastomers and rubbers. There are several scales of durometer, used for materials with different properties. The A scale is for softer plastics, while the D scale is for relatively harder ones. In present research D scale was used.

## 3. Results and discussion

### *3.1 Effect of Fibre Loading*

*3.1.1 Tensile Properties.* Tensile properties of the composite samples were measured for each fiber content (5, 10 and 15 wt%) with the help of stress/strain curves. The variation of these properties for different percentage weights of raw sheep wool and jute fiber (wool: jute=1:1) reinforced hybrid polypropylene composites are shown in figure 1. Tensile strength of prepared composites increased with fiber loading, as shown in figure 1 (a). Young's modulus values of jute and wool fiber reinforced polypropylene composites for different fiber loadings are shown in figure 1 (b). It is observed that Young's modulus increased with an increase in fiber loading [6, 7]. This is because with an increase in fiber content, the brittleness of the composite increased and stress/strain curves became steeper. Poor interfacial bonding creates partially separated micro spaces which obstruct stress propagation between the fiber and the matrix. As the fiber loading increases, the degree of obstruction increases, which in turn increased the stiffness [8].

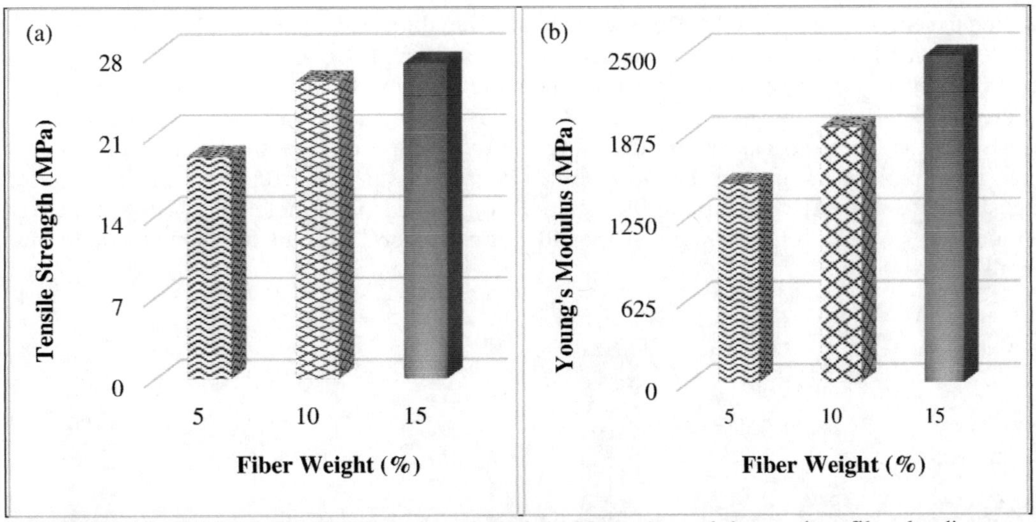


**Figure 1.**Variation of (a) tensile strength and (b) Young's modulus against fiber loading.

*3.1.2 Flexural Properties.* Flexural properties were measured for samples of each fiber content with the help of flexural stress/strain curves and respective equations. Flexural strength and flexural modulus of raw sheep wool and jute fiber reinforced hybrid polypropylene composites at different fiber loadings (wool: jute=1:1) are shown in figure 2. Both of these properties increased with an increase in fiber loading [6, 7]. The reason behind the increase in flexural strength may be due to the favourable entanglement of the polymer chain with the filler which has overcome the weak filler matrix adhesion with increasing fiber content [8]. The flexural modulus values increased with an increase in fiber loading. Since both jute and wool are high modulus material, higher fiber concentration demands higher stress for the same deformation [9]. So the coalition of the filler (rigid jute and wool fiber) with the soft polypropylene matrix results into the rise in the modulus.

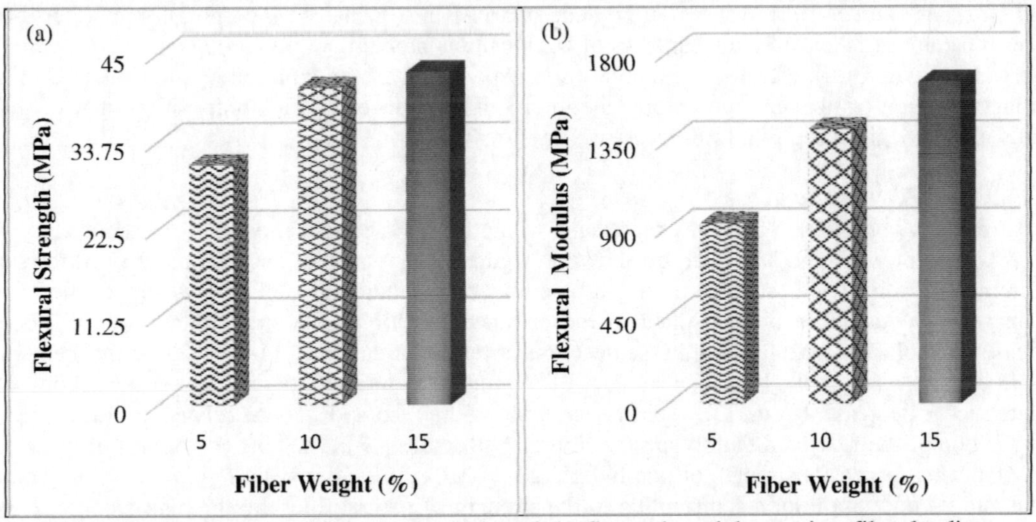


**Figure 2.**Variation of (a) flexural strength and (b) flexural modulus against fiber loading.

*3.1.3 Impact Properties.*Variation of Charpy impact strength with different fiber loading for raw jute and sheep wool fiber (jute: wool= 1:1) reinforced hybrid composite is shown in figure 3 (a). Impact strength increased with fiber loading [8, 10]. Impact strength of a material provides information regarding the energy required to break a specimen of given dimension, the magnitude of which reflects the material's ability to resist sudden impact. The impact strength of the fiber reinforced polymeric composites depends on the nature of the fiber, polymer and fiber-matrix interfacial bonding [11]. As presented in the figure, impact strength of all composites increased with fiber loading. This result suggests that the fiber was capable of absorbing energy because of favourable entanglement of fiber matrix. Fiber pull out is found to be an important energy dissipation mechanism in fiber reinforced composites. One of the factors of impact failure of a composite is fiber pulled out. With the increase in fiber loading, stronger force is required to pull out the fibers. This in turn increased the impact strength.

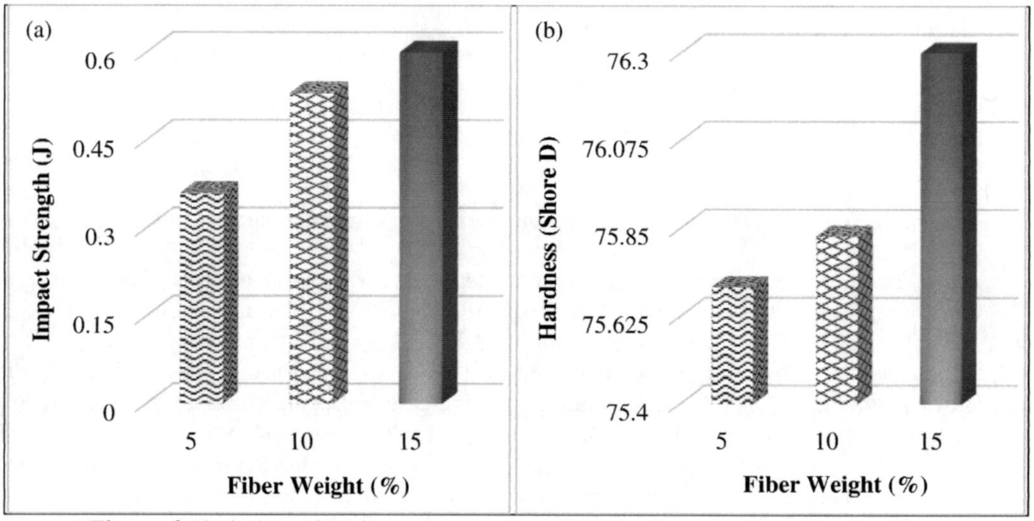


**Figure 3.**Variation of (a) impact strength and (b) hardness against fiber loading.

*3.1.4 Hardness Test Results.* The polypropylene matrix is a soft matrix. Incorporation of fiber into it makes the composite hard. So the hardness of composite is higher than polypropylene matrix which is observed in figure 3 (b). Hardness of a composite depends on the distribution of filler into the matrix and incorporation of fiber into polypropylene matrix. This reduces the flexibility of the matrix which results in more rigid composite [10].

*3.2 Effects of Fibre Ratio*

*3.2.1 Tensile Properties.* Variations of tensile strength and Young's modulus of composites with different ratio of wool and jute fiber are shown in figure 4. The value of tensile strength is different at different fiber ratio. The best value is shown by composite containing wool and jute fiber ratio of 1:3. An increase in jute fiber ratio resulted in the enhancement of tensile strength by 3.32%, whereas increase in wool fiber ratio resulted in the decrease of tensile strength by 14.44%. The enhancement of jute fiber ratio also resulted in enhancement of Young's modulus by 5.16% and that of wool fiber resulted in a decrease by 6.31%. Tensile strength of any composite are related to the chemical composition of the fiber and their respective internal structure [12]. The tensile strength of wool is 1-1.7 MPa, whereas tensile strength of jute fiber is 20-25 MPa. Jute fiber has also high cellulose content which plays important role in contributing to the strength of natural fiber .So the high value of tensile strength of composites containing more jute fiber is related to its single fiber properties according to which the tensile strength of single jute fiber is higher compared to sheep wool fiber. As jute fiber has higher modulus (13-26 GPa) compared to wool fiber (1.4 GPa) so higher concentration of jute fiber incorporation demands higher stress for the same deformation and results into the enhancement of Young's modulus [13].

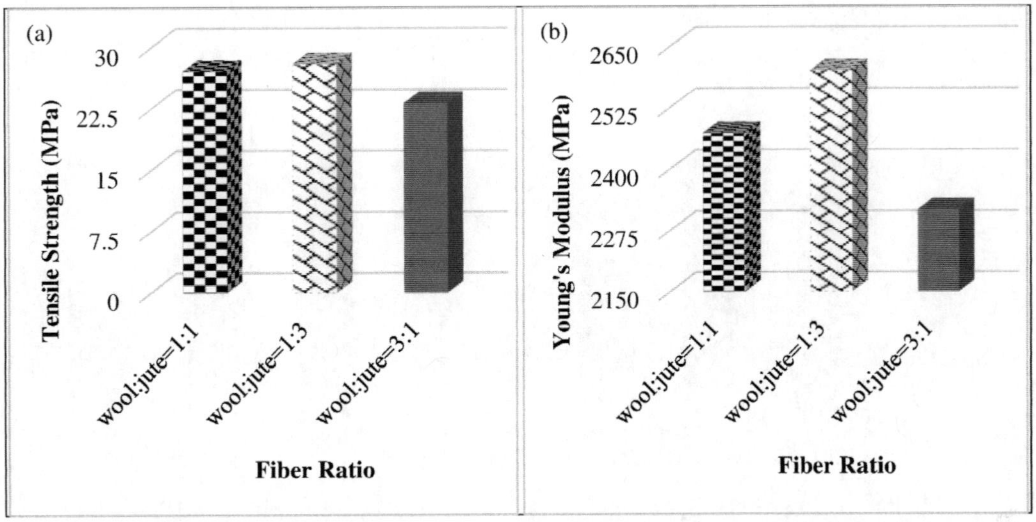


**Figure 4.**Variation of (a) tensile strength and (b) Young's modulus against fiber ratio.

*3.2.2 Flexural Properties.* The flexural strength was changed with different fiber ratio. Enhancement of wool and jute ratio from 1:1 to 1:3 resulted in enhancement of flexural strength by 7.17%, whereas enhancement of wool content resulted in a decrease in flexural strength by 11.66%. The higher cellulose content and the smaller percentage of hemicellulose and lignin may be the reason behind the higher bending properties of high jute containing composite in comparison with higher wool carrying one. The incorporation of 75% higher modulus fiber jute resulted in an enhancement of flexural modulus of 9.29% as compared to 50% jute incorporated composite. On the other hand, 75% wool fiber in the composite resulted in a decrease in flexural modulus by 11.33% as compared to 50% jute incorporated composite. Jute is a plant fiber. So high cellulose content may be the reason of the higher bending properties of high jute containing composite in comparison with higher wool carrying one [2].

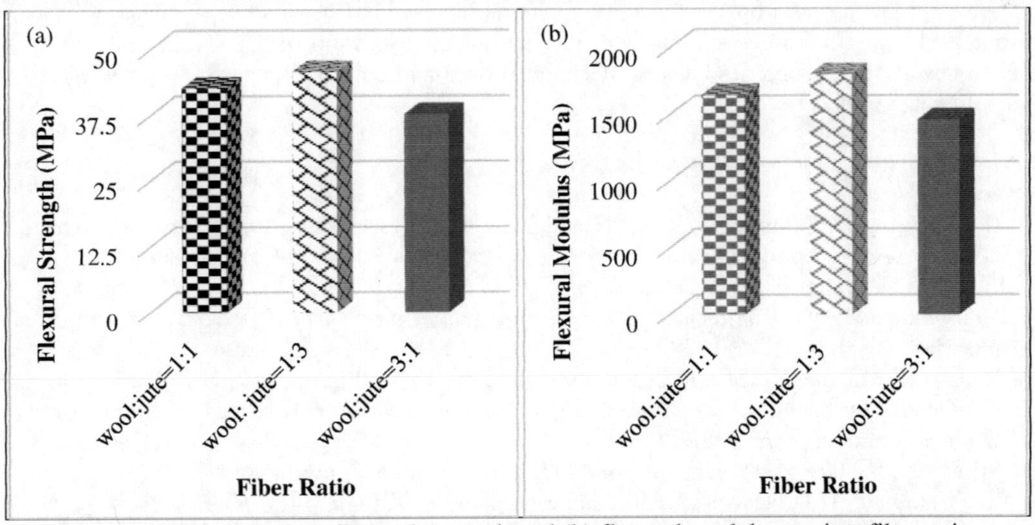


**Figure 5.**Variation of (a) flexural strength and (b) flexural modulus against fiber ratio.

*3.2.3 Impact and Hardness Properties.* The composites with high volume fraction of jute fiber showed higher impact strength as compared to the composite with high volume fraction of wool fiber. Jute increases the bonding capability and the area under stress-strain curve and so produces greater impact strength [14]. The change in fiber ratio did not affect the hardness data much as expected. Jute percentage enhancement to 50% and 75% from 25% resulted in enhancement of hardness value by 7.16% and 5.48% respectively.

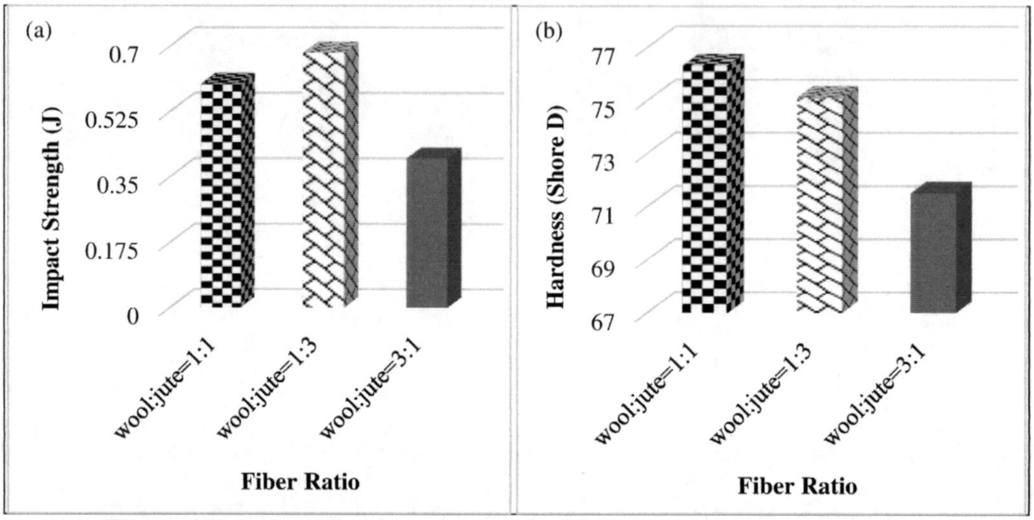


**Figure 6.**Variation of (a) impact strength and (b) hardness against fiber ratio.

## 4. Conclusion

Development of new composite products from the easily renewable natural materials has a strong potential to deliver novel biodegradable and recyclable materials suitable for the automotive, packaging industry and all other applications. Polypropylene is soft as compared to jute and wool fiber. In present research, jute fiber and sheep wool fiber reinforced hybrid polypropylene composites were prepared by varying fiber loading and fiber ratio. Prepared composites were subsequently characterized. Mechanical properties increased with an increase in fiber loading. Best set of properties were obtained from 15% fiber loading with jute and wool fiber ratio of 3:1. These tests could help further to unveil other properties besides mechanical ones and help in comparison of raw fibers with chemically treated fibers in future.

## 5. References

[1] Ku H, Wang H, Pattarachaiyakoop N and Trada M 2011*A review on the tensile properties of natural fiber reinforced polymer composites,* Composites B, Vol. 42(2), pp. 856-873.

[2] Siddika S, Mansura F and Hasan M 2014*Effect of reinforcement and chemical treatment of fiber on the properties of jute-coir fiber reinforced hybrid polypropylene composites*,Fibers and Polymer, 15(5), pp. 1023-1028.

[3] Rana A K, Mandal A, Mitra B C, Jocobson R, Rowell R and Banerjee A N 1998 *Short jute fiber reinforced polypropylene composite: Effect of compatibilizer,* Journal of Applied Polymer Science, Vol.69, pp.329-338.

[4] Srinivasan K 2009 *Composite Materials* (UK: Alpha Science International Ltd).

[5] Sathishkumar T P, Naveen J and Satheeshkumar S 2014 *Hybrid fiber reinforced polymer composites – a review,* Journal of Reinforced Plastics and Composites, Vol. 33(5), pp. 454-471.

[6] Rana A K, Mandal A and Bandyopadhyay S 2003 *Short jute fiber reinforced polypropylene composite: Effect of compatibilizer, impact modifier and fiber loading,* Composites Science and Technology, Vol. 63, pp.801-806.

[7] Joseph S, Sreekala M, Oommen Z, Koshy P and Thomas S 2006 *A comparison of mechanical properties of phenol formaldehyde composites reinforced with banana fibers and glass fibers,* Composites Science and Technology, Vol. 62, pp.1857-1868.

[8] Yang H S, Kim H J, Son J, Park H J, Lee B J and Hwang T S 2006 *Water absorption behavior and mechanical properties of lignocellulosic filler polyolefin biocomposites,* Composite Structure, Vol. 72, pp. 429-437.

[9] Mansour R, Osmani H, Abdellatif I and Noureddine B 2011 *Effect of chemical treatment on flexure properties of Natural Fiber-Reinforced Polyester Composite,* Procedia Engineering, Vol. 10, pp. 2092-2097.

[10] Jamil Md S, Ahmed I and Abdullah I 2006 *Effect of rice husk filler on the mechanical and thermal properties of liquid natural rubber compatibilized high-density polyethylene natural rubber blends,* Journal of Polymer Research, Vol.13, pp.315-321.

[11] Jayaraman K 2003 *Manufacturing sisal-polypropylene composites with minimum fiber degradation,* Composites Science and Technology, Vol.63, pp. 367-374.

[12] Maschinenwesen D F 2006 *Investigation on Jute Fibers and their Composites based on Polypropylene and Epoxy Matrices* (Vietnam: Technischen Universitat Dresden)

[13] Feughelman M and Robinson M S 1971 *Some mechanical properties of wool fibers in the 'Hookean' region from zero to 100% relative humidity,* Textile Research Journal, Vol. 41(6), pp. 469-474.

[14] Kumar M and Singh S 2013 *Analysis of mechanical properties and cost of glass/jute fiber-reinforced hybrid polyester composites*, Journal of Materials: Design and Applications, Vol. 229(3), pp. 202-208.

**Acknowledgment**
The authors are thankful to the department of Materials and Metallurgical Engineering, Bangladesh University of Engineering and Technology. I t gives authors great pleasure in acknowledging the support and technical assistance of all laboratory in-charges and other staffs of the department.

# Size Dependent Magnetic and Optical Properties of Mn Doped $Bi_{0.9}Ho_{0.1}FeO_3$ Nanoparticles

**Fabia Farlin Athena[1], Tilottoma Saha[1], Md Rafiqul Islam[1], M. A. Matin[2] and Ahmed Sharif[1]**

[1]Department of Materials and Metallurgical Engineering, Bangladesh University of Engineering and Technology, Dhaka 1000, Bangladesh
[2]Department of Glass and Ceramic Engineering, Bangladesh University of Engineering and Technology, Dhaka 1000, Bangladesh

**Abstract.** In this work, $Bi_{0.9}Ho_{0.1}Fe_{1-x}Mn_xO_3$ (x=0.0,0.05,0.10,0.15) nanoparticles have been successfully synthesized by sol–gel technique and subsequent annealing process. The variation of Mn doping in $Bi_{0.9}Ho_{0.1}FeO_3$(BHFO)on the properties such as morphological, structural, magnetic, and optical have been investigated in this study. X-ray diffraction (XRD) patterns demonstrated that Mn substitution in BHFO led to a structural transition from rhombohedral (R3c) to orthorhombic (Pn21a) phase. The field emission scanning electron microscope (FESEM) showed that Mn doping decreased particle size and consequently higher percentages of Mn resulted agglomeration of particles severely. In this case particle sizes reduced from 60 nm to 30 nm. For Mn doping up to 5% in BHFO nanoparticles, the saturation magnetization (Ms) was enhanced, implying suppression of space modulated spin structure by structural transition. However, for a further increase of doping up to 15%, the Ms was started to decrease again. UV–Visible absorption spectra revealed that Mn substitution in BHFO led to strong reduction of band gap energy significantly. Band gap decreased from 2.9 to 2.38 eV on Mn doping, which allows the absorption of light extended to visible range.

**Keywords:** Nano-particles, X-ray diffraction, Orthorhombic, Magnetization.

## 1. Introduction

Perovskite based bismuth ferrite $BiFeO_3$(BFO), is one of the most promising multiferroic materials and has received special attention in recent years because of its simultaneous ferro-electricity and antiferromagnetism above room temperature ($T_c$=1102K and $T_n$=643K) [1-3]. The limitation of miniaturization and speed which is being faced in conventional devices can be solved by its benchmark properties. These materials with their multi properties can successfully terminate the "Si Age" with next generation miniature high speed devices. For next generation devices including nonvolatile memories, spintronics, high density microacutators and piezoelectric sensors etc. these outstanding properties have aroused superior candidate [4,5]. However, it is very difficult to observe the ferromagnetic loop of a bulk $BiFeO_3$ sample at room temperature due to its low spiral spin modulation which is caused by defects and nonstochiometry in the bulk $BiFeO_3$ sample. Hence, it is essential to improve the ferromagnetic properties of $BiFeO_3$ without much disturbing its ferroelectric properties before its actual use in device application. The weak magnetization is attributed to the superimposition of a spiral spin structure on BFO antiferromagnetic order. In this spiral spin structure, the antiferromagnetic axis rotates through the crystal with an incommensurate long wavelength period of 62nm [6], which cancel the macroscopic magnetization and also inhibit the observation of the linear magnetic effect in bulk $BiFeO_3$.

To overcome the problems of BFO many techniques have been followed (eg. doping). A-site substitution by lanthanide ions (Gd, Sm, La, Dy, Ho, Sm) is known to be an effect way to overcome these obstacles [7]. These lanthanide ion substitutions would effectively suppress the formation of impure phases and oxygen vacancies. Suppression of spiral modulated spin structure occurred by substitution due to different ion radii between lanthanide and $Bi^{3+}$ ion. Kang et. al. reported $Ho^{3+}$ substitution for Bi site is an effective way to increase the magnetization. Transitional metal doped at Fe site are thought to be an effective way to improve the ferromagnetic properties at room temperature. BFO may find application as photo catalytic material due to its small bandgap. It allows carrier excitation by laser pulse, hence enables the researcher to develop optoelectronic devices. Moreover, bandgap value of pure BFO can be tuned by substitution of heterovalent ion [8]. Therefore, it is widely expected that the strategies based on nanotechnology and aliovalent ion doping will have a great impact on their magnetic, optical properties as a result of the suppression of the helical order and quantum confinement effect [8].

## 2. Methodology

Modified pechini sol-gel method was used to synthesize 0, 5, 10, 15% Mn doped BHFO to make the co-doped bismuth ferrite nanoparticles using citric acid as the chelating agent. The reagents used in the present work were analytical grade bismuth nitrate pentahydrate ($Bi(NO_3)_3 \cdot 5H_2O$), iron nitrate nonahydrate ($Fe(NO_3)_3 \cdot 9H_2O$), citric acid ($C_6H_8O_7$), ethylene glycol ($C_2H_6O_2$) and glacial acetic acid ($CH_3COOH$) (purity>99.98%). The final product was obtained by auto-combustion. Further the gel was dried at 120°C for 24 hrs. in a drier to obtain fluffy green xerogel. The xerogel samples were then ground into powders and annealed with 3°C/min at 500°C for 2 hrs. to get $BiFeO_3$ nanoparticles. Subsequently these samples were structurally characterized by XRD. Field emission scanning electron microscopy (FESEM) was used to determine surface texture and nanoparticles size. Magnetic property was measured using a vibrating sample magnetometer (VSM). Finally, the optical property was measured using a UV/Vis spectroscopy.

## 3. Results and Discussion

### *3.1 Structural and micro structural characteristics*

Powder diffraction (PXRD) patterns of the samples BHFO, $Bi_{0.9}Ho_{0.1}Fe_{0.95}Mn_{0.05}O_3$(BHFMO-5), $Bi_{0.9}Ho_{0.1}Fe_{0.90}Mn_{0.1}O_3$(BHFMO-10), and $Bi_{0.9}Ho_{0.1}Fe_{0.85}Mn_{0.15}O_3$(BHFMO-15) were recorded at room temperature and are shown in Figure 1(a) According to literature pure BFO structure possess rhombohedral (R3c) symmetry. Although observed patterns are slightly different but in agreement with the pattern of literature suggested BFO, so observed pattern confirms the peaks of doped BFO nanoparticle and shows polycrystalline behavior with good crystallinity.

According to the literature some impurity peaks are observed in pure BFO but in this XRD pattern impurity peaks are scarcely observed due to some reactions. In pure BFO:

$$2Bi^{3+} + 3O^{2-} \rightarrow \mathbf{Bi_2O_3\ (evp)} + 2\,V_{Bi} + 3\,V_o \quad (1)$$

$$2Fe^{3+} + O^{2-} \rightarrow 2Fe^{2+} + \mathbf{0.5O_2\ (gas)} + 2V_o^{\cdot\cdot} \quad (2)$$

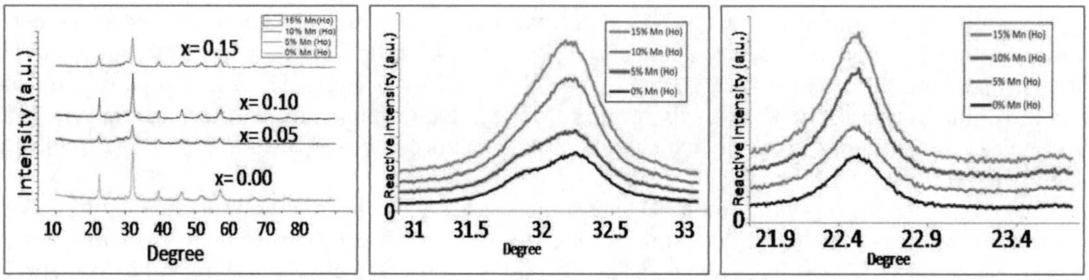


(a) (b) (c)

**Figure 1: (a)** XRD patterns of 0,5,10, 15 mol% Mn doped BHFO nanoparticles (from bottom to top). Enlarged version of XRD in the range of 2θ from **(b)** 31-33 and **(c)** 21.9-23.4 for 0, 5, 10, 15 mol% Mn doped BHFO sample sintered at 500$^0$C (from bottom to top).

When $Bi^{3+}$ is partially replaced by $Ho^{3+}$ which will suppress the $Bi^{3+}$ volatile because melting point of $Ho_2O_3$ (2367$^0$C) is much higher than $Bi_2O_3$(820$^o$C). Vacancies become smaller amount for BHFO than BFO. When Mn is added then co-doping completely hinders the formation of impurity phase [9]. Figure 1(b), 1(c) shows magnified graph from 2θ(31 to33°) ranges for BHFO having slightly valley of dual peak which was merged later due to Mn substitution. This transition suggests a structural transformation from the rhombohedral structure (pure BFO) toward coexistence of tetragonal and orthorhombic symmetry structure. Due to slighter ionic radii difference between host ions ($Bi^{3+}$1.17Å and $Fe^{3+}$0.645Å) and dopant ions ($Ho^{3+}$1.015Å and $Mn^{2+}$0.58 Å)[10], which makes it easier that $Ho^{3+}$ (or $Mn^{2+}$) enters the lattice to replace $Bi^{3+}$(or $Fe^{3+}$) respectively. Structural transformation of nanoparticles can be explained using the Goldschmidt tolerance factor (t).

$$t= \frac{(\langle rA\rangle+rO)}{(\sqrt{2}\langle rB\rangle+rO)} \quad (3)$$

where, ⟨rA⟩ and ⟨rB⟩ are the average ionic radii of A and B site cations, respectively; and rO is the ionic radius of oxygen. By substituting the dopants RA, and ¯RB into the modified t is shown:

$$t = \frac{(1-x)RBi + xRHo + RO}{\sqrt{2}(0.95RFe + 0.05RMn + RO)} \quad (4)$$

Where, $R_{Bi}$, $R_{Ho}$, $R_{Fe}$, $R_{Mn}$, $R_O$ are 1.17, 1.015, 0.645, 0.58, 1.38 Å, respectively. Calculated t values of NP's for BFO 0.8904, for BHFMO 0.8864 ($Bi_{0.9}Ho_{0.1}Fe_{0.95}$ $Mn_{0.05}O_3$)[11]. The slight decrease of t proves the lattice distance of these planes gradually decreased leading to crystal distortion of the samples takes place. Further increase of Mn content causes more changes of Fe-O-Fe bond angle and Fe-O bond length so ratio of rhombohedral to orthorhombic (Pn21a phase) is changed [12].

Figure 2 shows the FESEM micrographs of $B_{0.9}H_{0.1}F_{1-X}M_XO$ powders annealed at temperature 500°C for 2 hrs. Particle sizes are average 60, 45-50, 45, 30nm for x= 0.0, 0.5, 0.10, 0.15 respectively. Substitution of B-site $Fe^{3+}$(0.645 Å) with Mn leads to smaller particle size. Mn can present in $Mn^{2+}$ (0.83), $Mn^{3+}$ (0.645 Å), $Mn^{4+}$ (0.53 Å) ions which lead to a lattice distortion and hinder crystallite nucleation and thereby reduces particle size [13]. Grain size is also related to oxygen vacancies which is a source of Bi losing and the transformations of $Fe^{3+}$to $Fe^{2+}$. Mainly co-doping causes suppression of oxygen vacancies hence slower oxygen ion motion and slower grain growth rate. When 15% Mn was used then particle sizes have been reduced greatly to 30nm which has higher surface area and higher energy, hence highly reactive [14] and the ultra-fine BFO nanocrystals due to high surface energy form neck by solid state diffusion and evaporation-condensation process which leads to agglomeration and particle growth [15]. Thus, severe agglomeration was observed.

**Figure 2:** SEM micrographs of $BiHo_{0.1}Fe_{1-x}Mn_xO_3$ (x=0.00,0.05,0.10,0.15) samples (50,000x).

*3.2 Magnetic characteristics*

The magnetization vs. applied field (M-H) measurements of the BHFO and BHFMO powders was carried out at 300K in order to investigate the magnetic behavior and M–H loops are presented in Figure 3. From the plotted graphs, it can be seen that magnetization value is higher than pure BFO from literature.

According to literature BFO nanoparticles have weak ferromagnetic behavior because no spin canting is considered due to spin compensation into its two sublattices. However, here the SEM micrographs reveal that nanoparticles (NP's) size decreased with doping. Enhancement of the magnetization of doped BFO occurred because of the uncompensated surface spins induced by the size effect of the NP's. The magnetic parameters extracted from the magnetic hysteresis loops measured at 300K are presented in Table 1.

**Table1:** Magnetic property analysis results.

| Sample | Remnant magnetization, Mr(emu/gm) | Average Coercive Field,Hc(Oe) | Maximum Magnetization, (Ms)(emu/gm) |
|---|---|---|---|
| BHFO | 0.0898 | 64.676 | 1.08 |
| BHFMO-5 | 0.0177 | 12.476 | 1.77 |
| BHFMO-10 | 0.02277 | 20.966 | 1.06 |
| BHFMO-15 | 0.0043 | 115.305 | 0.304 |

BHFO exhibits a narrower hysteresis loop (Mr 0.089emu/g and Hc (64.6 Oe) in Figure 3, which reflects weak ferromagnetic behavior arising from canting of the antiferromagnetic sublattices. These features can be accounted due to the difference in ionic radius, ($Ho^{3+}$-0.901Å, $Bi^{3+}$-1.03Å) it is also expected that this distortion improves the magnetic properties[16]. Also, Ho causes increase magnetism by modifying exchange interaction due to decreasing of Fe-O-Fe bond angle that is the result from the distortion of lattice and destruction of spiral modulated spin structure (SMSS) due to tilting of $FeO_6$ octahedra or structural phase transition [16]. The substitution of non-magnetic

$Bi^{3+}$([Xe]$4f^{14}5d^{10}6s^{2}6p^{0}$) ion by magnetic $Ho^{3+}$([Xe] $4f^{10}$, theoretical magnetic moment 7.5μB) ion may be another reason of this increased Ms.

Among all, BHFMO-5 showed the maximum remnant magnetization and coercive field (Hc). There are many possible factors such as decrease in the average particle size as observed due to codoping. Destruction of spin arrangement has occurred due to structural transition. $Mn^{3+}$(0.46Å) is smaller than $Fe^{3+}$(0.645Å) hence structural transition is accelerated this change in bond angle and bond length hence tilting angle of $FeO_6$ octahedron and thereby suppress SMSS[17,18]. By Dzyaloshinskye Moriya interaction in the orthorhombic phases the spin canting can originated which may introduce magnetization[19,20]. In addition, Mn generally stays at multivalent (+2,+3and+4) states in BFO and the substitution of $Fe^{3+}$([Ar] $3d^5$,5μB) by $Mn^{2+}$([Ar] $3d^5$,5μB), $Mn^{3+}$([Ar] $3d^4$,4μB) and $Mn^{4+}$ ([Ar] $3d^3$,3μB) may causes incomplete compensation of SMSS. Also, the redox coupling between Fe and Mn ions or may be due to incorporation $Mn^{3+}$ (4.89 μB) at Fe site (5.92 μB) that develops local

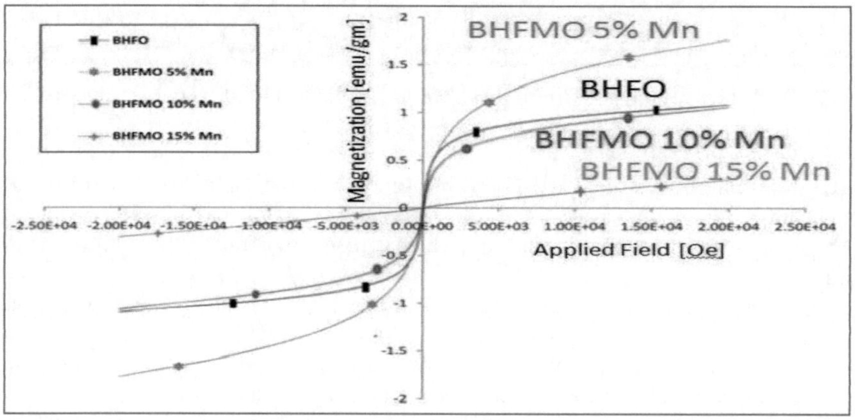


**Figure 3:** Comparison of magnetic moment of BHFO and after various % Mn doping where compositions are $BiHo_{0.1}Fe_{1-x}Mn_xO_3$ (x = 0.00, 0.05, 0.10, 0.15).

ferromagnetic spin structure [21,22]. Increase of %Mn to higher level causes a decrease of saturation magnetization this may be due to thermal effects. Saturation magnetization is inversely proportional to coercivity as:

$$Hc= \frac{\sigma * K}{Ms} \quad (5)$$

Where, K=anisotropy constant, σ is constant which depends upon crystal structure and degree of alignment. The reason for the decrease in saturation magnetization for 10% Mn doping might be due to $Fe^{2+}$ ion that act as inhibitor and weakens the super-exchange interaction existing between $Fe^{3+}$-O-$Fe^{3+}$[23]. Differences in coercive field values on either side of x-axis which signify the occurrence of both ferromagnetism and anti-ferromagnetism simultaneously and hence might result in unidirectional anisotropy shift. Also due to substitution of $Mn^{3+}$(4.89μB) on $Fe^{3+}$(5.92μB) site I increased and Ho content remain the same thus lower Bohr magnetron and increase of anisotropy constant nullify the effect of Ho as a result saturation magnetization value is decreased. Another reason is the Fe-O-Mn super exchanges is of antiferromagnetic nature and Ho Mn exchange interaction may compete with the ferromagnetic interaction resulting from co-doping and hence further reduced magnetization in BHMFO samples has been observed[12].

*3.3 Optical property analysis*

*3.3.1 Spectral analysis:* Figure 4 shows absorbance vs. wavelength plot of BHFO and BHFMO samples derived from diffuse reflectance data. BHFO shows absorption peak at 430 nm. In addition of Mn, the absorbance of BHFO increases and change in peak position occurs for 5% Mn. As %Mn is increased to 10%, absorption peak of BHFO is gradually suppressed and absorption peak of Mn appears.

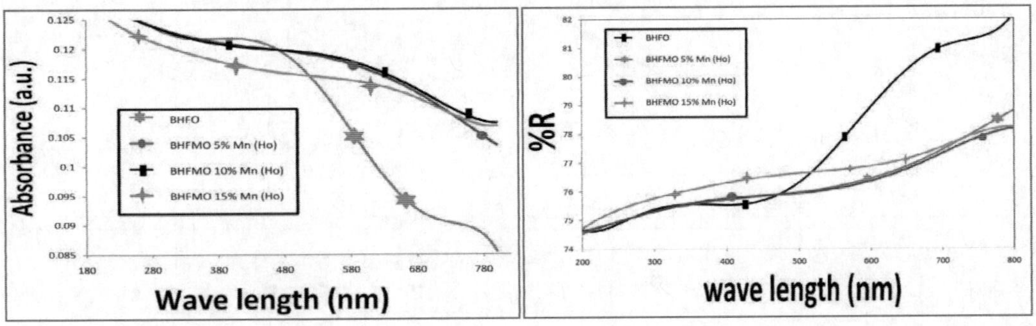


**Figure 4:** UV-Vis **(a** absorption **(b)** reflectance spectra of different percentage Mn doped BHFO.

The Mn addition causes an increment of absorbance value. On the addition of Mn on BHFO, the absorption peak shifts towards right or higher wavelength position around 520 nm. Since visible light ranges from 400-700 nm, it can absorb visible light. Another peak occurs towards lower wavelength below 300nm. UV light have wavelengths ranges from 200-400nm. So, the Mn and Ho doped structure can absorb visible as well as wave that pure BHFO alone can't do. So, Mn and Ho codoping increases visible light response and narrower bandgap which has a great application as photocatalyst. In Figure 4(b) the reflectance graph BHFO peak is found at above 320 nm another at around 680 nm and with the addition of Mn reflectance value is decreased. Also, from morphological analysis it has been found that higher the concentration of Mn darker the sample color.

*3.3.2 Band Gap Measurement:* To determine $E_g$ (band gap energy), $[F(R)*hv]^2$ vs. hv graph has been plotted for the synthesized powders. Here the intersection of the tangent line with $[F(R)*hv]^2=0$ represents the optical band gap energy, $E_g$. Here, F(R) is given by Kubelka–Munk function.

$$F(R) = \frac{\{(1-R)^2\}}{2*R} \quad (6)$$

Where, R=diffused reflectance value. This absorption can correspond to the electronic transition from the top of the valance band, which is mainly composed of O-2p hybridized with Bi-6s, to the bottom of the conduction band composed of Fe-3d. Figure 5 demonstrates that Ho single doping has bandgap of 2.9eV and for Mn doping percentages of 5%, 10% and 15% band gap values are 2.8, 2.47 and 2.4 eV respectively. This is due to reduced degree of hybridization associated with a stable electronic configuration (half occupation $4f^{10}$ $5d^0$ $6s^0$) of $Ho^{3+}$ ion, in between Fe 3d and O 2p a unique energy level is formed and which causes this effective band gap of Ho doped BFO[24]. Cation doping changes the Fe-O and Fe-O-Fe bond angle which plays a critical role in modifying one-electron bandwidth (W) and resulting change in bandgap energy [25]. The empirical formula relating electron bandwidth with bond length and angle is:

$$W \approx \frac{\cos\omega}{d_{Fe-O}^{3.5}} \tag{7}$$

where, ω = ½ [(pi)-(Fe-O-Fe)] and $d_{Fe-O}$ = Fe-O bond length The band gap is related with W as follows: Eg = Δ-W. where Δ is the charge-transfer energy [27,28].

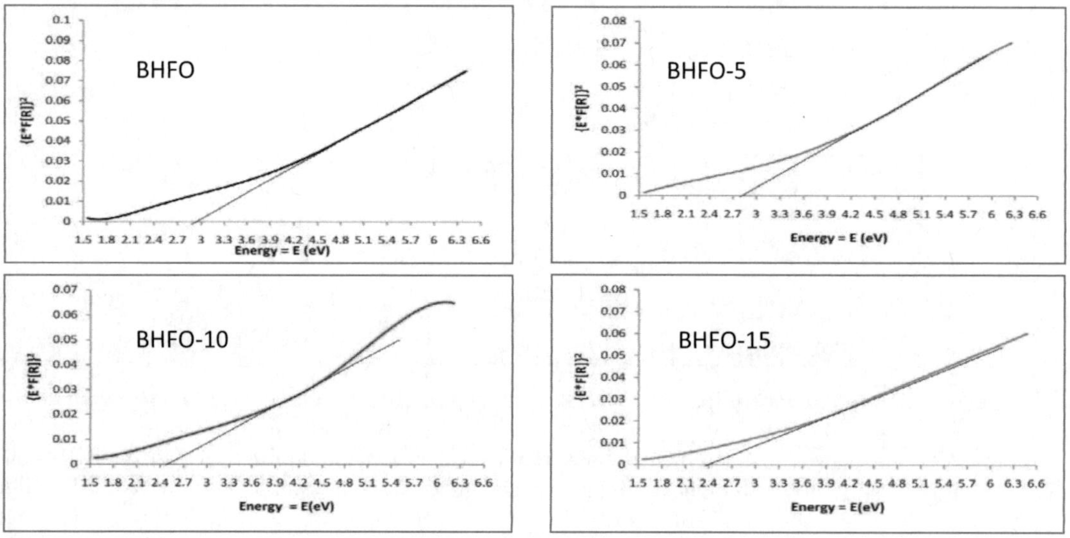


**Figure 5:** $[F(R)*h\nu]^2$ vs. hν plot for band gap calculation of BHFO and after various % Mn doping.

As the Fe-O bond length of rhombohedral phase is larger than orthorhombic phase [28], Mn-doping decreases the Fe-O bond lengths hence increases orthorhombic phase, which results in an increase of the value of W and decrease of effective band gap energy. In addition, the 3d conduction band-edge ($E_{cb}$) of $Mn^{4+}$(Ecb~5.83 eV) is lower than $Fe^{3+}$ (Ecb~4.78 eV) state [25] and brings the holes in the d band [26] which may reduce the effective energy gap between 3b conduction band and O2p valence band and resulting decrease in nanoparticles optical band gap.

## 4. Conclusion

The sol-gel method followed by thermal treatment has been used to prepare $Bi_{0.9}Ho_{0.1}Fe_{1-x}Mn_xO_3$(x= 0, 0.05, 0.10, 0.15) NP's system and their structural, magnetic and optical properties have been studied using standard experimental techniques. Studies of basic structural parameters using XRD reveals that all samples were crystallized perovskite form. Structural transition occurred toward orthorhombic and tetragonal biphasic form. Mn has smaller ionic radius than Fe, so decrease in bond length, bond angle and distortion of structure is confirmed by peak merging. Very small amount of $Bi_2FeO_9$ impurity phase is observed in BHFO. A highly dense SEM microstructure recorded on the sample has confirmed that NP's size was decreased significantly therefore finally found to be around 30nm.Using VSM studies of magnetic property reveals that maximum magnetization Ms increased on 5% Mn addition in BHFO, due to tilting of octahedral. However, increasing Mn% to 15 % causes a decrease of magnetization value due to anisotropy. Still all the samples showed greater than literature value of pure $BiFeO_3$. The UV-vis exhibits that all samples has lower bandgap energy. Normally $TiO_2$ based photo catalysts have a bandgap energy of 3.2eV while we have found values decreasing from 2.9 to

2.38eV. Also, from absorption spectra it is clear that photocatalytic activity under visible light, 48% of total solar energy, is prominent and $TiO_2$ based photocatalyst is only active for UV light, 4% of solar energy, is not very efficient. Therefore, this Mn doped BHFO can serve as a promising photocatalyst either for water splitted hydrogen generation or for degradation of organic pollutants.

## 5. Reference

[1] Charkrabarti K, Das K, Sarkar B and De S K 2011 *Enhanced magnetic and dielectric properties of Eu and Co co-doped $BiFeO_3$ nanoparticles* J APPL. Phy.110, 1035-5

[2] Ye W, Tan G Q, Dong G H, Ren H J, Xia A 2016 *The effects of grain size on electrical properties and domain structure of $BiFeO_3$ thin films by sol–gel method.* J Matter sci. Matter electronCeram.Int.41 (2015) 4668-4674

[3] Bai Y, Chen J, Zhao S, and Lu Q 2016 *Magneto-dielectric and magneto electric anisotropies of $CoFe_2O_4/Bi_5Ti_3FeO_{15}$ bilayer composite hetero structural films* RSC Adv. 6(57), 52353–52359

[4] Raghavan C M, Kim J W and Kim S S 2014 *Effects of Ho and Ti doping on structural and electrical properties of $BiFeO_3$ thin films* J. Am. Ceram.soc, 97235-240

[5] Pradhan S K, Roul B K, Sahu D R 2012 *Enhancement of ferromagnetism and multiferroicity in Ho doped Fe rich $BiFeO_3$* Solid State Commun.1521176-1180

[6] Chauhan S, Kumar M, Chhoker S, Katyal S C, Singh H, Jewariya M and Yadav K L 2012 *Multiferronic, magnetoelectric and optical properties of Mn doped $BiFeO_3$ nanoparticles.* Solid State Commun, pp152525-529

[7] Song G L, Song Y C, Su J, Song X H, Zhang N, Wang T X and Chang F G, 2017 *Crystal structure refinement, ferroelectric and ferromagnetic properties of Ho modified $BiFeO_3$ multiferroic material.* J. Alloys and Compound.503-509

[8] Chauhan S, Kumar M, and Katyal S C 2016 *Band-gap tuning and magnetic properties of heterovalent ions (Ba, Sr and Ca) substituted $BiFeO_3$ nanoparticles* AIP Conference Proceedings 1731, 130029

[9] Hongfeng, Guoqiang H, Li B C, Zhu Y, Yang J and Pradhan S K 2010 *Effect of Ho, Mn co-doping on the structural, optical and ferroelectric properties of $BiFeO_3$* Journal of Physics and Chemistry of Solids 71 1557–156.

[10] Hasan M, Basith M A, Zubair M A, Hossain M S, Mahbub R, Hakim M A, and Islam M F, 2016 *Saturation magnetization and band gaptuning in $BiFeO_3$ nanoparticles via co-substitution of Gd and Mn* J. Alloys Compd. 687, 701–706.

[11] Zhang Z, Wu P, Chen L and Wang J, 2010 *Systematic variations in structural and electronic properties of $BiFeO_3$ by A-site substitution*, Appl. Phys. Lett. 96 012905.

[12] Hasan M, Islam M F, Mahbub R, Hossain M S and Hakim M A, 2016 *A soft chemical route to the synthesis of $BiFeO_3$ nanoparticles with enhanced magnetization,* Mater. Res. 179e186.

[13] Hernandez B A, Chang K S, Fisher E R and Dorhout P K 2002 *Solgel template synthesis and characterization of $BaTiO_3$ and $PbTiO_3$ nanotubes,* Chem. Mater. 14, 480

[14] Lobo R, Moreira R L, Lebeugle D and Colson D 2007 *Infrared phonon dynamics of a multiferroic $BiFeO_3$ single crystal* Physics Review B, Vol. 76, pp. 172105.

[15] Kumar P, Shankhwar N, Srinivasan A and Kar M 2015 *Oxygen octahedra distortion induced structural and magnetic phase transitions in $Bi1xCaxFe1xMnxO_3$ ceramics* J. Appl. Phys. 117 194103

[16] Kumar P and Kar M 2014 *Tuning of net magnetic moment in $BiFeO_3$ multiferroics by co-substitution of Nd and Mn,* Phys. B Condens. Matter 44890e95.

[17] Khomchenko V A, Pereira L C J and Paix~ao J A 2011 *Substitution-driven structural and magnetic phase transitions in Bi0.86(La, Sm) 0.14FeO*$_3$ *system*, J. Phys. D Appl. Phys. 44185406.

[18] Sergienko I A and Dagotto E 2006 *Role of the Dzyaloshinskii-Moriya interaction in multiferroic perovskites*, Phys. Rev. B 7.

[19] Majetich S A, Artman J O, and Tanaka C 1994 *Calculation of magnetic moments in* $Ho_2C_3$ *nanocrystal* J. Appl. Phys. 76, 10

[20] Kittel C, 1967 *Introduction to solid state physics* American Journal of Physics,Vol.35, pp547-548

[21] Fiebig M 2005 *Revival of the magnetoelectric effect*, Journal of Physics D: Applied Physics, Vol 38, pp. R123.

[22] Gao N, Chen W, Zhang R, Zhang J, Wu Z, Mao W, Yang J, Li X A and Huang W, 2016 *First principles investigation on the electronic, magnetic and optical properties of* $Bi_{0.8}M_{0.2}Fe_{0.9}Co_{0.1}O_3$ Com. Theor. Chem. 108436e42.

[23] Zhang Z, Wu P, Chen L and Wang J, 2010 *Systematic variations in structural and electronic properties of* $BiFeO_3$ *by A-site substitution* Appl. Phys. Lett. 96 012905.

[24] Xu Y, and Schoonen M A A, 2000 *The absolute energy positions of conduction and valence bands of selected semiconducting minerals* Am. Mineral. 85 543556.

[25] Rodríguez-Carvajal J, Hennion M, Moussa F, Moudden A H, Pinsard L and Revcolevschi A 1998 *Neutron-diffraction study of the Jahn-Teller transition in stoichiometric* $LaMnO_3$ Phys. Rev. B 57 R3189eR3192.

[26] Medarde M, Mesot J, Lacorre P, Rosenkranz S, Fischer P and Gobrecht K 1995 *High pressure neutron-diffraction study of the metallization process in* $PrNiO_3$ Phys. Rev. B 52

[27] Radaelli P G, Iannone G, Marezio M, Hwang H Y, Cheong S W, Jorgensen J D and Argyriou D N 1997 *Structural effects on the magnetic and transport properties of perovskite* $A1_xA0_xMnO_3$ *(x¼0.25, 0.30)* Phys. Rev. B 56

[28] Tütüncü H M and Srivastava G P 2008 *Electronic structure and lattice dynamical properties of different tetragonal phases of* $BiFeO_3$ Phys. Rev. B 78 235.

**Acknowledgments**

This research was endorsed by the Department of Materials and Metallurgical Engineering, BUET. We are also grateful to the Department of Glass and Ceramic Engineering, BUET and Bangladesh Atomic Energy Commission for their support.

# Synthesis and Characterization of $Ba^{2+}$, $La^{3+}$ and $Zr^{4+}$ Co-Modified Multiferroic $BiFeO_3$

**F Khan[1], S Mondal[2], H M Usama[3], A Sharif[4], M A Matin[5]**

[1,2,3,4]MME Department, Bangladesh University of Engineering and Technology, Dhaka-1000, Bangladesh
[5]Department of Glass and Ceramics, Bangladesh University of Engineering and Technology, Dhaka-1000, Bangladesh

***Abstract :*** The objective of this research project was to reduce the electrical leakage, enhance the magnetic and electric property as well as to stabilize the $BiFO_3$ structure by co-doping La, Ba and Zr. At first the solid state sintering process was done to synthesize the $Bi_{.8}La_XBa_{0.2-x}Fe_{.95}Zr_{.05}O_3$ (where, x = 0.05, 0.10, 0.15, 0.20). X-ray diffraction data revealed a stabilized $R_3C$ structure with the dopants. The amount of zirconia in B-site resulted in the reduction of vacancy which ultimately reduced leakage current. For the higher percentage of $Ba^{2+}$ the oxygen vacancy increased which led to higher leakage current. Thus for increasing the $La^{3+}$ up to 15 mole % with decreasing $Ba^{2+}$ dielectric constant increased For further addition of $La^{3+}$ (i.e. 20%) the dielectric constant decreased again. From the SEM analysis the coherent grain size was confirmed for the temperature at 950°C. With the increase of $La^{3+}$ and decrease of $Ba^{2+}$ the grain size increased. For the composition with 5% of $La^{3+}$, the p-loop was found to be satisfactory with the remnant polarization of 0.25 (μC/cm$^2$) and the coercivity of 330V. However the p-e loop for the other compositions was found to be unsatisfactory.

## 1.Introduction

Multiferroics are multifunctional materials which display more than one ferroic order in the same phase. In recent years, multi-ferroics have been drawing the attention of researchers due to their unique behavior of coupling between two or more properties of ferroelectricity, ferromagnetism and ferroelasticity and also because of their potential technological applications in devices in spintronics, information storage, sensing and actuation. There are many magnetic and ferroelectric materials, there are relatively few multi-ferroic materials. $BiFeO_3$ (BFO) is perhaps the only material that is both magnetic and a strong ferroelectric at room temperature. Bismuth ferrite is one of the most promising multi-ferroics and at room temperature, bulk $BiFeO_3$ presents a rhombohedrally distorted perovsklte structure belonging to the space group R3c. Compared to other multi-ferroics, BFO exhibits a higher ferroelectric Curié temperature (TC ~830 °C) and a high G-type antiferromagnetic ordering (TN ~370 °C) temperature. Bulk BiFeO3 suffers from a high leakage current, thought to be the result of vacancies invariably produced due to the volatility of $Bi_2O_3$. High leakage current translates into loss of stored charged. However, it is very difficult to observe the ferromagnetic loop of a bulk $BiFeO_3$ sample at room temperature due to its low spiral spin modulation which is caused by defects and non-stoichiometry in the bulk $BiFeO_3$ sample. Hence, it is essential to improve the ferromagnetic properties of $BiFeO_3$ without much disturbing its ferroelectric properties before its actual use in device application. And the formation of oxygen vacancies induces impurity energy level in the band gap and increases the free carrier density by hopping of electrons to these defect levels. The weak magnetization is attributed to the superimposition of a spiral spin structure on BFO antiferromagnetic order.The main purpose of this research was to determine the effects of Ba and La along with co-modification by Zirconia on bismuth ferrite, with a view to improving its dielectric and magnetic properties. Solid state process was the route of choice at first. Sintering was done at 950°C to obtain coherent grain size. The room temperature magnetization improved remarkably compared to pure BFO after co-modification, more work needs to be done before its implementation in real life applications. The increased amount of $Zr^{4+}$ resulted in the reduction of leakage current. For the higher percentage of $Ba^{2+}$ the oxygen vacancy increased which led to higher leakage current. Thus for increasing the $La^{3+}$ up to15 mole % with decreasing $Ba^{2+}$ content, dielectric constant increased. For further addition of $La^{3+}$ (i.e. 20%) the dielectric constant decreased again.

## 2. Experimental

At first, constituent powders were mixed at proper amount depending on the doping level. Next, the mixture was ball milled at 120 rpm in acetone for 18 hrs. Finally, the mixture was pressed into pellets using a hand pressing unit and subsequently sintered. The basic raw materials used are powders of $Bi_2O_3$, $Fe_2O_3$, $La_2O_3$, $ZrO_2$ and $BaCO_3$. These powders were taken according to the target composition of the batch to be prepared. . Four compositions were chosen for this thesis:
a) $Bi_{0.8}La_{0.20}Fe_{0.95}Zr_{0.05}O_3$ b) $Bi_{0.8}La_{0.15}Ba_{0.05}Fe_{0.95}Zr_{0.05}O_3$ c) $Bi_{0.8}La_{0.10}Ba_{0.10}Fe_{0.95}Zr_{0.05}O_3$
d) $Bi_{0.8}La_{0.05}Ba_{0.15}Fe_{0.95}Zr_{0.05}O_3$

Then ball milling was done for proper mixing and homogenization of the raw materials. The powders of different compositions were taken into different high density polyethylene (HDPE) pots serving as the milling chamber. $Y_2O_3$ stabilized $ZrO_2$ balls, together with acetone, were used as the grinding medium. All the compositions were milled for 18 hours. Then the dried product was taken into a mortar and grinded to form fine powders. After milling, the mixture taken into suspension in acetone and then carefully extracted into a glass beaker (which was ultrasonically cleaned). The beaker was covered with a perforated aluminum foil and kept in a drier at around 100°C. It was important to keep the mixtures in the drier for a sufficiently long time, so that a completely dry powder mixture was obtained. Then the powder was calcined at 800°C for 2 hours. As we used $BaCO_3$, we need to calcine to form BaO. That's why calcination is done. Around 0.6 gm of powder was taken and pressed into disk-like pellets using a manual hydraulic pressing machine. 2 tons of load was applied for 60 seconds. The pressed pellets were then dried in an oven for at least 24 hours to enhance their green strength. Then sintering is done at 950°C to get $BiFeO_3$ pellet of required composition. Then density measurement, SEM Dielectric properties measurement by impedance analyzer , Ferroelectric properties measurements were done.

## 3. Results and Discussion

### 3.1 XRD Analysis:

XRD analyses of the samples were done at room temperature. The diffraction patterns obtained are shown in Figure 1 for $Bi_{0.8}La_xBa_{0.2-x}Fe_{0.95}Zr_{0.05}O_3$ compositions. All the samples were sintered at 950°C. The pattern of x = 0.05 sample showed that BFO structure was indeed formed. The peaks are not identical to pure BFO; there are some differences in peak positions and some split peaks seem to merge upon modification with 20% La. This is in accordance with previous reports [1, 2] and is assumed to be the result of space group modification, i.e. the space group is no longer R3c. The diffraction pattern is nearly the same for x = 0.1. And there is no identical 2[nd] phase present in any of the sample

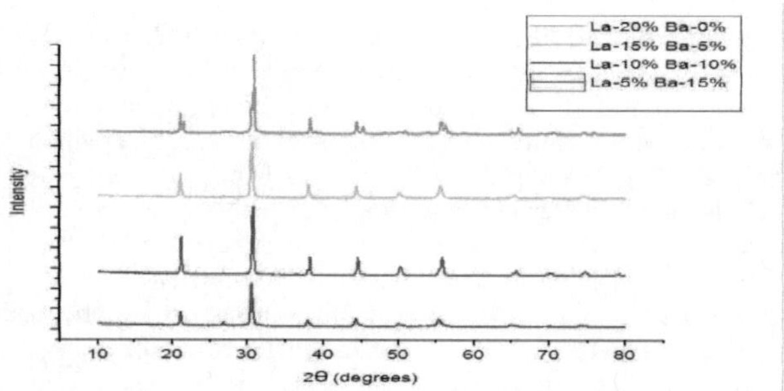


**Figure 1** : Room temperature XRD patterns for $Bi_{0.8}La_xBa_{0.2-x}Fe_{0.95}Zr_{0.05}O_3$ sintered at 950° C.

Figure 2 shows the peak around 29.5-31.5 degrees more closely for different compositions. It can be readily seen that the peak gradually shifts towards higher theta values with increasing $La^{3+}$ content & decreasing $Ba^{2+}$ content. Since the theta value (θ) and the planar spacing (d) for a given peak are related by the equation,

$$\lambda = 2d_{hkl} \sin(\theta_{hkl})$$

Where λ is the wavelength, it follows that higher theta values correspond to smaller d values, i.e. lattice spacing. So it can be said with increasing $La^{3+}$ content & decreasing $Ba^{2+}$ content results in change of the lattice parameter, and this is expected because differences in ionic size. Looking now at the patterns for La-15% Ba-5% & La-20% Ba-0%, we see that the peak in question is quite prominently split indicating that whatever structural modification was set in by La-substitution. The peaks at this position belong to the [100] family of planes in the pseudo-cubic cell. All six [100] planes have the same d-spacing in cubic and rhombohedral symmetry and so this peak splitting indicates that the symmetry has been lowered to at least tetragonal, possibly even lower.

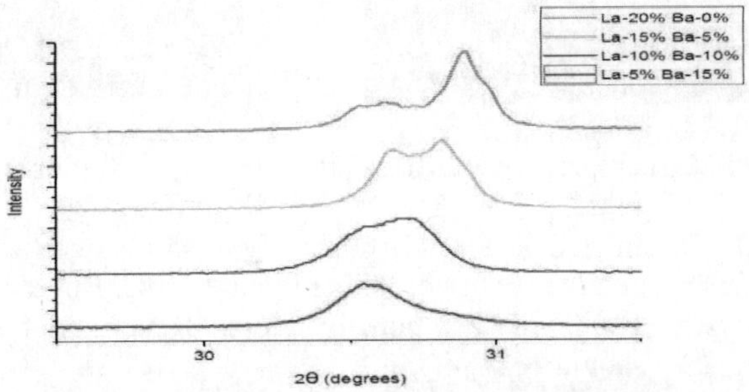


**Figure 2** : Change in peak position with increasing La $^{3+}$content & decreasing Ba $^{2+}$content

3.2Microstructure development:

Obtaining dense samples is one of the important parts of the optimization process. The best properties will be obtained for those samples that have both high %TD and desirable grain size. Figure 3 shows that $Bi_{0.8}La_{0.05}Ba_{0.15}Fe_{0.95}Zr_{0.05}O_3$ sintered at 900°C contain large amount of porosity. And average grain size is ~ 0.2 μm. It is evident that densification had not occurred to any respectable degree and higher temperature would be required for obtaining proper microstructure. This situation is improved in the case of sintering at 950°C. Porosity decreases sufficiently and the grain size also increased to ~ (0.6-0.7)μm.

Literature review reveals that the $Ba^{2+}$ ion substitution in $BiFeO_3$ has a strong grain growth inhibiting effect in the microstructure [20]. Hence, sintering temperature at 950°C is seemed to be optimum for $Bi_{0.8}La_{0.05}Ba_{0.15}Fe_{0.95}Zr_{0.05}O_3$. The %TD and grain size achieved in this temperature range was satisfactory.

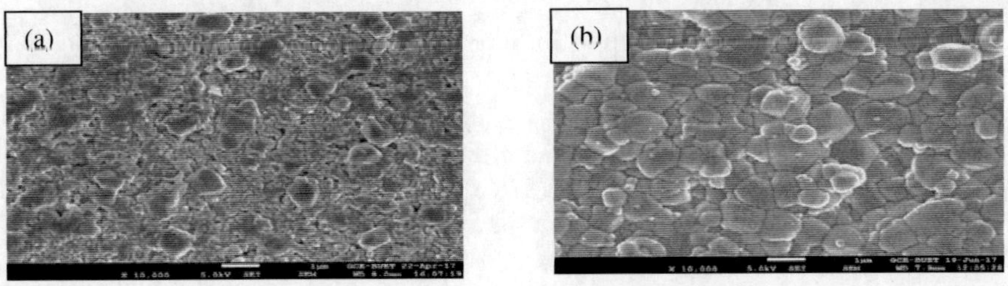


**Figure 3**: FESEM micrograph of $Bi_{0.8}La_{0.05}Ba_{0.15}Fe_{0.95}Zr_{0.05}O_3$ sintered at (a) 900°C (b) 950°C

Figure 4 shows the micrographs of $Bi_{0.8}La_{0.10}Ba_{0.10}Fe_{0.95}Zr_{0.05}O_3$ sintered at various temperatures. Sintering of $Bi_{0.8}La_{0.10}Ba_{0.10}Fe_{0.95}Zr_{0.05}O_3$ at 850°C shows the similar effect of porosity as obtained in the case of $Bi_{0.8}La_{0.05}Ba_{0.15}Fe_{0.95}Zr_{0.05}O_3$ sample sintered at 900°C. Here sintering is not up to the mark. The situation is improved at 950°C where porosity decreases to a significant amount and grain size increases. Average grain size is found to be ~1 μm.

**Figure 4**: FESEM micrograph of $Bi_{0.8}La_{0.10}Ba_{0.10}Fe_{0.95}Zr_{0.05}O_3$ sintered at (a) 900°C (b) 950°C

$Bi_{0.8}La_{0.15}Ba_{0.05}Fe_{0.95}Zr_{0.05}O_3$ sintered at 900°C shows Figure 5(a) significant amount of porosity which on increasing sintering temperature gives better densification at 950°C. But the grain size here is much larger (~3 μm). And there is some amount of porosity too. Hence, optimum sintering temperature for $Bi_{0.8}La_{0.15}Ba_{0.05}Fe_{0.95}Zr_{0.05}O_3$ should be 950°C.

**Figure 5**: FESEM micrograph of $Bi_{0.8}La_{0.15}Ba_{0.05}Fe_{0.95}Zr_{0.05}O_3$ sintered at (a) 900°C (b) 950°C

Figure 6 shows the micrographs of $Bi_{0.8}La_{0.20}Fe_{0.95}Zr_{0.05}O_3$ sintered at various temperatures. Sintering of $Bi_{0.8}La_{0.20}Fe_{0.95}Zr_{0.05}O_3$ at 850°C shows the similar effect of porosity as obtained in the case of $Bi_{0.8}La_{0.20}Fe_{0.95}Zr_{0.05}O_3$ sample sintered at 900°C. Here sintering is not up to the mark. The situation is improved at 950°C where porosity decreases to a significant amount and grain size increases. Average grain size is found to be ~1 μm.

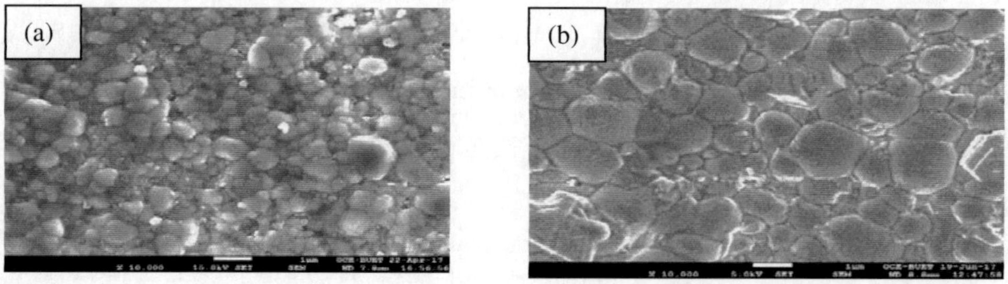


**Figure 6**: FESEM micrograph of $Bi_{0.8}La_{0.20}Fe_{0.95}Zr_{0.05}O_3$ sintered at (a) 900°C (b) 950°C

La and Ba have important effect on grain growth. La and Ba have dissimilar identity regarding the grain growth effect. Generally, the increased amount of La results in the increase of grain size as well as helps grain growth. As when amount of La is increased the round grain transformed to elongated grain and root mean square surface roughness is decreased as well. On the other hand, Ba impedes or retards the grain growth.
From the figure for La (5%) & Ba (15%) in 7(a), As the amount of La is less in compared to other compositions available for La , the grain growth increased effect is least and at the same time the increased percentage of Ba results the impede of grain growth. Thus the overall grain size in the microstructure is smallest.
From the figure for La (10%) & Ba (10%) in 7(b), Here the amount of La% is increased which results in the increase of grain size to a small extent compared to the former one. As the amount of Ba is decreased thus retardation characters of Ba decreased thus the combined effect results in increase of grain growth. From the figure for La(15%) & Ba (5%) in7(c), Here the La% is increased to a further extend which enhances the grain growth characters at the same time the decreased quantity of Ba lowers the impeding characters and their combining effect at this composition results in the increase of grain growth than the previous two compositions.
From the figure for La (20%) & Ba (0%) in 7(d), unusual effect is shown. The aim of our research is the analysis of the combining effect of both Ba & La. When both the dopants are present then it shows simultaneous continuous properties. But when only one of the dopant is used in the absence of another one then completely identical different properties shown which is not the follow of the previous one. Exactly same situation happens in our research. The absence of Ba and the presence of high La% should tend to develop high grain size but the fact from the experiment found that the grain size lowers than the La (15%) & Ba (5%).

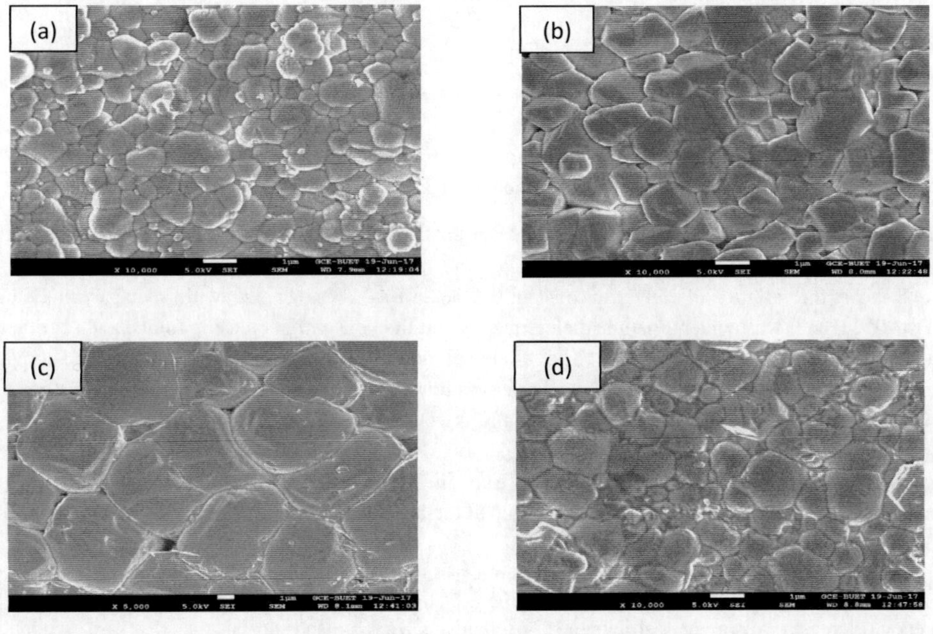


**Figure 7**: FESEM micrograph of $Bi_{0.8}La_xBa_{0.2-x}Fe_{0.95}Zr_{0.05}O_3$ sintered at 950°C. (a) X=.05, (b) x=0.10, (c) x=0.15, (d) x=0.20

3.3 Dielectric study:

The dielectric constant measures the polarizability of a medium. Upon application of an electric field, the medium which polarizes more and does not conduct electricity will be the one to show a higher dielectric constant. Already present polarization, if any, of the material does not contribute to the dielectric constant; it is the change in polarization with electric field which contributes. Thus if a material contains dipoles in its structure, but the dipoles do not change direction or change in magnitude appreciably with applied electric field the dielectric constant will be low.

As we know, the crystal structure of bismuth ferrite is non-centrosymmetric, and the unit cell does have a polarization in the pseudo-cubic [111] direction. The ferroelectric nature of the material allows the dipoles to reverse direction in response to an oscillating electric field. At low frequencies these dipoles can follow the electric field and contribute to the dielectric constant. But as the frequency is increased a point is reached where, the dipoles cannot reorient fast enough and so a drop in dielectric constant is expected. In this range only electronic polarization contributes to the dielectric constant.

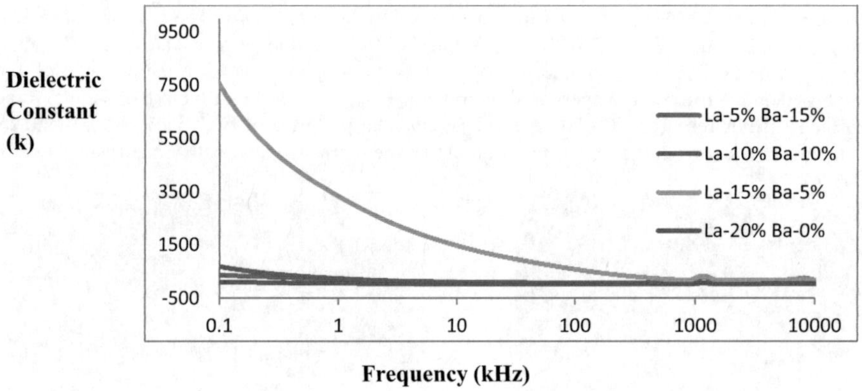


**Figure 8**:Frequency dependence of dielectric constant for Bi0.8LaxBa0.2-xFe0.95Zr0.05O3 sintered at 950° C

Figure 8 shows the variation of dielectric constant of $Bi_{0.8}La_xBa_{0.2-x}Fe_{0.95}Zr_{0.05}O_3$ with frequency in the range 100 Hz –10 MHz. It was evident that the dielectric constant falls off with increasing frequency, settling to a nearly constant value at high frequencies. The reason behind this phenomenon has been described above. Something that's not explained by the above discussion however, is the abnormally high dielectric constant for BLFO at low frequencies. This can be attributed to Maxwell-Wagner type contribution to the dielectric constant. This effect is related to space charge relaxation at the interface. The space charges can arise due to vacancies in the bismuth or oxygen site i.e$VO_2^+$,$VBi^{3-}$. These space charges can follow the electric field at low frequencies but do not have enough time to build up at the interface at high frequencies. Thus that contribution steadily decreases as frequency is increased.

At lower frequencies, the net contribution from various polarizations viz. space charges, ionic defects, permanent dipoles, induced atomic and ionic dipoles, contribute dielectric resulting in high values [3]. As frequency increases, different participating mechanisms for dielectric constant decrease gradually and subsequently. In the present scenario, $Ba^{2+}$ substitution in $Bi^{3+}$ions increases the oxygen vacancy in the lattice of $Bi_{0.8}Ba_{0.2}FeO_3$ resulting in the increase of space charge [4]. Thus, for $Bi_{0.8}Ba_{0.2}FeO_3$ samples at low frequency show a high value of dielectric constant.

The observed dielectric constant decreases rapidly with the frequency, the decrease of dielectric constant with increase in frequency can be explained by a dipole relaxation phenomenon which is due to Maxwell-Wagner type of interfacial polarization contribution [5]. A high dielectric constant value in $Bi_{0.8}Ba_{0.2}FeO_3$ could also be due to structural distortion in the sample. The $Zr^{4+}$substitution decreases the oxygen vacancy by charge compensation.

For increasing Ba content, leakage current is increasing. So the dielectric constant is lower. For La-15% Ba-5%, leakage current lower. So, the dielectric constant is highest. But unusual results have been found in only La-20% sample. Here, the effect of Ba is absent. So, the leakage is higher. So, the dielectric constant is lower.

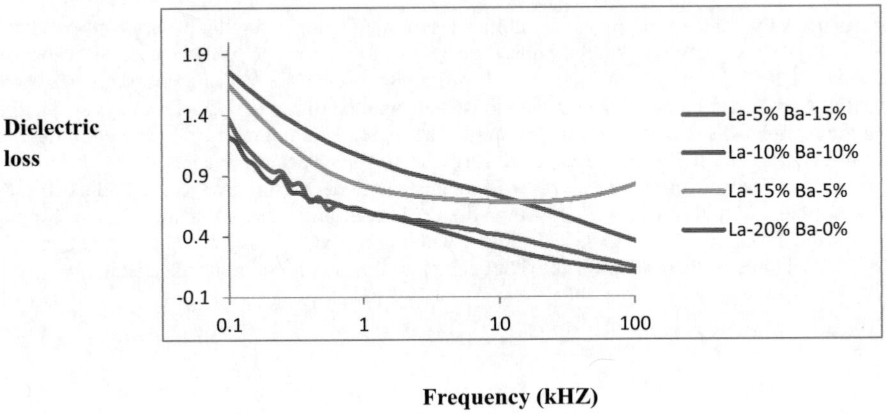


**Figure 9**: Frequency dependence of dielectric loss (D-factor) for $Bi_{0.8}La_xBa_{0.2-x}Fe_{0.95}Zr_{0.05}O_3$ sintered at 950°C

Figure 9 represents the frequency dependence of dielectric loss. Here same results have been found in the frequency range of (0.1-100 kHz). We know that higher dielectric loss means higher leakage current. For Ba-0%, dielectric loss is minimum. Increasing Ba content means increasing dielectric loss. But for Ba-15%, dielectric loss is lower for lower frequencies.

3.4 Ferroelectric property analysis

When Ba is doped in the BFO then it results in the creation of oxygen vacancy. As a result of the increase of the oxygen vacancy it generates electrons which ultimately increase leakage current. Thus increased amount of Ba result in the decrease of the remnant polarization. From the composition of La (5%) and Ba (15%) the p-loop found was satisfactory. It gives a decent identical curve on the polarization Vs electric field axes. And from that curve the precise data of remnant polarization and the coercively can be found. The comparison with the other composition cannot be done in our research. As the P-Loop obtained from the other variable compositions is not satisfactory as well as improper. As the P-Loop obtained from other compositions are unsatisfactory, this may happen due to the lack of sample preparation, improper surface area calculations or mishandling during technical operations.

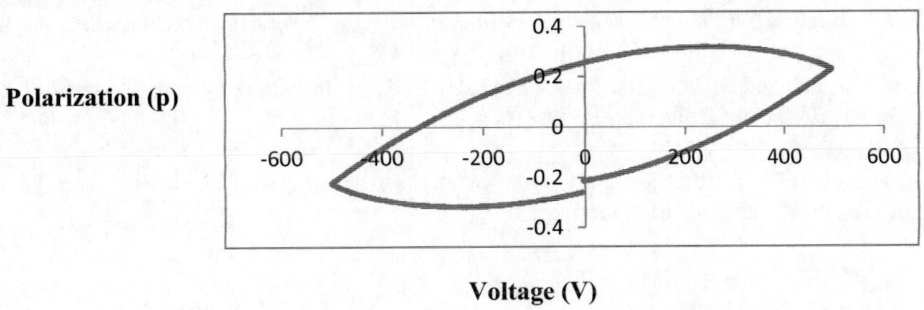


**Fig 10**: P-E loop for $Bi_{0.8}La_xBa_{0.2-x}Fe_{0.95}Zr_{0.05}O_3$ La-5% Ba-15%

## 4. Conclusion:

The main purpose of this research was to determine the effects of Ba and La along with co-modification by Zirconia on bismuth ferrite, with a view to improve its dielectric and magnetic properties. Solid state process was the route of choice at first. Sintering was done at 950C to obtain coherent grain size. The room temperature magnetization improved remarkably compared to pure BFO after co-modification, more work needs to be done before its implementation in real life applications. The followings are in brief the highlighted findings from this research work.

Pure BFO produced by solid state process exhibited less amount impurity which was confirmed by XRD analysis. When dopants were used unknown impurities were not found in the XRD analysis. It also found that with the increase of $La^{3+}$ the grain size increased. $Bi_{0.8}$ $La_x$ Ba. $_{02\text{-}x}$ $FeO_{0.95}$ $ZrO_{0.05}$ (x=0.05, 0.10, 0.15, 0.20) were successfully produced by solid state methods which was confirmed by XRD analysis that indicated formation of the main BFO structure in all the samples. The increased amount of $Zr^{4+}$ resulted in the reduction of leakage current. For the higher percentage of $Ba^{2+}$ the oxygen vacancy increased which led to higher leakage current. Thus for increasing the $La^{3+}$ up to15 mole % with decreasing $Ba^{2+}$ content, dielectric constant increased. For further addition of $La^{3+}$ (i.e. 20%) the dielectric constant decreased again. For the compositions with 5% of $La^{3+}$ the p-loop was found to be satisfactory with the remnant polarization of 0.25 (μC/cm$^2$) and the coercivity of 330V. However the p-loop for the other compositions was found to be unsatisfactory.

## 5. References:

[1] H Singh and K L Yadav 2011 *Dielectric,magnetic and magnetoelectric properties of La and Nb codoped bismuth ferrite* Journal of Physics:Condensed Matter vol 23 pp 385901:1-6

[2] M Cazayous, D Malka, D Lebeugle, D Colson 2007 *Electric field effect on $BiFeO_3$ single crystal investigated by Raman spectroscopy* Applied Physics Letters vol 91 pp 071910-071910

[3] Z M Tian, S L Tuan, X L Wang, X F Zheng, S Y Yin, C H Wang and L Liu 2009 *Size effect on magnetic and ferroelectric properties in $Bi_2Fe_4O_9$ multiferroic ceramics* Journal of Applied Physics vol 106 pp 103912:1-4.

[4] S Cheong and M Mostovoy 2007 *Multiferroic:a magnetic twist for ferroelectricity* Nature Materials vol 6 p 13-20.

[5] N A Hill 2000 *Why are there so few magnetic ferroelectrics?* Journal of Physical Chemistry vol 104 p 6694-6709.

[7] C Ederer and N A Spaldin 2005 *Weak ferromagnetism and magnetoelectric coupling in bismuth ferrite* Physics Review B vol 71 pp 060401.

[8] I Sosnowska, T P Neumaier and E Steichele 1982 *Spiral magnetic ordering in bismuth ferrite* Journal of Physics C:Solid state Physics vol 15 pp 4835.

[9] Y K Jun and S H Hong 2007 *Dielectric and magnetic properties in Co and Nb-substituted $BifeO_3$ ceramics* Solid State Communicationvol 144 p 329-333

[10] M E Lines and A M Glass 2004 *Principles and Applications of Ferroelectris and Related Materials* Oxford University Press.

**Acknowledgements**

The authors are thankful to the Department of Materials and Metallurgical Engineering, BUET; the Department of Glass and Ceramics Engineering, BUET.

# Effect of Concentration of Precursor Solution on Surface Morphology and Optical Properties of Titania Thin Film

**N. Barua, J.F. Moushumi, A.K.M.B. Rashid**

Department of Materials and Metallurgical Engineering, Bangladesh University of Engineering and Technology, Dhaka 1000, Bangladesh

**Abstract.** Titania thin film has become a great interest for the researchers due to its amazing photocatalytic activities and anti-bacterial effect. In present paper, films were deposited on normal silica glass substrate by dynamic spin coating using stable titania precursor solutions of different concentrations. Stable titania solution were synthesized by aqueous sol-gel route using titanium tetra isopropoxide.The films were calcined at $500^0$C were characterized with thermo gravimetry anddifferential ccanning calorimetry to determine phase transition point, X-ray diffraction for phase analysis, UV-Visible spectrometry for optical properties analysis, surface profilometer for thickness measurement, Optical Microscope and Scanning Electron Microscopy (SEM) to determine surface morphology. The objective was to analyze the effect of concentration of titania precursor solution on the surface morphology and optical properties of titania thin films. Solution of comparatively lower concentration produced uniform and stable titania film with better optical properties. With increasing concentration, the adhesion of the film degraded along with its surface morphology and optical properties.

## Introduction

Interest in titania thin film has grown rapidly in recent years due to its photocatalytic activity, availability as well as chemical inertness which has led to a global spark of interest into several applications such as water and air purification by photo catalysis, self-cleaning or anti-fogging materials (due to its superhydrophylicity) and solar energy conversion.Many investigations have focused on its production and applications in solar energy conversion and environmental purification since Fujishima and Honda discovered the photocatalytic splitting of water on the $TiO_2$ electrodes in 1972[1]. $TiO_2$ is transparent in the visible range and has a wide optical band gap [2]. This, together with its chemical and thermal stability, makes $TiO_2$ an excellent candidate for optical coatings. Many researchers have been developing various ways to apply $TiO_2$ coatings on various substratesto avoid the use of powder, which must be separated from the water in a slurry system after photocatalytic reaction.

There are several methods for the fabrication of titania thin films including hydrothermal methods, electronbeam evaporation, DC and RF magnetron sputtering, solvo-thermal synthesis and sol–gel methods [3-7]. However, some of these methods have some disadvantages for industry applications [8] like vacuum evaporation, sputtering, and chemical vapor deposition methods require special apparatuses for deposition of films. The sol–gel process provides new approaches and a better control in the preparation of materials compared with the other methods[9].It offers several advantages: fine powders and films with high homogeneitycan be prepared using simple equipment, the thickness and the porosity of the films can easily be controlled by changing the concentration of the sol and large surface areas can be coated by spin or dip coating.Spin coating is a simple process for rapidly depositing thin coatings onto relatively flat substrates.This technique offers many advantages over other deposition techniques due to the use of very simple large area thickness uniformity, good reproducibility, controlled growth and low temperature deposition[10].

The photocatalytic activity of $TiO_2$ thin films strongly depends on the preparing methods and process parameters like solution concentrations,calcination temperature as they have decisive influence on chemical and physical properties of titania thin films[11]. It was our interest to analyze the effect of solution concentration on the overall quality of titania thin film as not many works have been done on this.In this work both the aqueous and non-aqueous sol-gel routes for facile preparation of titania thin film were explored to find an inexpensive production route of uniform and transparent $TiO_2$ thin film via spin coating method.With this, the effect of concentration of the precursor solution on the surface morphology, quality and optical properties of titania thin film was analyzed which was our major concern.

## 2. Experimental

### *2.1 Materials*

Titanium(IV) isopropoxide (TTIP) ($Ti[OCH(CH_3)_2]_4$ ) (98.0%) were purchased from Acros Organics and acetylacetone ($C_5H_8O_2$ )(99.0%), acetic acid($CH_3COOH$) (98.5%) were purchased from Merck Chemical Industries, Ltd. All reagents were used as received, and all the experiments were carried out under ambient conditions without eliminating the water from the atmosphere.

### *2.2 Preparation of Aqueous Titania Sol*

Titanium (IV) isopropoxide was added to acetyleacetone ($\eta_{acac}/\eta_{TTIP}=1$) slowly under continuous stirring at room temperature. The mixture was stirred for 10 minutes. The resulting yellow solution of TTIP/ Acac (acetyle acetone) precursor was added dropwise to an aqueous solution of AcOH under continuous stirring for 1 hour at room temperature, yielding yellow transparent solutions(20ml).The water and AcOH ratio were determined via trial and error to obtain the desired pH.5 batches of solutions of different concentrations (0.3M ,0.5M,1M,1.25M,1.5M) with pH ranging from (3.5- 3.6) were produced to study the effect of concentration on the stability, uniformity, and properties of the films.

### *2.3 Preparation of Titania Film*

Thin films of $TiO_2$ were prepared by dynamic spin-coating method. The substrate used for coating was a silica glass substrate with 25*25*1.15mm dimension.The substrates were cleaned with soft bristled brush to avoid scratches; detergent was used to get rid of oily or greasy substances. The cleaning is followed by ultrasonic bath in ethanol, acetone and water respectively, each for 10 minutes. The clean and dry substrates were then dropped with precursor solution while the substrate was rotating at 2500 rpm for 20 seconds in the spin coater (MIDAS, Model 1200T). The coated film was dried at room temperature for 10 min and then at 60°C for 10 minutes. The set of spin coating and drying was further repeated twice. The as-prepared film was then heated at 500 °C for 2hours to yield stable and Transparent$TiO_2$ thin films.

### *2.4 Characterization of Titania Film*

The surface morphology of the$TiO_2$films were observed using anoptical microscope, field emission scanning electron microscope (FeSEM, JEOL-7600F). Surface roughness and thickness were determined using surface profilometer (BRUKER, Model :DektakXT).Changes in the crystal phase and crystallite size were monitored using X-ray diffraction (Model EMMA) rotating anode diffractometer (45 kV, 40 mA) with CuK$\alpha$ radiation (lKa ¼ 1.5406 A). UV–Vis Spectrometer (Cary Win UV 50, Australia) and (Cintral 2020, GBC corporation,Australia) were used to measure absorbance of the thin film over an extended spectral range.

## 3. Result and discussion

### *3.1 Characterizations of $TiO_2$ films prepared by spin-coating*

Preparation of titania film was attempted by both static and dynamic spin coating at different spinning speed, spinning time and different concentration. It was observed that dynamic spin coating at 2500 rpm produces more uniform film than static spinning. It was also observed that film produced from solution with lower concentration produced more stable, uniform and transparent film than the film produced from higher concentration precursor solution. A visual difference between the films made from different concentration precursor solution can be seen from Fig. 1.

**Figure 1** : Effect of concentration on the uniformity and transparency of the films (1) C= 0.5M (2)C=1M (3) C= 1.25M (4) 1.5M

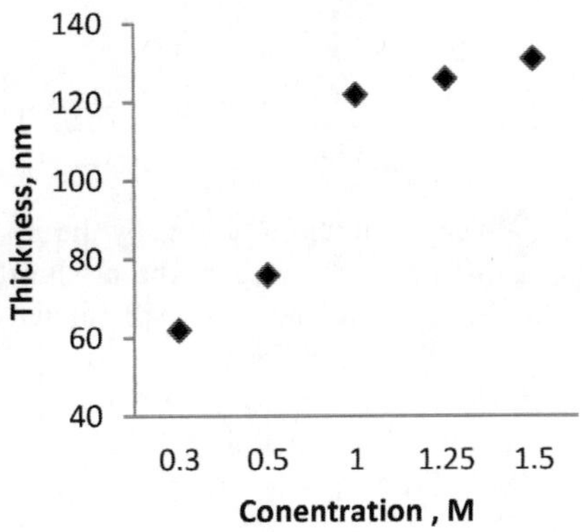


**Figure 2**: Effect of concentration on film thickness

The thermo-gravimetry and differential thermal analysis was conducted to determine the calcination temperature. Because it was difficult to analyze precisely the changes accompanied by the crystallization and/or phase transition owing to the quite small amount of $TiO_2$ loaded on the glass substrate, the TG-DTA analysis was conducted on the powder samples that were obtained by the simple dryness of the precursor sol solutions. No clear peaks were observed for titania powder from 500 to 900 °C, whereas exothermic peaks corresponding to phase transition from anatase to rutile are generally observed at around 800 °C for titania powder obtained from other precursor solutions[12].The TG-DTA result indicates the interesting nature for acac/AcOH for preventing the phase transition of anatase to rutile at high temperatures. It has been reported that the addition of acac and/or AcOH into the organic solvents containing the titania precursor effectively suppress the phase transition[12] and crystal growth[13].However, it has also been found that the crystal growth increases with increasing calcination temperature [2] so 500°C calcinations temperature were selected for this experiment.

Figure 2 shows the thickness of $TiO_2$ thin film with different concentration from 0.3, 0.5M, 1M, 1.25M and 1.5M.The thickness are increasing gradually. It is found that as the thickness increase it will affect the electrostatic contact between solute particles [14]. The particles become larger and resulted in increasing the probability of more solute particles to be grouped creating agglomerate particles. This result was supported by structural analysis.

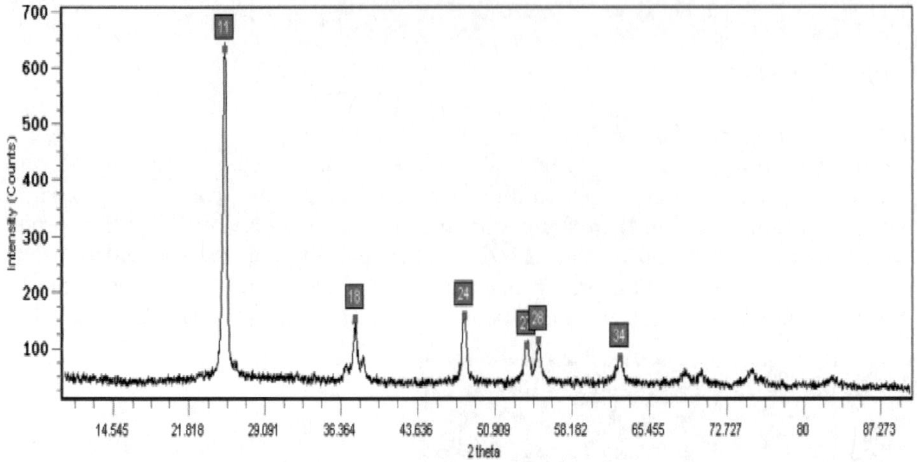


**Figure 3**: XRD spectra of Titania thin film (C=0.3M)

The X-ray diffraction pattern of the synthesized titania film is shown in Figure 3.The absence of spurious diffractions indicates the crystallographic purity [7]. The experimental XRD pattern agrees with the JCPDS card no. 21-1272 (anatase $TiO_2$) and the XRD pattern of $TiO_2$ nanoparticles other literature [8]. Strong diffraction peaks at 25° and 48° indicating $TiO_2$ in the anatase phase [10]. The 2θ peaks at 25.27° and 48.01° confirm its anatase structure. The X-ray diffraction peaks of the powders show a significant broadening due to the small size of crystallites. The average crystallite sizes (D) were calculated from the most intense diffraction peak (101) using the Scherrer formula. It was observed that crystallite size increases with increasing concentration.

**Table 1**: Different Crystallite size with varying Concentration.

| Sample No | Concentration | wavelength, λ | FWHM, 2Θ | Diffraction angle at 101 peak 2Θ | Crystallite Size |
|---|---|---|---|---|---|
| 1 | 0.3M | 0.154 nm | 1.47 | 25.12 | 486nm |
| 2 | 0.5M | 0.154 nm | 1.42 | 25.2 | 502nm |
| 3 | 1.0M | 0.154 nm | 1.2 | 25.32 | 595nm |
| 4 | 1.25M | 0.154 nm | 0.95 | 25.18 | 751nm |
| 5 | 1.5M | 0.154 nm | 0.87 | 25.3 | 820nm |

### *3.2 Surface morphology of the Titania films*

In Fig. 4(a)the substrate showssmooth continuous coating with few inclusion type matter that has a broken glass type morphology.It is assumed that inclusion particles during drying or heating breaks the continuity of the coating. Figure 4(b) shows absolute broken glass morphology. When observed closely a smooth and continuous layer beneath the broken glass like layer can be seen which posits a very important conclusion that with increasing concentration the continuity/uniformity of the coating decreases. This may occur due to grain coarsening or particle-agglomeration resulting in degradation of other properties too.

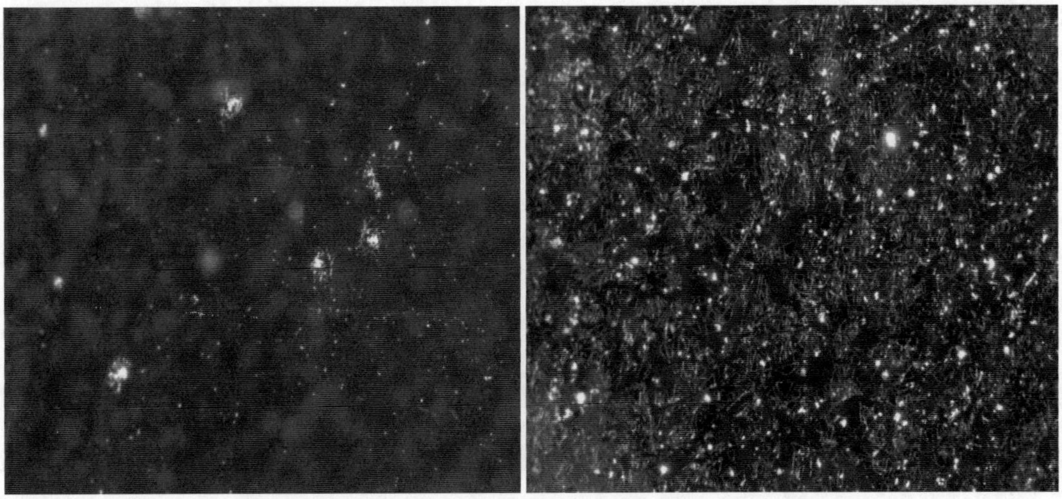

**(a)** **(b)**

**Figure 4** : Optical micrographs showing surface morphology of titania thin film.
(a) C= 0.3M,(b) C= 1.5M.

Figure 5 shows SEM images of $TiO_2$films at low and high concentration of the precursor solution. As described earlier, all the films were confirmed to have anatase phase after calcination (Figure 3 for the XRD patterns). The SEM image demonstrates that grain sizes of $TiO_2$ particles gradually increased with increasing concentration. The difference in crystal growth was also confirmed in the average sizes of $TiO_2$ particle calculated from the full width of half maximum intensity of the (101) diffraction peaks by thc Schcrrcr formula (Table 1).

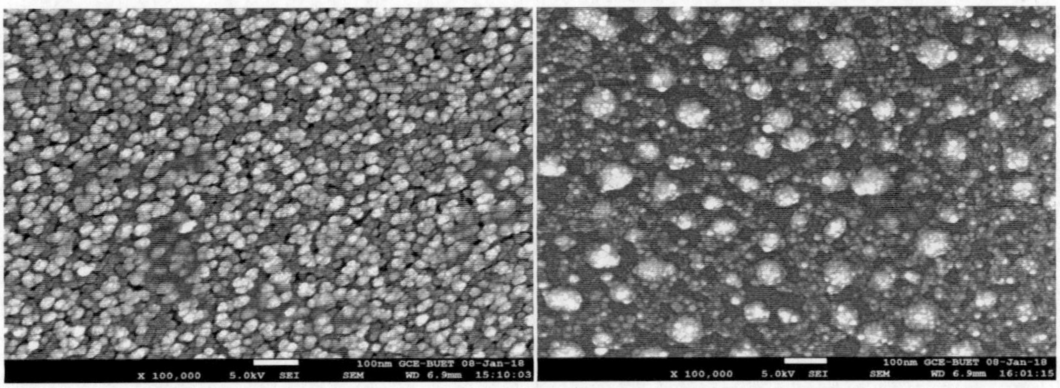


**(a)** **(b)**

**Figure 5**:SEM micrograph of titania thin film surface. (a)0.3M,(b) 1.25 M (at X=100k).

It is found that scarcity of sufficient amount of acid and water larger titania nano particles forms in precursor solution from alkoxide [14].This degrades the surface property like smoothness, adhesion with substrate and causes cracking on the surface.Sufficient amount of acetic acid hinders crystal growth and phase transformation from anatase to rutile during calcinations[4]. Lower volume of acid in causes crystal growth during calcinations and degrades surface morphology.

*3.3 Optical properties of Titania films*

Figure 6 shows the absorbance spectra of $TiO_2$. The thin films from all the solutions show absorbance in UV range and it is from 290 nm to 350 nm. But they show different peak value for absorbance at different wavelength. With increasing concentration, the wavelength for the highest peak value shifts towards left side of the wavelength. That means with increasing concentration the wavelength for the highest absorbance decreases.

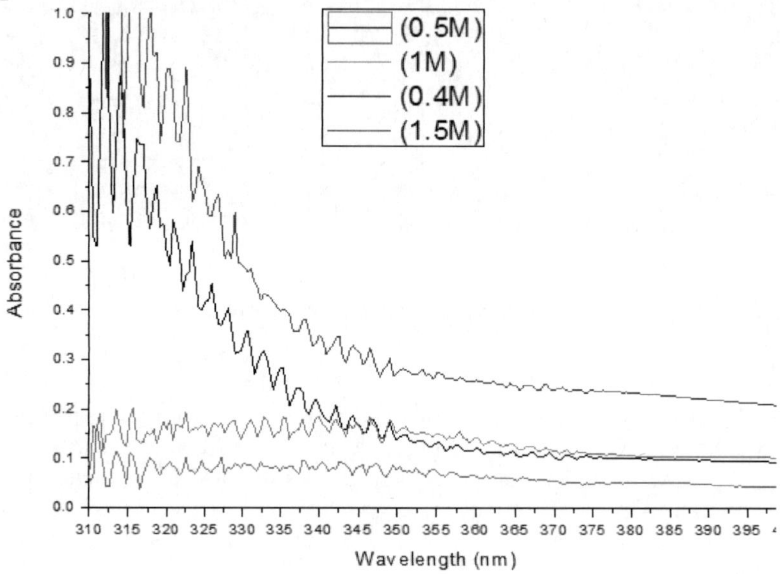


**Figure 6:** Absorbance spectra of titania thin film produced from different concentration precursor solution.(Wavelength range: from 310nm to 400nm)

It is evident from this research that precursor's concentration affects the optical properties of the thin film to a great extent. It is because it also affects the crystallization behavior, the size, stability, and morphology of the produced sol from titanium alkoxides [3].These are also strongly affected by the water titanium molar ratio ($r=[H_2O]/[Ti]$) [10].The formation of colloidal $TiO_2$ at high r ratio is of great interest, because the small size of particles of titania is formed under this condition. The smaller the nanoparticles of titania the more stable the precursor's colloidal solution of titania [7].And this type of solution produce film of smooth, transparent surface with higher surface area due to very small titania particle size. Film with higher surface area absorbs higher amount of energy.

With increasing concentration amount of water reduces that reduces the value of r. This enlarges the particle size of titania that causes film of lower quality. So that the value of peak absorbance reduces with increasing concentration.On the other hand, with increasing concentration, TITP to acetic acid ratio decreases, Table 2.

Now,Although the TIPT molecules had undergone hydrolysis and condensed in the initial period into amorphous titania nano colloids with Ti–O–Ti networks, the surface Ti species were effectively protected from further hydrolysis and condensation via the chelation of acac molecules to Ti cations[16]. This stabilization mechanism in the aqueous system seems reasonable because it is known that the chelated acac molecules can effectively suppress the hydrolysis and further condensation of various alkoxides, such as TIPT, in organic solvents [21].As for the roles of AcOH, it is probably that AcOH molecules also inhibit the condensation of titaniacolloids by affecting the surface Ti cations and/or hydroxyl groups[18–21].In addition, it has been reported that the added AcOH molecules can work as peptizer of $TiO_2$-sols in organic solvents[11].Although the sole addition of acac or AcOH is effective to suppress the hydrolysis and condensation of TIPT in organic media, neither can suppress the rapid hydrolysis of TIPT in aqueous media. The co-existence of acac and AcOH in an appropriate amount specially can stabilize the amorphous titania colloid even in aqueous media through a type of concerted mechanism.Decreasing the amount of acid with increasing concentrations couldn't supply the sufficient amount of acid for the peptizing and chelate action. That causes larger titania particle and this degrades the film quality. It has been reported that the addition of acac and/or AcOH into the organic solvents containing the titania precursor effectively suppressed the phase transition and crystal growth during the calcinations which causes higher surface area and higher absorbance[13].

**Table 2**: Molar Ratio of TTIP to Acid with varying Concentration

| Concentration of solution | TITP to acetic acid volume ratio |
|---|---|
| 0.3 M | 1:0.4 |
| 0.5 M | 1:0.37 |
| 1 M | 1:0.27 |
| 1.25 M | 1:0.21 |
| 1.5 M | 1:0.33 |

## 4. Conclusion

We developed a very simple method for preparing stable titania solution by aqueous route employing three common and cheap chemicals, TIPT, acac and AcOH, as raw materials and by mixing them with appropriate ratio and sequence to prepare solutions wth different concentration. The present titania sols possess desirable properties as a precursor for fabricating $TiO_2$ thin films, such as quite high stability (for more than one year), no use of organic solvents, mild acidity (pH 4), and can indeed be easily transformed into homogeneous and transparent $TiO_2$ thin films on various substrates post calcination. Surface quality and adhesion degraded with increasing concentration of precursor solution. The value of absorbance peak reduced and shifted towards lower wavelength (UV portion) with increasing concentration. Solution with 0.3 M (pH 3.2) yielded the most stable solution and the best thin film.Effect of pH of solution on the film quality was also studied. With increasing the pH of solution surface quality and adhesion of the films degraded but the effect wasn't as noticeable as the effect of concentration. Thin film produced from lowest pH (3.2) yielded least bandgap width. The $TiO_2$ thin films prepared from the lower concentration titania sols showed considerably higher photocatalytic activity under UV light irradiation, certainly due to the considerably smaller size of $TiO_2$ particle and the higher content of highly active anatase phase. These superior features of the newly developed aqueous titania sols will widen the application of $TiO_2$ thin films in the practical use on large scales.

**Acknowledgements**
The authors would like to express their deepest appreciation to all the staffs of Materials and Metallurgical department and Glass and Ceramic Engineering departments for their assistance in completing this work. Special gratitude for Dr. Muhammad Shariar Bashar, Senior Scientific Officer, BCSIR for his guidance, co-operation and giving us permission to use the lab facilities for characterization.

## References

[1] Anpo, L., Yamashita, H., et al., 1995. Photocatalytic Reduction of $CO_2$ with $H_2O$ on Various Titanium Oxide. Electroanalyt. Chem. 396, 21-26.

[2] Asahi, R., Morikava, T., Ohwaki, T. et al., 2001. Visible-Light Photocatalysis in Nitrogen-Doped Dioxide. Science. 293, 269-271.

[3] Bu, S.J., Jin, Z.G., Liu, X.X. et al., 2005. Synthesis of $TiO_2$ Porous Thin Films byPolyetheleneGlycol Templating and Chemistry of the ProcessJ. Eur. Ceramic Soc. 25, 673-679.
[4] Ƙaploviƙová, M., Bellie, P., Ƙaploviƙ, L. et al., 2012. On the True Morphology of High Photoactive Anatase $TiO_2$ Nanocrystals. Appl. Cat. B.117-118, 224-235.
[5] S. Valencia, J.M. Martin, G. Restepo, Study of the bandgap of synthesized Titanium Dioxidenanoparticles using the sol-gel method and a hydrothermal treatment, The Open Mater. Sci.Journ., 4, 2010, 9-14.
[6] M.B. Casu, W. Braun, K.R. Bauchspieb, S. Kera, B. Megner, C. Heske, R. Thull, E. Umbach,A multitechnique investigation of $TiO_2$ flms prepared by magnetron sputtering,Surface Sci., 602, 2008,1599-1606.
[7] M. Sreemany, A. Bose, S. Sen, A study on structural, optical, electrical and microstructural properties of thin TiOx flms upon thermal oxidation: Effect of substrate temperature and oxidation temperature, Physica B, 405, 2010,85-93.
[8] N.M.Thuy, L.T.H.Hai, T.M.Duc and N.T.H.Thanh, "A visible light activity of $TiO_2$ basedphotocatalysts", *The 5th International Workshop on Advanced Materials Science and Nanotechnology (IWAMSN2010) - Hanoi, Vietnam* – Nov. 09-12, 2010.
[9] Chekhlova, T.K., Jivtsov, S.V., Grabowski, E.I., 2006. Temperature Dependence of the Sol-Gel Waveguides. J. of Com. Tech. and El. 51, 834-843.
[10] Fujishima, K. Hashimoto, T. WATANABE,$TiO_2$ Photocatalysis, Fundamentals andApplications, BKC, Tokyo 1999.
[11] Miao, L., Tanemura, S., et al., 2004. Microstructure and Bacterecidal Ability of Photocatalytic $TiO_2$ Thin Films. Appl.Surf. Sci. 238, 125-131.
[12] Mor, G.K., Carvalho, M.A., Varghese, O.K. et al., 2004. A Room-Temperature TiO2 –Nanotube Hydrogen Sensor Able to Self-Clean Photoactively from Environmental Contamination. J. Mater. Res. 19, 628-634.
[13] Myata, H., Fukushima, Y., Okamoto, K. et al., 2011. Remarkable Birefringence in a TiO2–Composite Film with an Aligned Mesoporous Structure. J. Am. Chem. Soc. 133,13539-13544.
[14] Negishi, N., Takeuchi, K., 2005. Preparation of Photocatalytic $TiO_2$ Transparent Thin Filmby Thermal Decomposition of Ti-Alkoxide with Ị- Terpinol as a Solvent. Thin Solid Films 479, 249-253.
[15] Guo, B., Liu, Zh., Hong, L. et al., 2005. Sol Gel Derived Photocatalytic Porous $TiO_2$ Thin Films. Surf. Coat. Techn. 198, 24-29.
[16] A. Kleimana, A. Marqueza, M.L. Veraa, J.M. Meichtrya, M.I. Littera, Photocatalytic activity of $TiO_2$flms deposited by cathodic arc, Appl. Catalysis B: Environm., 101, 2011, 676-681.
[17] W.Choi, A.Termin and M.R. Hoffmann, "The Role of Metal Ion Dopants in Quantum-SizedTiO2: Correlation between Photoreactivity and Charge Carrier Recombination Dynamics", *J.Phys. Chem.* **98** (1994)13669
[18] Sanni Ma,JianqiangMeng, JinghuaLi,YufengZhang,LeiNi, Synthesis of catalytic membranes enabling visible-light-driven photocatalytic degradation of dyes in water, Journal ofMembraneScience453(2014)221–229.
[19] Chen, X., Mao, S.S., 2007. Titanium Dioxide Nanomaterials: Synthesis, Properties, Modifications, and Applications. Chem Rev. 107, 2891-2959.
[20] I. H. Andersen, K. P. Almtoft, L. P. Nielsen, M. B. Sorensen, S. Johannsen, J. Bottiger, „Photocatalytic active and antibacterial TiO2 coatings deposited by industrial scale magnetron sputtering", paper C2, Photocatalytic and Superhydrophilic Surfaces Workshop, Manchester Metropolitan University.
[21] Van Popta, A.S., Cheng, J., Sit, J.C. et al., 2007. Thermal Annealing of Birefringent TiO2 Thin Films Formed by Oblique-Angle Deposition. Proc.SPIE 6647, Nanocoatings, 66470C.

IOP Conf. Series: Materials Science and Engineering **438** (2018) 012033 doi:10.1088/1757-899X/438/1/012033

# Al- & Al-Ca-doped ZnO Thin Films by Sol-Gel Method for Solar Cell Application

**S. Bhowmik, M.I. Raiyan and M. Moniruzzaman**

Department of Materials and Metallurgical Engineering, Bangladesh University of Engineering and Technology, Dhaka 1000, Bangladesh

**Abstract.** Zinc oxide (ZnO) thin films, due to their excellent photovoltaic properties, are gaining much attention to be used as n-type semiconductor material for modern day solar cell. The change in photoelectric properties of zinc oxide with doping is an area of much consideration. In this work, 1.17%Al, 1.17%Al-2%Ca and 1.17%Al-4%Ca doped ZnO thin films were produced on glass substrate by Sol-Gel spin coating technique. Zinc acetate dihydrate, triethanolamine and isopropanol were used for making precursor solution. For Al doped ZnO, aluminum nitrate nona hydrate was added to the precursor solution as source of Al. Aluminum nitrate nona hydrate and calcium chloride dihydrate were utilized as source of Al and Ca respectively for Al-Ca doped ZnO. The films were deposited from precursor solutions at 1300 rpm for 20 seconds. After deposition, annealing was done at $520^0$C for one hour. Properties of deposited films were characterized by X-ray Diffractometer (XRD), Scanning Electron Microscopy (SEM) and UV-Vis spectroscopy. From XRD spectra the theoretical grain size was measured which increased from 0.31 nm to 0.33 nm because of doping. From SEM micrograph the uniformity and homogeneity of thin film was observed. The band gap of deposited thin films was measured from absorbance spectra. The band gap decreased from 3.27 eV to 3.17 eV because of doping 4% Ca in addition to 1.17% Al. From transmittance spectra, it was found that all films were transparent with average percentage of transmittance of more than 86%.

## 1. Introduction

A solar cell or photovoltaic cell (previously termed "solar battery") is an electrical device that converts the energy of light directly into electricity by the photovoltaic effect, which is a physical and chemical phenomenon [1]. Photovoltaic (PV) systems provide the possibility to generate electricity with environmental advantages as zero carbon dioxide emission during operation as well as scale flexibility, simple operation, and low-cost maintenance [2]. Despite its numerous advantages, two principal problems are the intermittent nature of environmental conditions and the low energy yield in PV systems that increase area-relate costs [3].

Different strategies are proposed to address this low-efficiency problem. Of them thin film solar cells are the most promising ones because of their low materials cost and energy cost during manufacturing. The present PV industry is dominated by crystalline and multi-crystalline Si wafer based solar cells. Because of high production cost, threat to environment and low efficiency, its development is getting limited attention. In contrast, thin film solar cells, only need very small amount of semiconductor materials. Thin film solar cells can be deposited onto various kinds of substrates such as glass, stainless steel, polymer, which account for a large fraction of the materials cost and energy cost. The development of new materials, such as polymer, may help reducing the thin film solar cells cost. For example: CZTS (Copper Zinc Tin Sulpher), Zinc Oxide (ZnO) and it's doping, Perovskite materials etc.

In the past ten years we have witnessed a revival of, and subsequent rapid expansion in the research on zinc oxide (ZnO) as a semiconductor. The prospects of gaining control over its electrical conductivity have led a large number of groups to turn their research for electronic and photonic devices. While a number of research groups have reported achieving p-type ZnO, there are still problems concerning the reproducibility of the results and the stability of the p-type conductivity [4]. Even the cause of the commonly observed unintentional n-type conductivity in as-grown ZnO is still under debate [4]. Band-gap engineering of ZnO can be achieved by alloying like with MgO or CdO. Adding Mg to ZnO increases the band gap whereas Cd decreases the band gap, similar to the effects of Al and In in GaN [4].

Most of the current ZnO application have made use of polycrystalline films that are grown by a variety of deposition techniques, mostly on glass substrates. For optimized growth and processing, environments (partial pressures and temperature) are necessary. Current techniques that allow for this level of control include Pulsed Laser deposition (PLD), Chemical Vapor Deposition (CVD), Metal-Organic CVD (MOCVD) and Molecular-beam Epitaxy (MBE), and to a lesser extent sputtering [5]. But for making low cost thin film solar cells, the sol–gel method has emerged as one of the most promising processing route as it is particularly efficient in producing thin, transparent, homogeneous, multi component oxide films of many compositions on various substrates and it allows the tuning of the refractive index and thickness of the film by varying synthesis parameters [6]. The objective of this work is to find out the change in band gap of ZnO because of doping. The process, which is free from the requirement to use expensive equipment, minimum wastage generation, environment friendly and low temperature deposition method, is used.

## 2. Experimental

### *2.1 Making Precursor Solution*

Three basic methods were adopted to make precursor solutions. In method-1, $Zn(Ac)_2.2H_2O$ in Isopropanol, for Al doping Aluminum nitrate nona hydrate, was added. For increasing solubility of Isopropanol, TEA (Triethanolamine) was added at molar ratio of 1.5:1 with Zinc acetate dihydrate. After that the solution is stirred with magnetic stirrer at $60^0C$ by using hot plate for 2 hours. Then the solution was not aged before thin film making process. For 1.17% Al-2% Ca-doped ZnO & for 1.17% Al-4% Ca-doped ZnO this procedure was not continued because Isopropanol has boiling point $82.6^0C$ so isopropanol was boiling out from reaction place which was hindering the process of making thin films by increasing solutions viscosity because of lack of solvent.

In method-2, $Zn(Ac)_2.2H_2O$ was mixed with 20% (v/v) Isopropanol to make 0.5M solution. In that, for Al doping Aluminum nitrate nona hydrate was added. For Ca doping Calcium chloride dihydrate was used. They were added as they were calculated. After mixing for some time TEA was added at molar ratio of 1.5:1 with Zinc acetate dihydrate. After that the solution is stirred with magnetic stirrer at $60^0C$ by using hot plate for 2 hours. After that the Solution was aged for 0-24 hours before thin film making process.

In method-3 (Reflux method),Isopropanol was boiling out from reaction place which was hindering the process of making thin films by increasing solutions viscosity because of lack of solvent. So reflux method was applied to get the boiled Isopropanol back to the reaction place to complete the thin film making process. $Zn(Ac)_2.2H_2O$ was mixed with 20% (v/v) Isopropanol to make 0.5M solution. After that for Al doping Aluminum nitrate nona hydrate was added. After adding TEA at molar ratio of 1.5:1 with Zinc acetate dihydrate, the mixture was taken into round bottom flask. It was stirred with magnetic stirrer at $60^0C$ by using hot plate for 2 hours. Then the Solution was aged for 3 days before thin film making process.

### *2.2 Spin Coating Process*

Precursor Solution was used to make thin films. The coating was done by dynamic spin coating process. Each solution was dispersed onto the substrate surface, which was spinning at a 1300 rpm. The process continued for 20 seconds. After that the film was dried at 200$^0$C for 3-5 minutes. This whole process (from spin coating to drying) was continued for 2-6 times for making multiple layered thin film.

*2.3 Annealing Process*
After completing the whole thin film making process each films were annealed at 520$^0$C. The heating rate was 10$^0$C/min and the holding time was 1 hour. Annealing was done to evaporate the impurity and lowering the defects from films.

## 3. Results and discussion

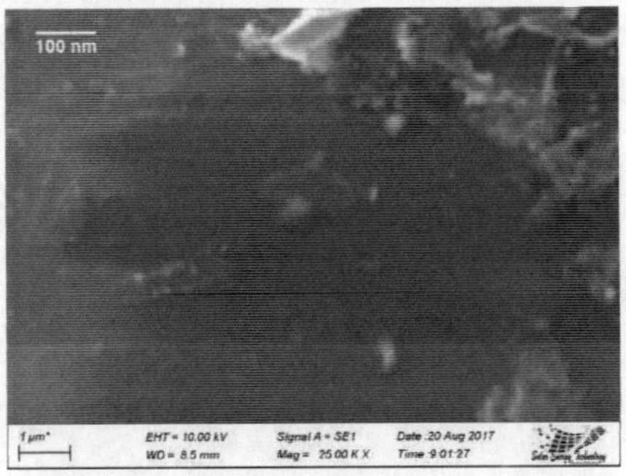


**Figure 1:** AZO from method 1 (Coated 1 time)

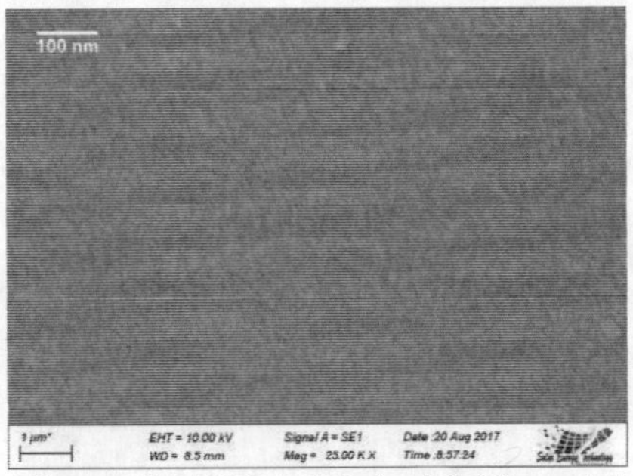


**Figure 2:** AZO in Method 3 (Coated 1 time)

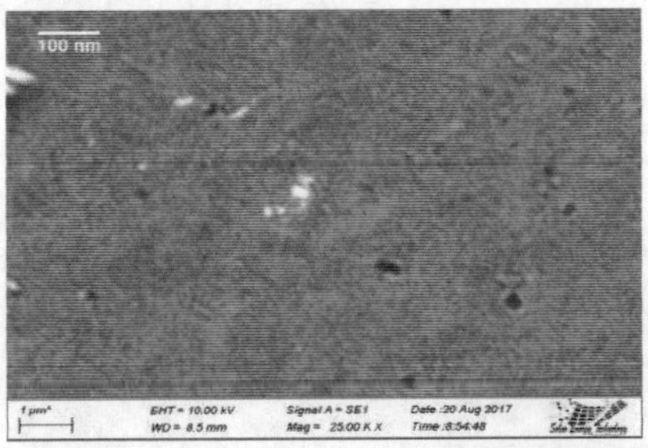


**Figure 3:** AZO in Method 3 (Coated 4 times)

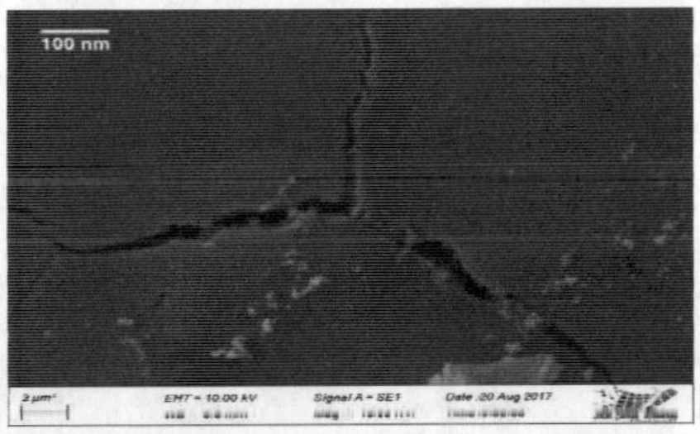


**Figure 4:** 1.17% Al- 2%Ca doped ZnO (Coated 1 time)

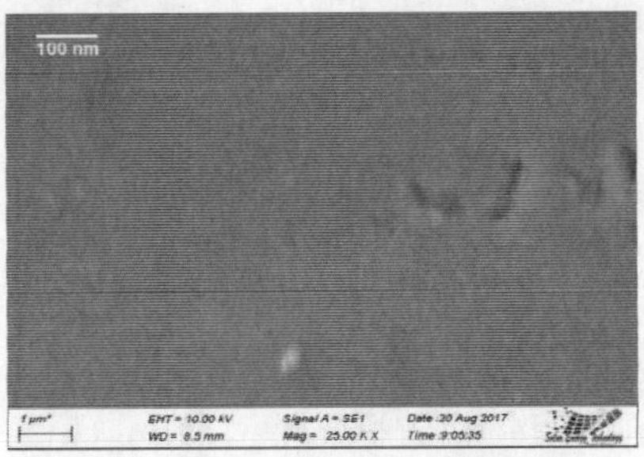


**Figure 5:** 1.17% Al- 4%Ca doped ZnO (Coated 1 time)

*3.1 Scanning Electron Microscopy (SEM)*
Figure 1, shows non uniform nature of the thin film. This might cause from change in viscosity because of still evaporating isopropanol. This hinders the spin coating process. The micrograph could not be taken properly because of reflection from glass slides. In figure 2, homogeneous and uniform film is found. The

film is also defect free for one time coating. In figure 3, uniform films were found but there were some nano level defects found on the films much likely pores. In figure 4, shows cracks due to solidification shrinkage in film and nano level precipitation. In figure 5, shows some non-uniformity in film which can cause from any fault in spin coating.

### *3.2 X-Ray Diffractometry (XRD)*

*3.2.1 Phase identification.* For Figure 6, the peak is at the position $2\theta = 36.2^0$ which corresponds to the crystal plane (002). The standard position should be $36.06^0$. This slight deviation from standard position could be because of residual stress. Plane (002) indicated that prevailing direction of orientation of the AZO film growth, a preferential orientation of c-axis is perpendicular to the substrate [7].No XRD peaks due to Al or $Al_2O_3$ phase could be observed, since Al had been doped [8]. The d-spacing is 2.474 $A^0$.

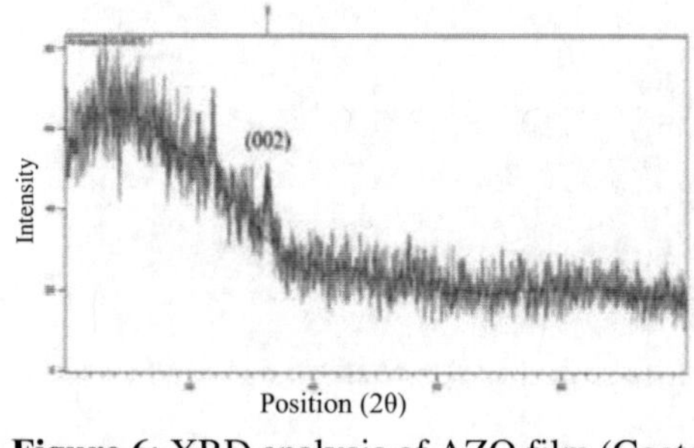

**Figure 6**: XRD analysis of AZO film (Coated 4 time) from method 1

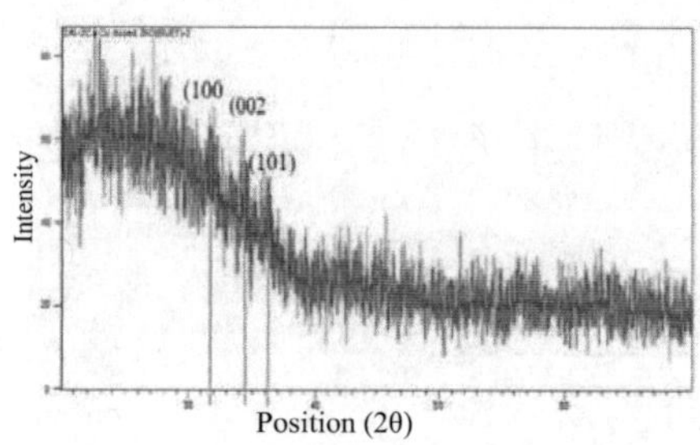

**Figure 7:** XRD Analysis of 1.17% Al - 2% Ca doped ZnO film

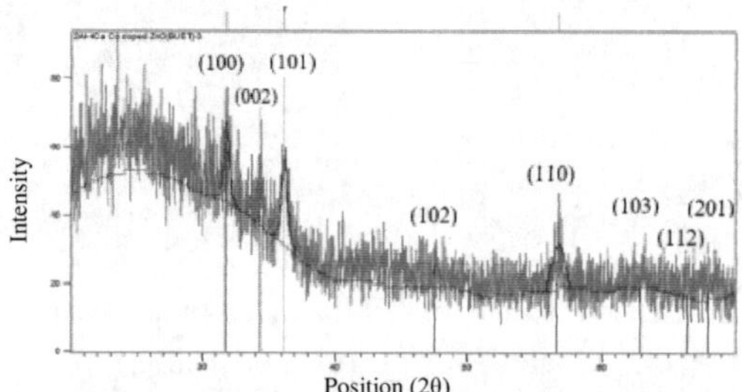

**Figure 8:** XRD Analysis of 1.17% Al - 4% Ca doped ZnO film

For Figure 7, film the XRD spectra did not show any specific peak. The reason could be having precipitation and not having continuous film from Figure 4.8. After observing closely weak peaks were found at the position $2\theta = 31.5^0$, $34.3^0$ and $36.1^0$ which corresponds to planes (100), (002) and (101) respectively. These are the planes of ZnO. So, doping of $Ca^{2+}$ shows no additional peaks, which confirms no additional impurity phase formation [9].

For Figure 8, the peaks are found at $2\theta = 31.8^0$, $34.3^0$, $36.2^0$, $47.5^0$, $56.7^0$, $62.9^0$, $67.3^0$ and $68.9^0$ which corresponds to the planes ZnO: (100), (002), (101), (102), (110), (103), (112) and (201) respectively. Of them three peaks are pronounced. They are originated from (100), (101) and (110) planes. Other peaks are weaker. No secondary additional phase was detected which proves the importance of the used process [10]. The d-spacing is for strong peaks which are 2.81 $A^0$, 2.47 $A^0$, and 1.62 $A^0$ for (100), (101) and (110) planes respectively.

3.2.2 *Grain size measurement.* If crystallites were in the form of elongated hexagonal cylinders, the average grain size can be calculated using the Debye-Sherrer's equation [10]

IOP Conf. Series: Materials Science and Engineering **438** (2018) 012033 doi:10.1088/1757-899X/438/1/012033

$$G = \frac{0.9\,\lambda}{B\cos\theta b} \quad (1)$$

Where λ is the X-ray wavelength (1.54060 $A^0$), $\theta_b$ is the maximum of the Bragg diffraction peak (in radians) and B is the line width at half maximum. From Figure 9, it is evident that, increased amount of doping has increased the grain size that consists of the film.

**Table 1:** Grain sizes according to planes

| Composition | Strong Peaks at planes | Grain Size, nm | Average Grain Size |
|---|---|---|---|
| AZO (4 times Coating) | (002) | 0.25 | 0.25 |
| 1.17% Al - 4% Ca doped ZnO | (100) | 0.305 | 0.316 |
| | (101) | 0.308 | |
| | (110) | 0.334 | |

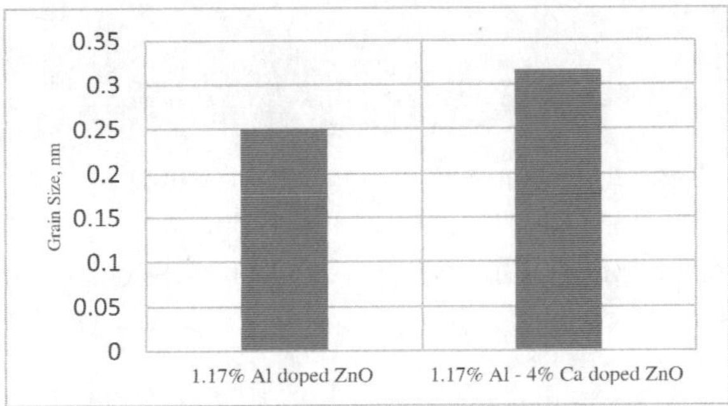


**Figure 9**: Comparison between average grain sizes

### *3.3 Ultraviolet-Visible Spectrometer*

*3.3.1 Absorbance.* From absorbance spectra, pattern of spectra are same for all compositions. AZO film from method-1 shows the edge of absorbance is between 312-385 nm. The maximum absorbance is 0.064. For AZO films from method-3 (Reflux method) edge of absorbance is between 312-394nm. The maximum absorbance is 0.303. For 1.17% Al - 2% Ca doped ZnO film edge of absorbance is between 314-395nm.

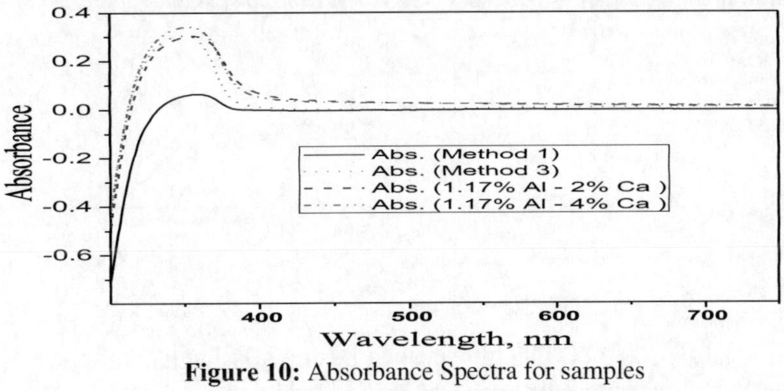


**Figure 10:** Absorbance Spectra for samples

But the increasing slope is between 314-360nm. The maximum absorbance is 0.3. For 1.17% Al - 4% Ca doped ZnO film, edge of absorbance is between 312-393nmThe maximum absorbance is 0.34. The

IOP Conf. Series: Materials Science and Engineering **438** (2018) 012033 doi:10.1088/1757-899X/438/1/012033

absorbance spectra is same for all films as all of them are a variation from ZnO. The absorbance edge did not change much for doping.

*3.3.2 Transmittance.* The optical transmittance is an important optical parameter for transparent oxides. The average optical transmittance for thin films for AZO film of method 1, AZO film of Method 2(R), 1.17% Al - 2% Ca doped ZnO film and 1.17% Al - 4% Ca doped ZnO film are 97.3%, 94.04%, 88.94% and 86.19% respectively. Transmission is decreased with increased doping. All films have an average optical transparency over 86% in the visible range. This implies that the AZO thin films transparent are largely preserved.

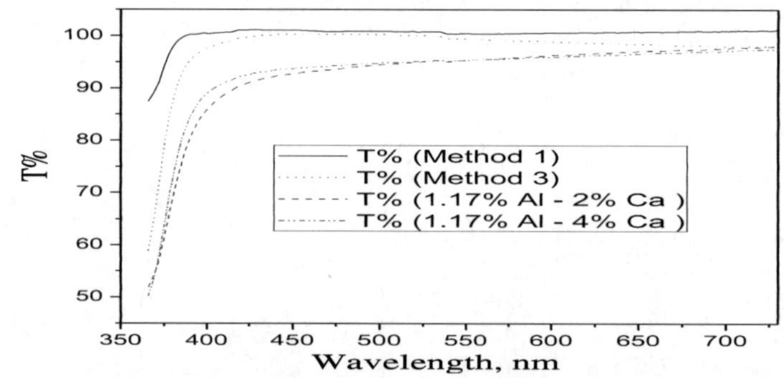


**Figure 11:** Transmittance spectra for samples

3.3.3 *Optical band gap*. From the optical absorption spectra, the band-gap $E_g$ of four samples could be calculated using the formula:

$$(\alpha h\vartheta) = A\,(h\vartheta - E_g)^{1/2} \qquad (2)$$

Where A is a constant, α is absorption coefficient and hϑ is the photon energy. The absorption coefficient can be calculated from this formula:

$$\alpha = 2.303\frac{A}{d} \qquad (3)$$

Where A is Absorbance and d is thickness of coating. Band gaps measured from graph $(\alpha h\vartheta)^2$ vs Binding energy,(hϑ). From Figure 13, it is observed that 4% Ca doping in addition to 1.17% Al decreases the band gap. But AZO film from method 1 shows unusual behavior having lowest band gap.

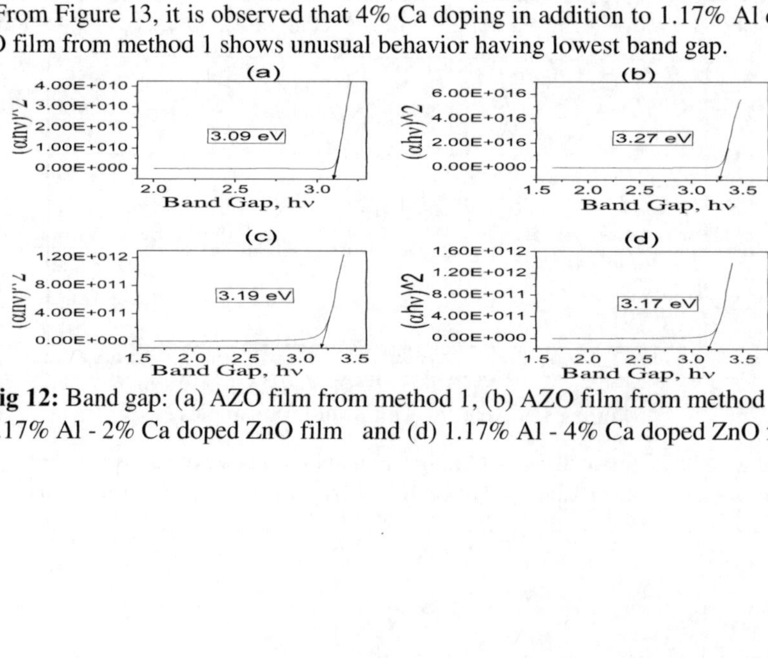


**Fig 12:** Band gap: (a) AZO film from method 1, (b) AZO film from method 3, (c) 1.17% Al - 2% Ca doped ZnO film and (d) 1.17% Al - 4% Ca doped ZnO film

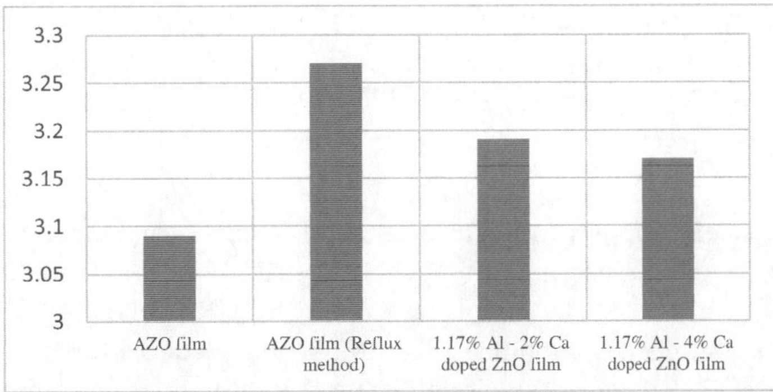


**Figure 13:** Change in band gap for different types of doping in ZnO

## 4. Conclusion

In this work 1.17% Al-doped ZnO (AZO), 1.17% Al-2% Ca-doped ZnO and 1.17% Al-4% Ca-doped ZnO thin films were fabricated by using low cost Sol-Gel spin coating technique. The effect of Al and Al-Ca doping (with different concentration of Ca) in ZnO is observed. The visual inspection proved that Al-doped ZnO have higher quality of films than Ca doped ZnO. The reflux method also provided better films than normal method for producing AZO. XRD analysis showed the bigger grains for with increasing doping. The SEM result showed defects in multiple layer thin films. The defects were not present in the case of single layered thin film. The band gap also decreased with increasing doping. The increased doping element was Ca. The films have optical transmittance more than 86%. So these conditions of fabrication can be used for making transparent thin films on glass substrate for solar cell application.

## 5. Reference

[1] URL http://www1.eere.energy.gov/solar/pdfs/solar_program_mypp_2008-2012.pdf
[2] URL http://www.ceramics.mmat.ubc.ca/introduction.html
[3] Shockley W and Queisser H J 1961 Detailed Balance Limit of Efficiency of p-n Junction Solar Cells *Journal of Applied Physics* 32(3) p 510
[4] Jagadish C and Pearton S J 2006 Zinc Oxide Bulk, Thin Films and Nanostructures (New York: Elsevier)
[5] Znaidia L, Touamb T, Vrela D, Soudeda N, Ben Yahiaa S, Brinzaa O, Fischerb A and Boudriouab A 2012 ZnO Thin Films Synthesized by Sol–Gel Process for Photonic Applications *International Congress on Advances in Applied Physics and Materials Science*
[6] Ratana T, Amornpitoksuk P and Suwanboon S 2009 The wide band gap of highly oriented nanocrystalline Al doped ZnO thin films from sol–gel dip coating Journal of Alloys and Compounds 470 p 408–412
[7] Kim D K, Kim H B 2009 Room temperature deposition of Al-doped ZnO thin films on glass by RF magnetron sputtering under different Ar gas pressure *Journal of Alloys and Compounds* 509 p. 421–425
[8] Karthikeyan B, Pandiyarajan T and Mangaiyarkarasi K 2011 Optical properties of sol–gel synthesized calcium doped ZnO nanostructures *Spectrochimica Acta Part A: Molecular and Biomolecular Spectroscopy* 82 p 97– 101
[9] Mir L E 2017 Luminescence properties of calcium doped zinc oxide nanoparticles *Journal of Luminescence* 186 p 98-102

[10] T. Fukumura, Jin Z, Ohtomo A, Koinuma H, and Kawasak M 1999 An oxide-diluted magnetic semiconductor: Mn-doped ZnO*Appl. Phys. Lett.* 75 p. 3366
[11] Masson G 2014 A snapshot of global PV (1992-2014) M. Brunisholz (Ed.) IEA-PVS Reporting Countries, EPIA, Becquerel Institute (BE), RTS Corporation (JP).

**Acknowledgments**
We acknowledge financial support from Department of materials and metallurgical engineering, Bangladesh University of Engineering and Technology (BUET). We thank Sajib Aninda Dhar, Engineer, Pilot Plant and Process Development Centre, BCSIR for helping us to use their SEM facilities. We express our deepest gratitude to Dr. Mahbubul Haque, Principal Scientific Officer, Atomic energy Centre for fiving us permission to the XRD facilities and support without which our research work could not be finished in time.

IOP Conf. Series: Materials Science and Engineering **438** (2018) 012034 doi:10.1088/1757-899X/438/1/012034

# Fire resistance of locally produced 500W structural steel bars

**M.R.N. Shaown, M.S. Shahriar, M.A. Islam**

Department of Materials and Metallurgical Engineering, Bangladesh University of Engineering and Technology, Dhaka 1000, Bangladesh

**Abstract.** Steel has been used for many decades in structural applications because of its good combination of mechanical properties. In recent years in Bangladesh the demand for TMT 500W steel bars has increased profoundly, for its higher strength and good ductility. But fire remains one of the serious potential risks to most buildings and structures. To understand the effect of fire on mechanical properties of steel bars at elevated temperature a locally produced 500W TMT bar of 8 mm diameter has been investigated. Both macroscopy and microscopy were carried out to reveal different zones such as hardened case, soft core and transition zone in the cross-section and microstructure respectively of the steel bars in the as received condition. Finally, high temperature tensile tests were conducted at 500$^0$C, 600$^0$C, 650$^0$C in steady state method at a strain rate of 0.5mm/min and the results at these 3 temperatures were noted. Data generated from the room temperature and high temperature tensile tests were compared. The test results demonstrated that the high temperature severely degrades the tensile strength of the 500W TMT steel bars and they can hardly bear any load at temperature above 600$^0$C.

**Keywords:** TMT steel bar, Fire resistance, High temperature mechanical properties, Coefficient of reduction

## *1. Introduction*

Concrete is a weak material in tension and cracks when the principle tensile stress exceeds its tensile strength. Consequently, concrete is reinforced to enable it to resist these stresses. Steel is the most popular reinforcement in concrete for its excellent mechanical properties. High strength structural steel (HSSS) usually refers to steels with nominal yield strength not less than 460 MPa. Compared with conventional mild structural steel, the application of HSSS could reduce the total steel consumption and emission of carbon dioxide in building industry.

Steel has excellent strength properties at ambient temperature, however, like other materials; steel loses its strength and stiffness with temperature. The yield strength of steel is reduced to about half at 550°C. At 1000 °C, the yield strength is 10 percent or less [1]. Near-total depletion of strength occurs at approximately 1,204$^0$C [2]. Because of its high thermal conductivity, the temperature of unprotected internal steelwork normally will vary little from that of the fire [1]. Young's Modulus does not decrease with temperature as rapidly as does yield strength [1].

If the duration and the intensity of the fire are large enough, the load bearing resistance can fall to the level of the applied load resulting in the collapse of the structure [3]. Thus, load bearing of steel decreases

when steel or composite structure is subjected to a continuous fire action. However, the failure of the World Trade Centre on 11th September 2001 and, in particular, of building WTC7 alerted the engineering profession to the possibility of connection failure under fire conditions [4]. The structural response of buildings to fire conditions has been the focus of intensive research activity in recent years. For composite steel/concrete buildings, this has been driven by the motivation to achieve more cost-effective designs and, more generally, by the need to attain a greater understanding of the underlying behavioural mechanisms that occur in fire.
Although sufficient data on the elevated temperature properties of concrete exist in the literature, the studies related to elevated temperature properties of steel bars are limited [5]which emphasises the need of more investigative work in this area. This paper presents the results of experimental testing which was conducted on the steel bars available in the local market in Bangladesh to determine their properties at high temperatures. The data of chemical composition, yield and tensile strength and strain and elongation of the tested bars are compared and the differences are discussed.
The presented data are essential to develop realistic models of the capacity of fire exposed buildings to the applied loads. The presented study is unique in its nature in that there is no published contribution to date (to the best of authors' knowledge) which has been carried out to assess the temperature dependent mechanical properties of steel bars available in Bangladesh. The parameters of the study included the local 500W rebar brands and heating temperatures.

## 2. Experimental

### 2.1Samples

3 local brands of 500WTMT of same diameter of 8 mm as raw material. The samples were marked as Sample 1, 2 and 3.The samples of necessary sizes were cut from them for composition, tensile and metallographic and high temperature tests.

### 2.2Composition Test

Three samples of about 150 mm were cut, each being of separate brand. Finally, they were tested on Optical Emission Spectroscopy (OES) and chemical composition was determined.

### 2.3Tensile Test

Three samples of about 260 mm were cut from each brands of rebar. 200 mm was selected as gauge length for each of the sample. Each sample was then weighed on weight taking machine and density was assumed as 7.8 gm /cc. Then from the equation given below, diameter for each of the sample measured. The samples were subjected to tensile test in UH-500kN SHIMADZU, tensile testing machine and after the test the two detached parts of each of the tested sample were placed in contact in the fractured portions to measure the change in gauge length, thereby to measure the % elongation. From the machine data, stress vs strain graph was plotted in excel for each sample.

### 2.4High temperature Tensile Test

For high temperature tensile test 3 samples of about 180mm length was cut from each brand of rebar. Then each rebar sample was made plate at each corner for holding in high temperature machine and 80mm length at the centre was selected as gauge length. In the machine test specimen was heated by the oven up-to the maximum expected temperature. Then the specimen was held at that temperature for a particular time (15 minutes). For my experiment the high temperature tensile tests were carried out at $500^0$C, $600^0$C and $650^0$C temperatures in steady state method with a constant strain rate of 0.5mm/min.

### 3. Results and discussion

*3.1Composition of the as received samples*

The following composition data were obtained in OES spectroscopy for the 3 samples:

**Table 1:** Chemical Compositions of re-bars

| Sample | Element proportion (%) | | | | | | | | | | |
|---|---|---|---|---|---|---|---|---|---|---|---|
| | Fe | C | Si | Mn | P | S | Cr | Mo | V | Nb | Cu |
| 1 | 98.46 | 0.21 | .19 | .71 | .01 | .02 | .08 | 0.011 | 0.002 | 0.001 | .22 |
| 2 | 98.60 | 0.244 | .25 | .81 | .008 | .02 | .01 | 0.002 | 0.001 | 0.002 | .001 |
| 3 | 98.60 | 0.24 | .20 | .81 | .03 | .01 | .01 | 0.002 | 0.001 | 0.001 | .001 |

The chemical analysis of the as received bars of all 3 samples was carried out by spectroscopy. These data of chemical composition of reinforcing bars may become helpful in explaining changes that take place in steel at high temperatures. About 30 elements were identified using OES machine. Of these, C, Mn, P, Cr, Ti and Co are the important elements which have a direct influence on strength, ductility and hardenability of steel. The proportions of the aforementioned elements in the control specimens are given in Table 1.

Carbon is the principal element which provides both strength and ductility to steel. [6-9]. According to ASTM A706 maximum carbon percentage in structural steel is 0.18%. All 3 sample we tested has greater carbon percentage which is the main reason for their greater strength. Manganese (Mn) was found to be in the range of 0.71-0.81 which is well within the range of ASTM A706 standard of 1.60% (maximum). The higher Mn content may partly responsible for better ductility of the samples. Phosphorous contents in steel are limited to a maximum .04 as higher phosphorus affects steels ductility [7]. It is noted in Table 1, that phosphorous contents in the bars comply with the requirements given in the respective standards. Addition of niobium up to 0.025 percent [8][10] enhances strength without any loss of ductility. The addition of 0.3-0.5 percent [9] copper improves strength and corrosion resistance of steel. Again the addition of sulphur is detrimental to the mechanical properties. The amount of Nb, Cu, S is well within the standards and are very small to have any significant effect on the steel properties. Molybdenum (0.04-0.06 per cent) [9-10] can increase elevated temperature strength and corrosion resistance of steel. But molybdenum in the TMT bars is significantly less than the recommended proportion in the above.

*3.2Mechanical Properties*

The mechanical properties of all 3 samples at deferent temperatures and there change with temperature are shown from table 2-6.

**Table 2**: Mechanical properties of Sample 1 different temperatures

| Temperature($^0$C) | Yield Strength (in MPa) | UTS(in MPa) | Young's Modulus (in MPa) | % Elongation |
|---|---|---|---|---|
| 25 | 526.35 | 609.3 | 210.52 | 12.6 |
| 500 | 265 | 303.2 | 51.96 | 16.3 |
| 600 | 102.1 | 109.5 | 40.3 | 22.7 |
| 650 | 70 | 72.3 | 35 | 27.1 |

IOP Conf. Series: Materials Science and Engineering **438** (2018) 012034 doi:10.1088/1757-899X/438/1/012034

**Table 3**: Mechanical properties of Sample 2 different temperatures

| Temperature($^0$C) | Yield Strength (in MPa) | UTS(in MPa) | Young's Modulus (in MPa) | % Elongation |
|---|---|---|---|---|
| 25 | 562.17 | 671.7 | 151.94 | 13 |
| 500 | 202 | 221.2 | 57.77 | 14.1 |
| 600 | 120 | 147.44 | 46.1 | 21 |
| 650 | 85.6 | 88 | 42.8 | 23.6 |

**Table 4**: Mechanical properties of Sample 3 different temperatures

| Temperature($^0$C) | Yield Strength (in MPa) | UTS(in MPa) | Young's Modulus (in MPa) | % Elongation |
|---|---|---|---|---|
| 25 | 590.8 | 703.6 | 252.17 | 11.4 |
| 500 | 175 | 322.3 | 76.1 | 16.4 |
| 600 | 77 | 98 | 48 | 27.4 |
| 650 | 88 | 89 | 44 | 38.1 |

**Table 5**: Average Mechanical properties of all 3 samples at different temperatures

| Temperature($^0$C) | Average Yield Strength (in MPa) | Average UTS(in MPa) | Average Young's Modulus (in MPa) | Average Elongation (in percentage) |
|---|---|---|---|---|
| 25 | 559.77 | 661.33 | 204.88 | 12.33 |
| 500 | 214 | 282.23 | 61.94 | 15.6 |
| 600 | 99.73 | 118.31 | 44.82 | 23.7 |
| 650 | 81.2 | 83.1 | 40.6 | 29.6 |

**Table 6**: Percentage change of mechanical properties of all 3 samples at elevated temperature compared to room temperature

| | Reduction in Percentage | | | Increase in Percentage |
|---|---|---|---|---|
| Temperature($^0$C) | Yield Strength (%) | UTS (%) | Young's Modulus (%) | Elongation (%) |
| 500 | 61.77 | 57.32 | 69.77 | 26.52 |
| 600 | 82.2 | 82.1 | 78.12 | 92.91 |
| 650 | 85.5 | 87.45 | 80.21 | 140.1 |

Figure 1 shows the stress versus strain curve for samples at high temperatures. The curves for as received samples are also shown for comparison purposes. No strain hardening is observed for the samples at high temperatures like for samples at ambient (25$^0$C).

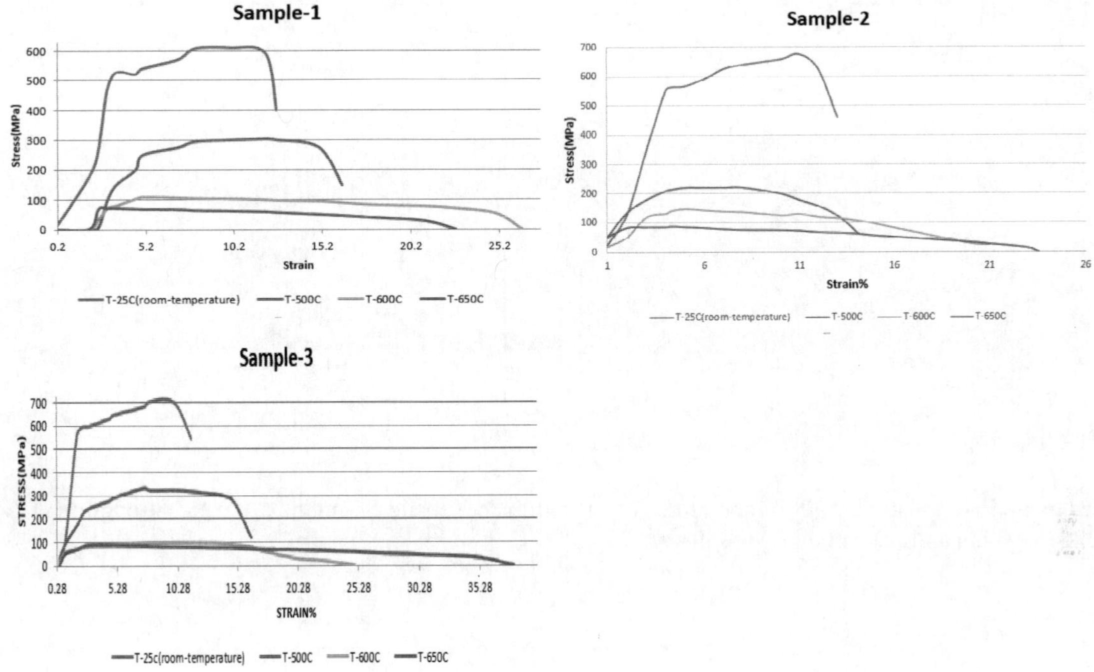


**Figure 1.** Stress vs Strain curves of tested rebars at elevated temperatures; (a) Sample 1, (b) Sample 2 and (c) Sample 3

### *3.3Yield Strength*

Figure 1 shows yield plateau for all three samples tested at room temperature. However, the yield plateau has disappeared at temperatures 500$^0$C, 600$^0$C and 650$^0$C. Moreover, in Figure 2(b), the yield strengths of all the samples decreased with increasing temperature. The reduction of yield strength occurs because of requiring reduced level of energy for movement of dislocation at high temperatures [1]. Figure 3 (b) illustrates the normalised yield strength (ratio of yield strength at high and room temperatures for a certain sample) which represents the coefficient for the reduction of yield strength. This is also called the reduction factor. The high temperature yield strength can be calculated by multiplying coefficient with the yield strength at room temperature. It is seen in Figure 3(b) that "Reduction Factor" is similar for all the three types of bars used in this study. Bar yield strength becomes nearly 15% yield strength at room temperature at 650$^0$C.

IOP Conf. Series: Materials Science and Engineering **438** (2018) 012034 doi:10.1088/1757-899X/438/1/012034

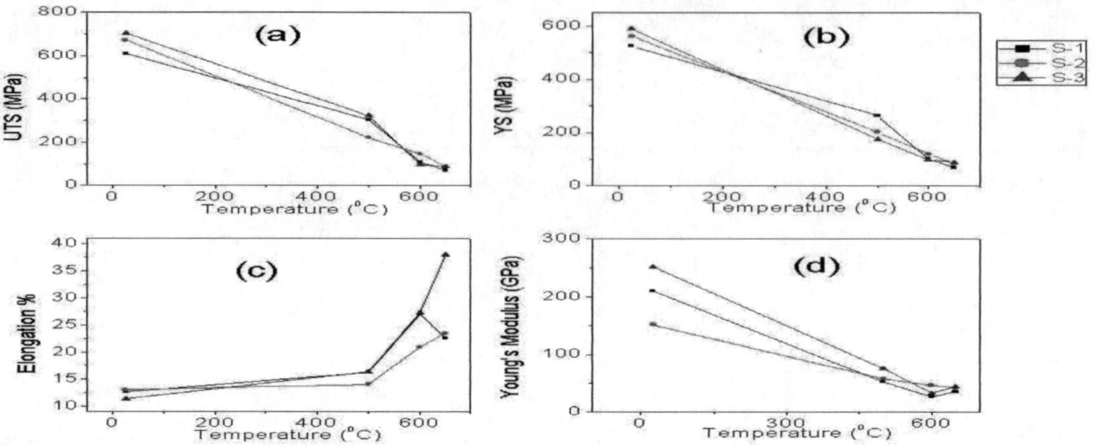


**Figure 2.**Relationships with Temperatures of (a) Ultimate Tensile Strength; (b) Yield Strength; (c) % of Elongation and (d) Young's Modulus

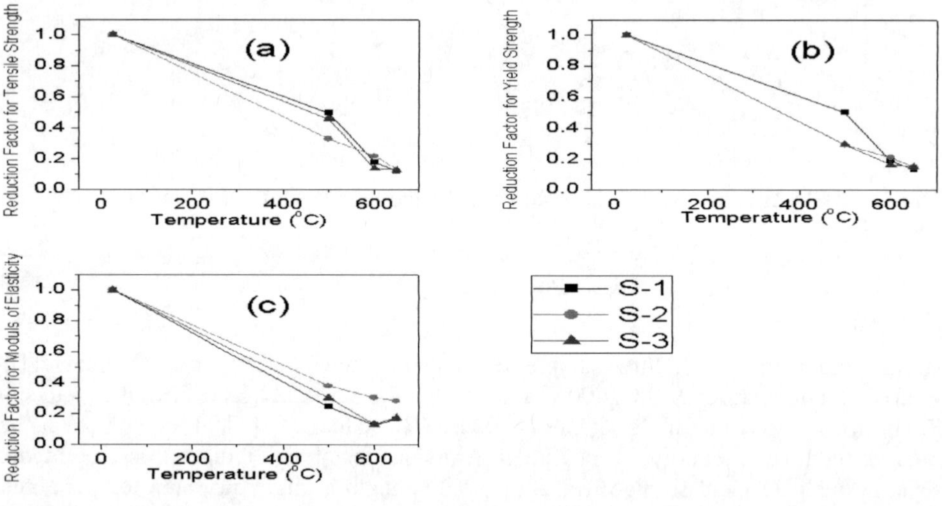


**Figure 3.** Relationships of Reduction Factor with Temperatures (a) Ultimate Tensile Strength; (b) Yield Strength and (c) Young's Modulus

### *3.4Ultimate tensile strength*

The ultimate capacity of the bar corresponds to maximum stress on the rebar stress–strain curve. Distinct differences in the behaviours of the bars in the strain hardening region are seen in Figure 1. Strain hardening in the bars became negligible after 500$^0$C. Figure 3(a) illustrates the normalised ultimate strength ratio of ultimate strength at high and room temperatures for a certain sample, which is the coefficient for the reduction or reduction factor of rebars ultimate strength. It is seen in Figure 3(a) that the curves for reduction factor of UTS follow a pattern similar to reduction factor of YS.

*3.5Modulus of Elasticity*

The elastic modulus of the bar at elevated temperature was calculated as tangent modulus of the initial linear elastic part of the stress–strain curve (Figure 2). Figure 3(a) illustrates the coefficient for the reduction or reduction for of re-bars elastic modulus which has been calculated as the ratio of modulus at high and room temperatures. The reduction of modulus from $500^0$C to $650^0$C is more rapid for sample 2 and 3 than sample 1.

*3.6Rebar ductility*

Ductility is often represented with percentage of elongation. From Figure 2 (c) the relationship between temperature and percentage of elongation can be observed. It has increased with increasing temperature for sample 1, 2 and 3 which characterises ductile behaviour of these bars at high temperatures. However, percentage of elongation had decreased for sample 1 when temperature was increased from $600^0$C to $650^0$C. The maximum ductility for sample 3 increased much higher than normal ductility at $650^0$C which was about 38%.

## 4. Conclusion

In this study we observed that the Yield strength, Tensile strength, Young's modulus of the re-bars decreased as the heating temperature increased (up-to $650^0$C). The weakening of the mechanical properties is very stiff after $500^0$C

The ductility or percentage of elongation for all 3 samples has increased profoundly with temperature and up to 38%. So the samples have become highly ductile at around fire temperatures.

It has been clearly observed that the samples lost ~62% yield strength and ~57% UTS at $500^0$C, whereas, at $650^0$C they lost almost all of their load bearing capability (~87% of UTS). So, it is clear that these local brands of 500W can hardly bear any load at temperature above $600^0$C. So, if these rods are used in concrete structures, they will be very vulnerable when exposed to extreme fire condition.

## 5. Acknowledgments

Authors wishing to acknowledge assistance of department of Materials and Metallurgical Engineering, Bangladesh University of Engineering and Technology for their tremendous support during the entire research. We would also like to thank Shubhrodev Bhowmik and Mahbuba Liza for their guideline in preparing this paper

## 6. Reference

[1] ASHI reporter," The Effects of Fire on Structural Systems" ,2007, *American Society of Home Inspectors*

[2] American Institute of Steel Construction, Inc, *Manual of Steel Construction, Facts for Steel Buildings-Fire*, 2003, AISC, Chicago, IL,2003

[3] Topcu, I B, Karakurt, C, "Properties of Reinforced Concrete SteelRebars Exposed to High Temperatures", *Research Letters in MaterialsScience*, vol 2008, Article ID 814137, p 1-4

[4] Wald, F, Da Silva, L S, Moore, D B, Lennond, T, Chladna, M,Santiagob, A,Benes, M, Borgesf, M, "Experimental behaviour of asteel structure under natural fire," 2006*Fire Safety Journal*, vol 41, no. 7,pp 509-522

[5] Zhong Tao, Xingqiang Wang, Brian Uy, Stress-Strain Curves of Structural and Reinforcing Steels after Exposure to Elevated Temperatures, 2013, *Journal of Materials in Civil Engineering*

[6] Avner, S H (1964), Introduction to Physical Metallurgy, *McGraw-Hill*, New York, NY.

[7] Basu, P C, Shylamoni, P and Roshan, A D (2004), "Characterization of steel reinforcement for RC structures: an overview and related issues", *Indian Concrete Journal*, Vol 78 No 1, p 19-30

[8] Callister, W D and Rethwisch, D G (2013), *Fundamentals of Materials Science and Engineering,Wiley*, New York, NY

[9] Fonstein, N (2015), Advanced High Strength Sheet Steels: Physical Metallurgy, Design, Processing, and Properties, *Springer*, Berlin
[10] Kan, M, Peppler, W and Stueck, G (2006), High strength steel, *Cross Reference to Related Applications*

# Microstructural changes under Fire in TMT 500W Structural Steel rebars

**M.S. Shahriar, M.R. Nawaz and M.A. Islam**

Department of Materials and Metallurgical Engineering, Bangladesh University of Engineering and Technology, Dhaka 1000, Bangladesh

**Abstract.** Reinforced cement concrete or RCC has become the most commonly utilized construction material for various buildings, bridges, flyovers, etc. As plain concrete can't easily withstand the stresses created by vibrations, wind or other forces; steel bars, plates or fibers are embedded as reinforcing agent, which amplify the load bearing capability. But as the temperature is increased, e.g. in case of unwanted fire the load bearing capability of steel decreases at an alarming rate, so the RCC structure becomes vulnerable at high temperatures. In our country, TMT 500W steel rebars are increasingly used as reinforcement in RCC, but there are no enough studies regarding the properties of this type of steel, especially for locally produced grades, at very high temperatures. For this study, TMT 500W steel rebars of 8mm diameter from three different local brands were collected and their composition & room temperature tensile properties were evaluated. Then microstructures at various high temperatures of 500, 600 and 650$^0$C along with room temperature had been investigated. The results show that, the microstructure of each of as received samples at room temperature consists of ferrite and pearlite at the core and tempered martensite at the edge. At high temperatures, martensite and bainite at the edge became virtually vanished due to high temperature exposure and converted to ferrite and pearlite. The grain coarsening with increasing temperature was also significant.

**Keywords:** TMT steel bar, Fire resistance, High temperature microstructures, Martensite.

## 1. Introduction

The performance of steel depends on the properties associated with their microstructures, that is, on the arrangements, volume fractions, sizes, and morphologies of the various phases constituting a macroscopic section of steel with a given composition in a given processed condition [1,2]. Because all the phases in steels are crystalline, steel microstructures are made up of various crystals[3], sometimes as three or four different types which are physically blended solidification, solid-state phase changes, hot deformation , cold deformation , and heat treatment. Each type of microstructure and product is developed into its characteristic property ranges by specific processing routes that control and exploit microstructural changes. Thus processing technologies do not only depend on microstructure, but are also used to tailor final microstructure [4]. Also the temperature and stress conditions largely dictate the performance of steel structures during operation [5, 6]. To evaluate the microstructural changes in high temperature and stressed condition, 3 samples of 8mm rods of different brands so of different composition but the same type were collected. All of the samples were 500 TMT steel Rebar and compostion of themshelves are shown in Table 1 . The samples were held at high temperature tensile tests for some more than 1 hour. The temperatures were 500, 600 and 650$^0$C. Tempering inevitably occurred at those very high temperatures in the TMT samples. During tempering a great numbers of phenomena occur such as C atoms dissolution, carbides evolution, austenite dissolution, cementite formation and spheroidization, ferrite recrystalization etc [7]. Hence combination of those type of phenomena induce complex changes in microstructure during tempering which will be carefully investigated. There are three stages of tempering which occur sequentially with temperature increases.

### *1.1. Three stages of tempering*

(a) The first stage of tempering is also referred to as low temperature tempering. This results in the formation of a low carbon martensite and a carbide by transformation of high carbon martensite. The low carbon martensite is usually referred as tempered-martensite. As far as mechanical properties are concerned, marginal decreases in hardness value takes place at this stage. Excellent wear resistance and reduced internal stresses are the characteristics of hardened steel tempered in this stage.

(b) The second stage of tempering consists of heating steels in the temperature range varying from 350-500$^0$C. During this stage, retained austenite transforms to tempered-troosite. Another name given to this stage is medium temperature tempering. Ductility and toughness increase by this treatment with a corresponding decrease in hardness and strength.

(c) The third stage of tempering is also popularly known as high temperature tempering. It consists of heating steel within a temperature range of 500-600$^0$C. Heating to such high temperature results in the formation of ferrite-cementite mixture which is referred as tempered sorbite.

The cooling rate has no effect on the microstructure of steels after being tempered.[8,9]

## 2. Methodology

Three rod samples of three different brands each were collected from local vendors. The rods were all 8 mm round 500 TMT steel rebars. Compositions of the samples were found out by OES spectrometer (Foundry-Master Pro, Oxford Instrument, Germany). In order to see the microstructures the necessarily cut samples was coarse grinded by wheel grinder (Shan, Taiwan), then rubbed with emery papers of 120, 320, 600, 800, 1200 and 1500 grades. Then they were wheel polished in metallurgical sample pre-grinder(YM-2A, China) , then etched with 2% Nital solution. Finally microstructures were carefully observed at 500X magnification with optical micrsoscope(B-600MET,Optika Inc. ,Italy). Necessarily cut standard samples were subjected to high temperature tensile testing by Testomeric M500-50CT at high temperature deformation laboratory of BUET .Then same procedures mentioned above were followed for microstructural observation after high temperature tensile tests.

**Table 1**. Composition of samples used(wt%)

| Comp. Brand | C | Si | Mn | P | S | Cr | Mo | Ni | Al |
|---|---|---|---|---|---|---|---|---|---|
| A | 0.24 | 0.21 | 0.81 | 0.026 | 0.021 | 0.011 | 0.002 | 0.004 | 0.001 |
| B | 0.21 | 0.18 | 0.70 | 0.011 | 0.020 | 0.077 | 0.010 | 0.044 | 0.001 |
| C | 0.26 | 0.24 | 0.79 | 0.009 | 0.018 | 0.009 | 0.002 | 0.004 | 0.002 |

(a)

(b)

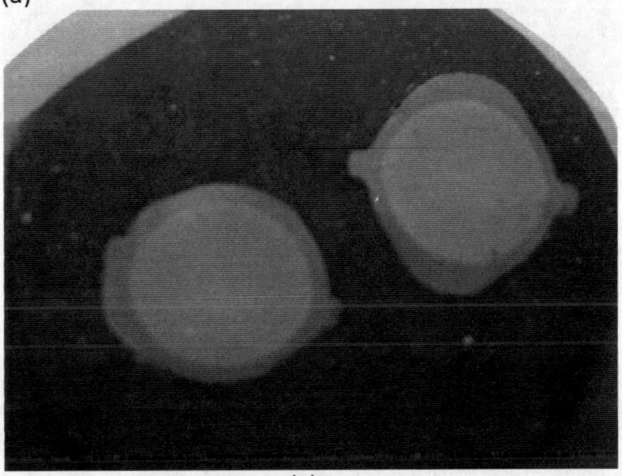

(c)

**Figure 1**. Macrostructures of samples from(a)Brand A, (b) Brand B & (c) Brand C all revealing martensite at the edges.

As the compositions revealed in Table 1. show that all the samples are mild steel and the macrostructures of the samples revealed martensite at the edges and core in the inner zones as expected in TMT 500W type steel rebars [10,11].

## 3. Results and discussion

The microstructures were found out by optical microscope and are shown in the figure 1, figure 2 & figure 3 for different brands separately. High temperature microstructures at 500, 600 & 650$^0$C along with room temperature microstructures are placed collectively in each figure for comparison purposes. The microstructures are all from the edges. The room temperature microstructures of as received sample all have martensitic structure in the edge which comprises needle shaped structure in the microstructures. Also there are tiny portion of retained austenite remained after quenching[12]. But in the high temperatures of 500, 600 & 650$^0$ C the microstructures have undergone extensive changes which is clear in the following figures. As the temperatures were very high(500$^0$ and above) , these microstructural changes can be termed as high temperature tempering phenomena which varied to some extent according to the compositional and process variations among the different brands mainly and also due to experimental variables to some extent. The soaking time and test time together took some more or less than 1 hour for each sample so the time effect on microstructure was ignored. The heating rate was constant for every sample so the microstructurual changes was mainly due to temperatures.

The cross sectional areas gradually decreased with increasing temperature and due to this , the grains were also contracted but this effect was neglected because the grains are so small and quantitative size variation with temperature was not needed to measure in this experiment, rather the grain shape changes with temperature was used as the basis to draw conclusions simply.

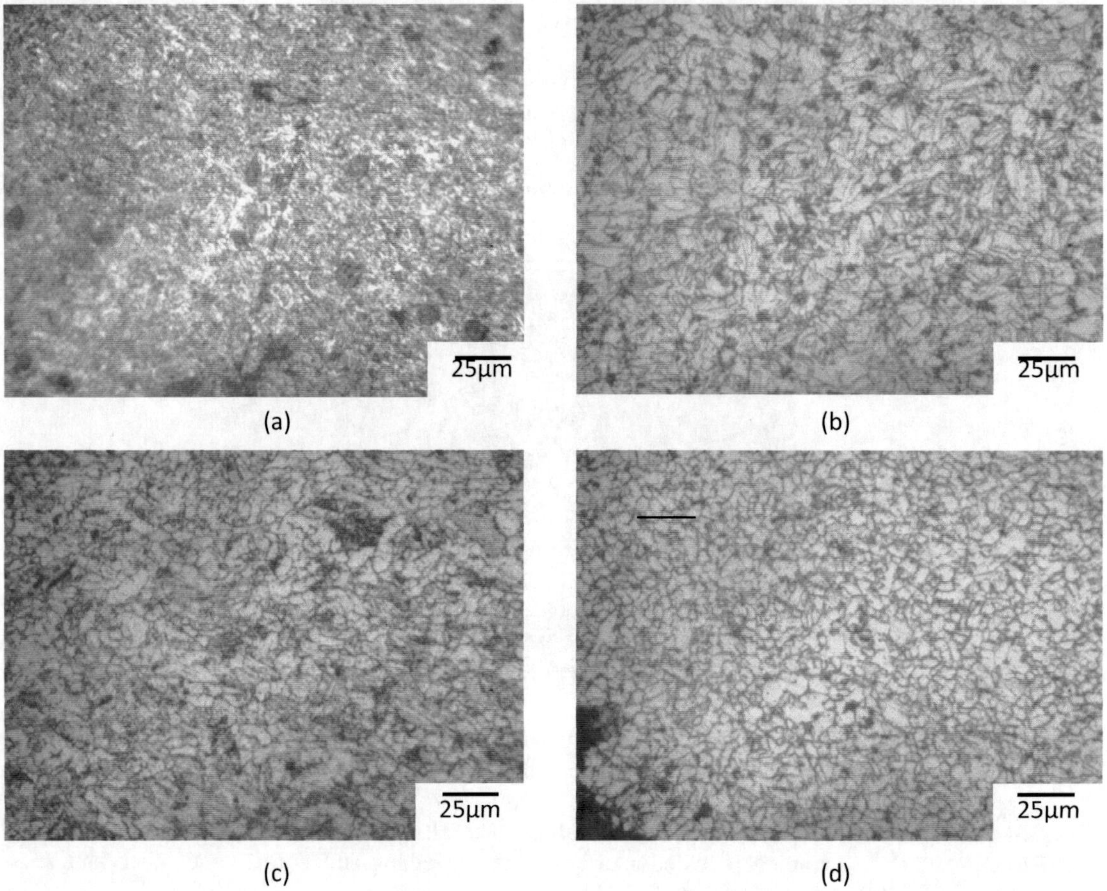


**Figure 2.** Microstructures of Brand A at(a) room temperature (b) $500^0$C (c) $600^0$C & (d) $650^0$C

From the microstructures of Brand A , this is apparent that high temperatures rendered clearer grains under microscope. This was due to the stress relief and recovery by dislocation movement, merging and thereby energy consumed[13,14]. The martensite structure completely losed its sharpness in high temperatures. The grains seemed elongated in one direction in $500^0$C because of the relatively recent dissolution of martensite into ferrite and cementite in this high temperature temperaing at $500^0$C; the needle shaped martensite rendered elongated grains after dissolution. At $600^0$C , the grains seemed less elongated as the elongated grains have higher energy , so to reduce the energy at this high temperature the grains tend to be equi- axial. At $650^0$C, the grains are virtually equi- axial.

IOP Conf. Series: Materials Science and Engineering **438** (2018) 012035 doi:10.1088/1757-899X/438/1/012035

The grain coarsening could be somehow realized by increased larger grains observed with temperature increase. The total number of grains seemly increased at 650$^0$C due to dynamic recrystallizatio

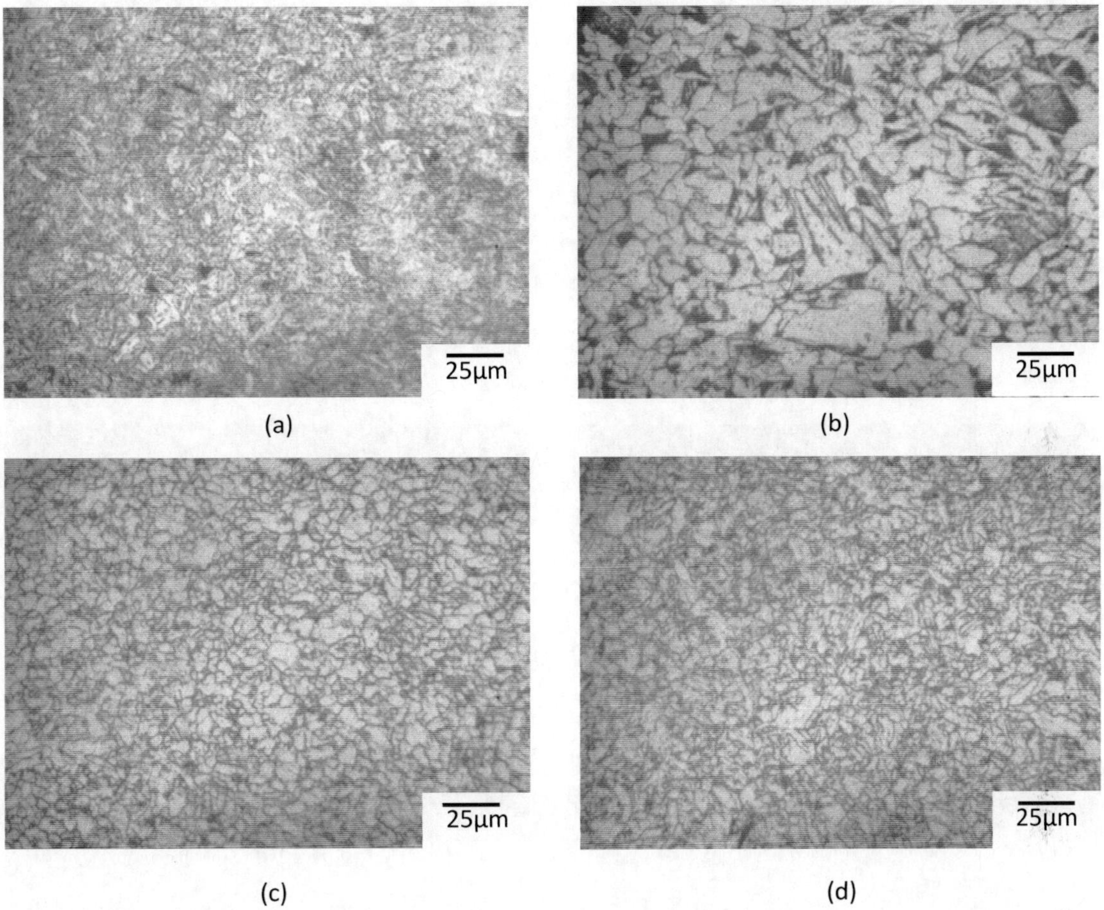


**Figure 3**.Microstructures of Brand B at (a) room temperature; (b) 500$^0$C; (c) 600$^0$C and (d) 650$^0$C

From figure 3, we can see that room temperature microstructure contains only martensite but high temperature microstructures don't process any martensite,, rather there are mainly ferritc grains found. As this is of mild steel composition, there should be very little carbide or pearlite in the equilibrium microstructure. As the very high temperatures mentioned are somehow sufficient for dislocation movement , so the microstructures tend to be like equilibrium structure . So there are more equi- axed grains with increasing temperature. The ferrite recrystallization seemingly occurred with increasing temperature as more grains are seen at 600$^0$C and 650$^0$C than at 500$^0$C. This phenomenon can be termed as dynamic recrystallization whereas mutual application of stress and heat accelerate recrystallization of grains [15]. The grain coarsening could not be easily realized from the increase in temperature from 600$^0$C to 650$^0$C; this may be due to little temperature change of 50$^0$C was not enough or may be the structure became stable till lower critical temperature.

IOP Conf. Series: Materials Science and Engineering **438** (2018) 012035 doi:10.1088/1757-899X/438/1/012035

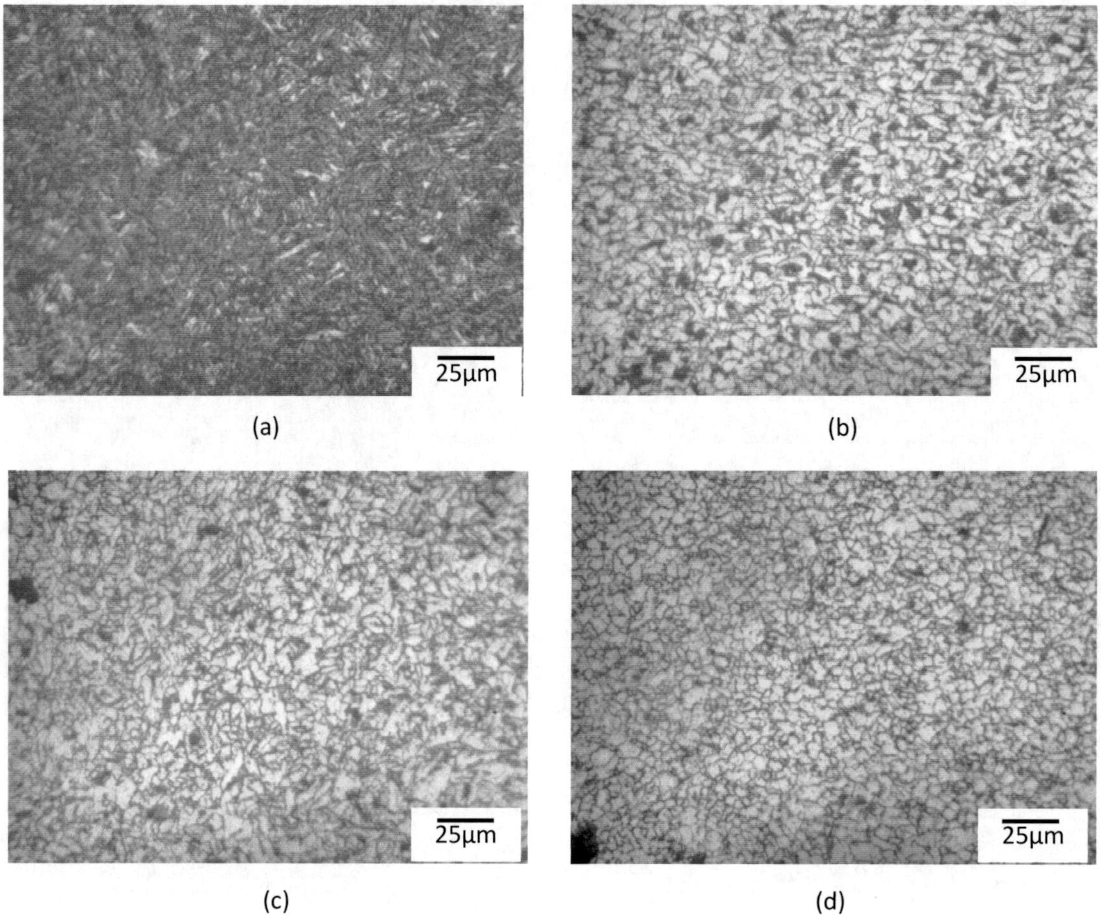


**Figure 4**.Microstructures of Brand C at (a) room temperature; (b) 500$^0$C; (c) 600$^0$C and (d) 650$^0$C

In figure 4, like the previous two brands, the martensite virtually decomposed into ferrite and pearlite in high temperatures. With temperature increase , the grains tended to be more and more equi-axed. There is also tendency for grain coarsening to occur but the opposite phenomenon ferrite recrystallization also occured, so grain coarsening couldn't be realized easily. This brand showed greatest tendency to reach equilibrium microstructure which indicated that dislocation movement was the easiest for this structure.

Among the three brands mentioned above, brand A showed the greatest change in microstructure, then brand B and brand C showed the least change at the high temperatures. But the relative ease of movement of dislocations was opposite for those brands. The easy martensite dissolution at high temperatures of 500, 600and 650$^0$ C to ferrite and pearlite indicated that there are severe lack of martensite stabililizer in the composition. The ferrite phase which occupies almost all of the equilibrium microstructure of mild steel [16] , is a very ductile phase. So all the samples of different brands showed dangerous vulnerability at fire temperatures for their respective brands.

## 4. Conclusions

Fire is a great concern for any structure, because at very high temperatures, almost all the materials degrade. So this is necessary to check and re-check the fire capability of TMT 500W steel rebars as they are increasingly used in this country. The investigation carried out in this experiment with local three brands helped to reach the following important conclusions;-

1. The martensite phase couldn't exist at the high temperatures indicating there was severe lack of martensite stabilizer in the composition. The resulting microstructure composed of almost ferrite which lacks necessary propreties for structural purpose.
2. The grains tended to be equi- axed with the increase in temperature.
3. In high temperatures, the grains tended to be coarsened but the opposite effect of ferrite recrystallization due to mutual application of stress and temperature was also significant.
4. Many of the most recent ferrite grains which formed due to martensite dissolution were elongated as evident in the lower temperatures.
5. Recovery and internal energy release due to the ease of dislocation movement at high temperatures, caused the microstructures become clearer.

## 5.References

[1] Keehan, E. (2004). *Effect of Microstructure on Mechanical Properties of High Strength Steel Weld Metals*. Ph.D. Chalmers University of Technology and Goteborg University.
[2] www.imetllc.com/manipulating-metal-properties.
[3] www.materials.ac.uk/resources/fe/heattreatment.ppt
[4] Karagöz, Ş., Atapek, Ş. and Yilmaz, A. (2010). Microstructural and Fractographical Studies on Quenched and Tempered Armor Steels. *Materials Testing*, 52(5), pp.316-322.
[5] French, H. (1922). *Effect of temperature, deformation, and rate of loading on the tensile properties of low-carbon steel below the thermal critical range, by H.J. French, ...* Washington: Government printing Office.
[6] Levings, J. and Sritharan, S. (2012). Effects of Cold Temperature and Strain Rate on the Stress-Strain Behavior of ASTM A706 Grade 420(60) Steel Reinforcement. *Journal of Materials in Civil Engineering*, 24(12), pp.1441-1449.
[7] https://www.slideshare.net/RakeshSingh125/f-tempering
[8] Jhen, S. (n.d.). Effect ofHeat Treatment on the Microstructure and Mechanical Properties of Steel.
[9] Talebi, S., Ghasemi-Nanesa, H., Jahazi, M. and Melkonyan, H. (2017). In Situ Study of Phase Transformations during Non-Isothermal Tempering of Bainitic and Martensitic Microstructures. *Metals*, 7(9), p.346.
[10] Ray, A., Mukerjee, D., Sen, S., Bhattacharya, A., Dhua, S., Prasad, M., Banerjee, N., Popli, A. and Sahu, A. (1997). Microstructure and properties of thermomechanically strengthened reinforcement bars: a comparative assessment of plain-carbon and low-alloy steel grades. *Journal of Materials Engineering and Performance*, 6(3), pp.335-343.
[11] Capitani, M., Ixtaina, V., Nolasco, S. and Tomás, M. (2013). Microstructure, chemical composition and mucilage exudation of chia (Salvia hispanicaL.) nutlets from Argentina. Journal of the Science of Food and Agriculture, 93(15), pp.3856-3862.
[12] www.heat-treat-doctor.com/.../RetainedAustenite.pdf
[13] www.heat-treat-doctor.com/.../Stress%20Relief.pdf
[14] Marcotte, M. (1973). Optimum Time and Temperature for Stress Relief Heat Treatment of Stainless Steel Wire. *Journal of Dental Research*, 52(6), pp.1171-1175.
[15] Huang, K. and Logé, R. (2016). A review of dynamic recrystallization phenomena in metallic materials. *Materials & Design*, 111, pp.548-574.
[16] Callister, W. and Rethwisch, D. (n.d.). *Fundamentals of materials science and engineering*

**Acknowledgments:** All thanks go to my thesis supervisor M.A Islam, for his charming teaching and insightful explanations. The contribution of my fellow thesis partner M.R. Nawaz was also vital. During the experiment, great help was received from staffs of laboratory, so I am also grateful to them.

IOP Conf. Series: Materials Science and Engineering **438** (2018) 012036 doi:10.1088/1757-899X/438/1/012036

# Stress Analysis of a Cylinder Subjected to Thermo-mechanical Loads by Using FEM

**Somnath Somadder, Md. Shahidul Islam**

Department of Mechanical Engineering, Khulna University of Engineering and Technology, Khulna 9203, Bangladesh

**Abstract.** Material science is the most emerging field in today's research and is getting widespread day by day. Numerical investigation of various kinds of products by using different kinds of material and changing boundary condition is getting popular because it is less time consuming and is beneficial from the cost benefit perspective. The behavior of material subjected to thermo-mechanical loads is of interest to researchers in a wide variety of discipline. Cylinders have wide use in industrial engineering like as containing various types of liquid. They often operate under static or fluctuating mechanical and thermal loads. Various methods are introduced for increasing strength of cylinders such as compound cylinder, increasing wall thickness. In this paper the analytical and numerical analysis of single and compound thick cylinder is given. The results of the analytical approach are validated to a finite element model.

**Keywords.** Stress analysis, Thermo-mechanical loads, Compound cylinders, Analytical and finite element analysis.

## 1. Introduction

The design of thick walled pressure vessels for operation at very high temperatures and pressure is a tough task with several concerns encompassing definition of the working and acceptable stress intensities, conditions of failure, material performance etc. Cylinders have been commonly used due to their significance for containing various types of fluids. Engineers are trying to improve the design of pressure vessel in order to ensure the safety and reliability of cylinder under various thermo-mechanical loads.For storing and bypassing fluids cylinders are broadly used in various kinds of industries like chemical, petroleum, pharmaceuticals, military industries as well as in power plants. When hot fluid passes through the cylinder it exerts a certain amount of pressure and heat transfer occurs between the surface of the cylinder and the fluid. Discontinuity in model and material properties causes fracture in ductile materials. High pressure cylinders used in industries, process plants, pneumatic tanks, hydraulic reservoirs, storage for gases are the general application of thick walled cylinders [1].

Earlier many investigation have been done for increasing the efficiency of the cylinders. Numerous research studies have been carried to determine acceptable stress range, conditions of failure and material performance at the interface of bimetallic cylinder. An analytical expression to determine the thermo-mechanical stresses of a multilayered compound cylinder considering the effect of closed ends was derived by Zhang et al. [2]. Vedeld et al. [3] derived the expression for two-layer cylinders exposed to pressure and thermal gradient for investigating stress and displacement induced in the body. It is noticeable that temperature has important influence in stress distribution. Plane stress condition was used by Sollund et al. [4] to develop analytical solutions for stress components of compound cylinders where the change along the length is assumed to be zero. The use of composite materials have widely been investigated. An overview of advanced materials was provided by David et al. [5] for high pressure hydrogen storage as storing high pressure hydrogen is important in many aspects. Design and reliability of bimetallic cylinder was inspected by Wilson et al. [6]. Kautor [7] investigated that for same thermo-mechanical loading using bimetallic pressure vessel is more convenient and the stress of a compound cylinder under pressure and temperature was also investigated. Minimization of material volume for the fabrication of multilayer cylinders subjected to pressure load was investigated by Patil [8].For reducing hoop stress shrink fit cylinder was analyzed and bursting pressure analysis was determined by Hareram [9]. For the compound cylinder with increasing inner pressure an elastic and plastic model is developed by Tazana for determining stress and displacement fields [10].In this paper, the resulting stress and displacement fields in both single and compound cylinders subjected to thermo-mechanical loads are presented. The analytical solutions are compared with finite element solutions.

## 2. Analytical model for determining the stress and displacement distribution of the cylinder

Displacement throughout the thickness of the cylinder can be obtained from following equation,

$$u_i(r)=\frac{(1+v_i)}{(1-v_i)}\left(\frac{\alpha_i}{r}\right)\int_{r_{in}}^{r}\rho T_{cyi}d\rho+C_{i1}r+\frac{C_{i2}}{r} \tag{1}$$

Where $C_{i1}$ and $C_{i2}$ are the constants of integration for cylinder $C_{yi=1,2}$

Using the displacement given by equation (1) and stress strain relationship, the radial and hoop stress can be calculated as,

$$\begin{aligned}\sigma_r^i &= E_i\left[-\frac{\alpha_i}{(1-v_i)r^2}\int_{rin}^{r}\rho T_{cyi}d\rho+\frac{C_{i1}}{(1+v_i)(1-2v_i)}-\frac{C_{i2}}{(1+v_i)r^2}\right]\\ \sigma_\theta^i &= E_i\left[\frac{\alpha_i}{(1-v_i)r^2}\int_{rin}^{r}\rho T_{cyi}d\rho+\frac{C_{i1}}{(1+v_i)(1-2v_i)}+\frac{C_{i2}}{(1+v_i)r^2}-\frac{\alpha_i T_{cyi}}{(1+v_i)}\right]\end{aligned} \tag{2}$$

The integration constants, $C_{ij}$, are determined using the boundary conditions in the compound cylinders. These boundary conditions are related to the application of the internal pressure, $p_i$, the continuity of the radial displacement and the radial stress at the interface layer between the two cylinders. They are summarized in the following equation:

$$\begin{aligned}&\sigma_r^1(a)=-p_i;\sigma_r^2(c)=0\\ &\sigma_r^1(b)=\sigma_r^2(b);u_1(b)=u_2(b)\end{aligned} \tag{3}$$

Substituting the temperature distributions, stresses and displacements of the inner and outer Cylinders into the boundary conditions, the equations system, given by Eq. (4), is obtained

$$\begin{bmatrix} 1 & k_{12} & -1 & -k_2 \\ 1 & k_{22} & k_{23} & k_{24} \\ 1 & k_{32} & 0 & 0 \\ 0 & 0 & 1 & k_{44} \end{bmatrix} \begin{bmatrix} C_{11} \\ C_{21} \\ C_{12} \\ C_{22} \end{bmatrix} = \begin{bmatrix} B_1 \\ B_2 \\ B_3 \\ B_4 \end{bmatrix} \tag{4}$$

With:

$$k_{22} = -\frac{(1-2\nu_1)}{b^2} = \frac{a^2}{b^2} k_{32} = \frac{c^2}{b^2} k_{44}$$

$$k_{12} = -k_{14} = \frac{1}{b^2}$$

$$k_{23} = -\left(\frac{E_2}{E_1}\right)\left(\frac{1+\nu_1}{1+\nu_2}\right)\left(\frac{1-2\nu_1}{1-2\nu_2}\right) = -k_{24}\left(\frac{b^2}{(1-2\nu_2)}\right)$$

$$B_1 = \left(\frac{1+\nu_1}{1-\nu_1}\right)\left(\frac{\alpha_1}{b^2}\right)\int_a^b \rho T_{cyi} d\rho = -B_2 \frac{1}{1-2\nu_1}$$

$$B_3 = \frac{(1-2\nu_1)(1+\nu_1)}{E_1} p_i$$

$$B_4 = \alpha_2\left(\frac{1+\nu_2}{1-\nu_2}\right)\left(\frac{1-2\nu_2}{c^2}\right)\int_b^c \rho T_{cy2} d\rho$$

By obtaining the integration constant we can calculate the stress and displacements at each radial position from equation (1) and (2).

## 3. Finite element model

The validation of the proposed analytical model is done by using finite element model with ABAQUS software. Simulation was done by considering the cylinder subjected to thermo-mechanical loads. As this is a thermo-mechanical analysis mesh element type is used C3D8T which is an 8-node thermally coupled brick, trilinear displacement and temperature. To examine the material properties effect on the behavior of the cylinder; steel is used for single cylinder and for compound cylinder steel and aluminum is used. Loads are changed to observe the effect of loads on the stress and displacement fields. For inner cylinder steel is used and for outer cylinder aluminum is used for the analysis. Coupled temperature displacement is used for stress analysis.

### 3.1 Physical Aspects of the Model

In this analysis two cylindrical element have chosen to analyze. The inner cylinder having inner radius $a = 0.01\ m$ and outer radius $b = 0.02\ m$. The outer cylinder having inner radius $b = 0.02\ m$ and outer radius $c = 0.03\ m$. The inner cylinder is made of steel and outer cylinder is made of aluminum. Bimetallic interface occurs at $r = b = 0.02\ m$.

IOP Conf. Series: Materials Science and Engineering **438** (2018) 012036 doi:10.1088/1757-899X/438/1/012036

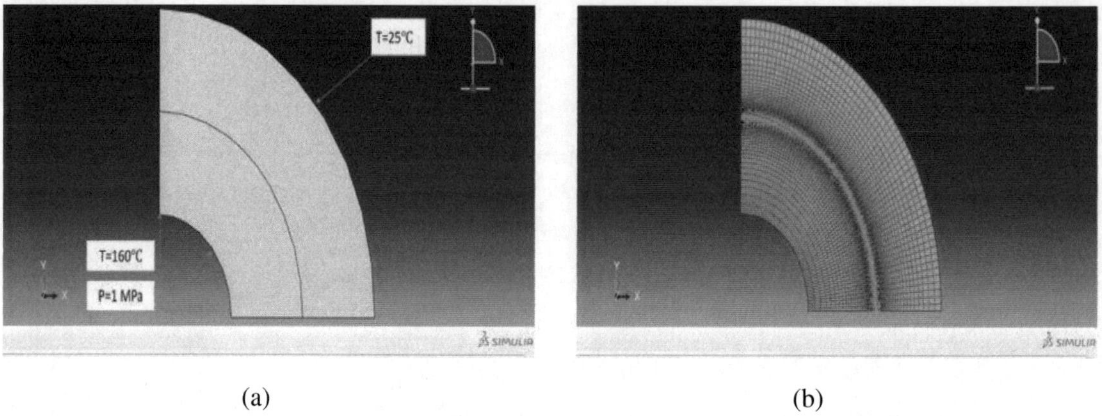


**Figure 1.** Model showing the (a) boundary conditions (b) mesh picture of the model.

Since at the interface material property changes so bias has been used so there is more element near the interface than the inner and outer side of the cylinder. As this is a thermo-mechanical analysis mesh element type is used C3D8T which is an 8-node thermally coupled brick, trilinear displacement and temperature.

3.2. Material properties
In this analysis, material Steel and Aluminum are used. Here aluminum is used as upper material and steel as lower material. The mechanical properties of aluminum and steel are:

**Table 1.** Properties of the two materials used for the analysis.

| Material | $E(GPa)$ | $v$ | $\alpha\,(°C)$ | $K\left(W.m^{-1}.K^{-1}\right)$ |
|---|---|---|---|---|
| Steel ASTM A564 H1150 | 210 | 0.3 | $11.6\times10^{-6}$ | 19.5 |
| Aluminum 1050A-H9 | 72 | 0.33 | $24\times10^{-6}$ | 234 |

## 4. Results and discussion
In this present work, ABAQUS software was used to analyze the stress and displacement variation of a thick walled cylinder under pressure and temperature. All stress components are plotted by changing loading conditions. Firstly the model was analyzed under pressure and then the model was analyzed under combined pressure and temperature loading. Symmetric boundary conditions are used as the geometry is symmetric and the boundary conditions are also symmetric. So, quarter of the model is used for analysis, resulting in fewer degrees of freedom and computational costs. The graphical presentations and the contours are shown below. All simulated results are plotted against r to investigate their stress and displacement characteristics. The graphical representations are shown below.

4.1 Mesh sensitivity analysis of the work

In case of numerical analysis mesh sensitivity analysis is the most important issue. The main objective of numerical analysis is to obtain the most accurate result in the shortest possible time. Here different mesh sizes and different number of elements were imposed on the model to verify the accuracy of work.

IOP Conf. Series: Materials Science and Engineering **438** (2018) 012036 doi:10.1088/1757-899X/438/1/012036

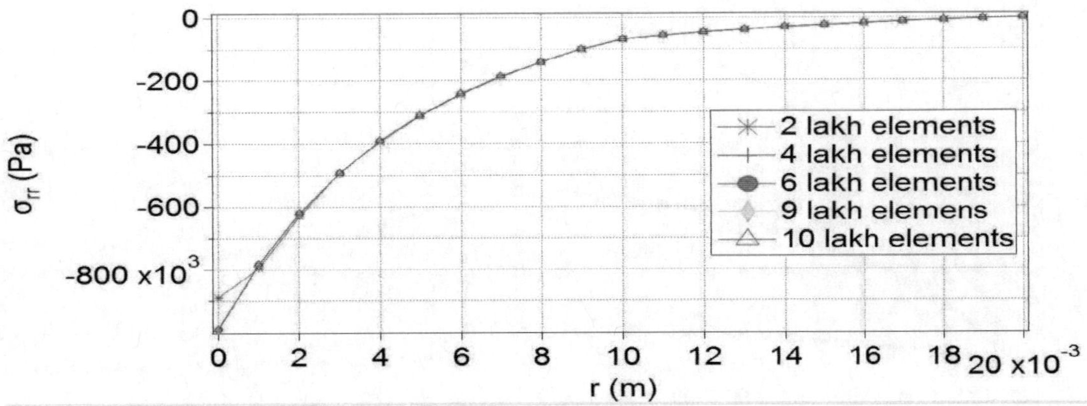


**Figure 2.** Mesh sensitivity analysis of the work showing variation of radial stress along the true distance of the radius.

From the figure it is clear that results are not so accurate using 2 lakh elements. Accuracy of the result is increased as element number is increased from 2 lakh to 10 lakhs. It was observed that the results obtained using 9 lakh elements is same as the result obtained 10 lakh elements. So, there is no reason to use 10 lakh elements and increase the computational time. So the optimum number of elements is found 9 lakh. Further analysis was done using this optimum number of elements.

4.2 Variation of stress for a 2D single cylinder

When fluid passes through cylinders it exerts uniform internal pressure on the inner surface of the cylinder. Here various magnitude of pressure was applied in the inner surface of the cylinder and the variation of radial and hoop stress throughout the radius was investigated.

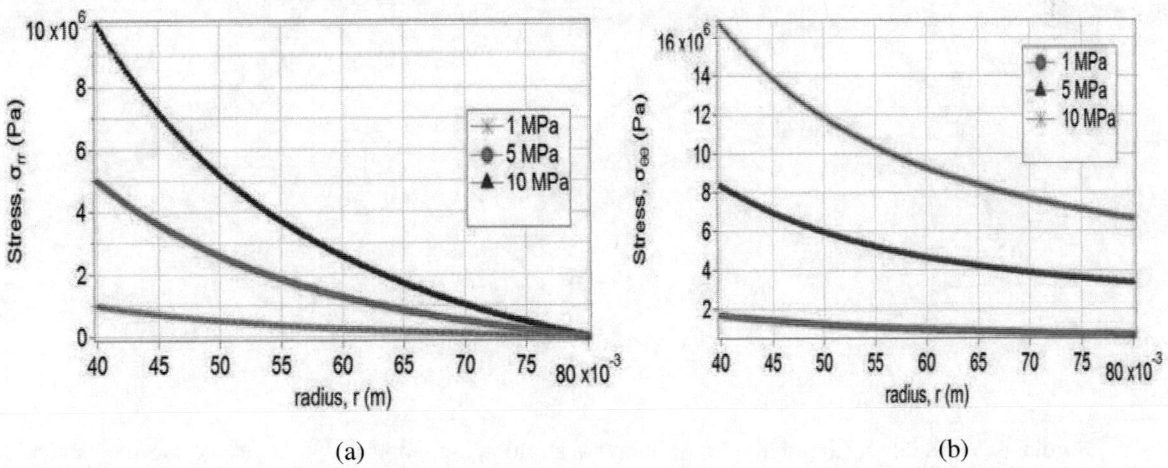


**Figure 3.**Variation of (a) radial and (b) hoop stress along radius subjected to internal pressure only.

Figure 3 indicates the variation of radial stress and hoop stress along the radius subjected to internal pressure only. Both radial and hoop stress is maximum at the inner radius and minimum at the outer radius as pressure is applied on the inner surface of the cylinder only. It is also noticeable from the graph that increasing internal pressure causes increase in radial stress and hoop stress.

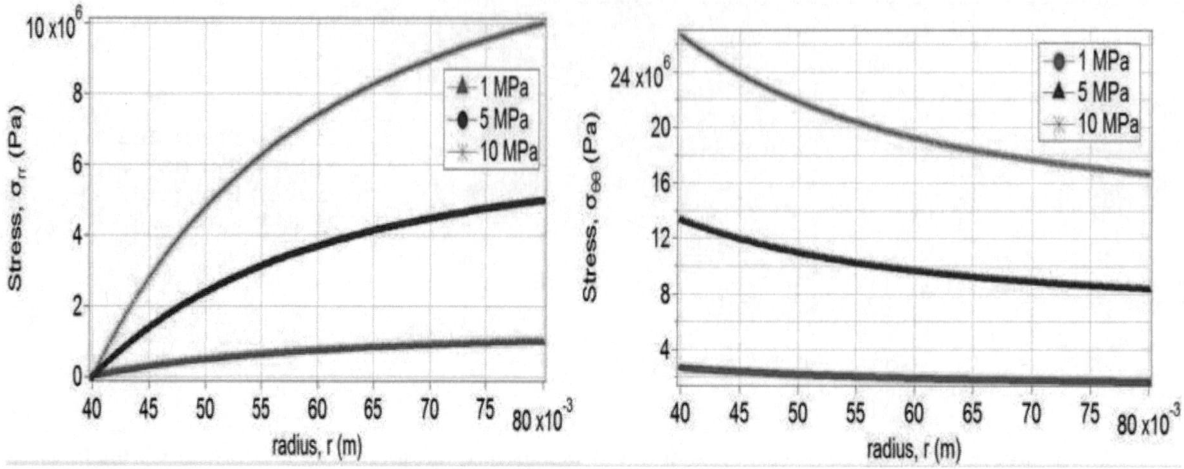


(a) (b)

**Figure 4.** Variation of (a) radial and (b) hoop stress along radius subjected to external pressure only.

In this graph it is seen that as the direction of applying pressure load changes to external surface of the cylinder only. Radial stress is nearly zero at the inner radius that satisfies the boundary condition. By increasing the magnitude of pressure, stress developed is also increased.

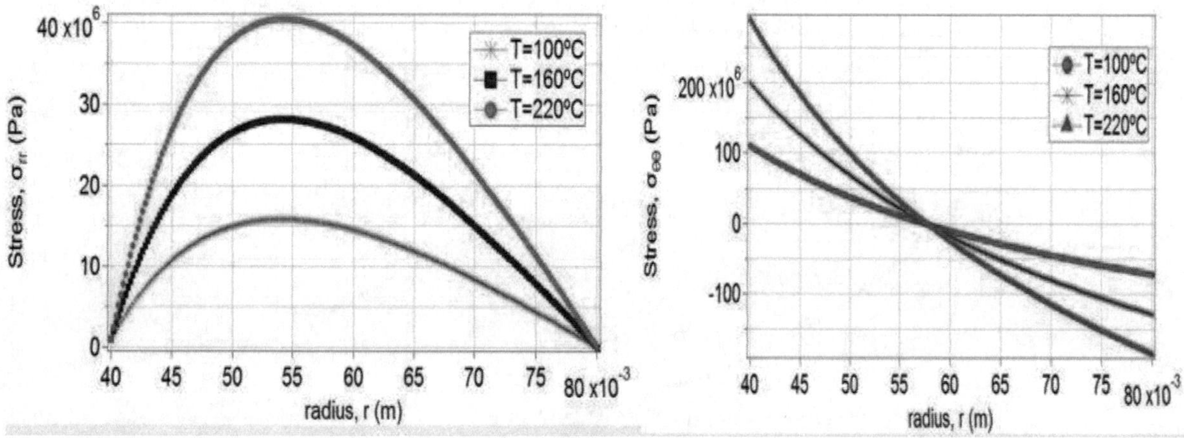


(a) (b)

**Figure 5.** Variation of (a) radial and (b) hoop stress along the radius subjected to thermo-mechanical loads.

In this graph the variation of radial and hoop stress subjected to thermo-mechanical loads is represented. The temperature gradient between the inner and outer surface of the cylinder creates radial displacements in each radial positions. When the temperature difference between inside and outside is large, the radial stress is more significant for all radial positions as illustrated.

4.3 Variation of stress for a 2D compound cylinder

**Figure 6.** Variation of radial stress along the radius subjected to pressure and temperature.

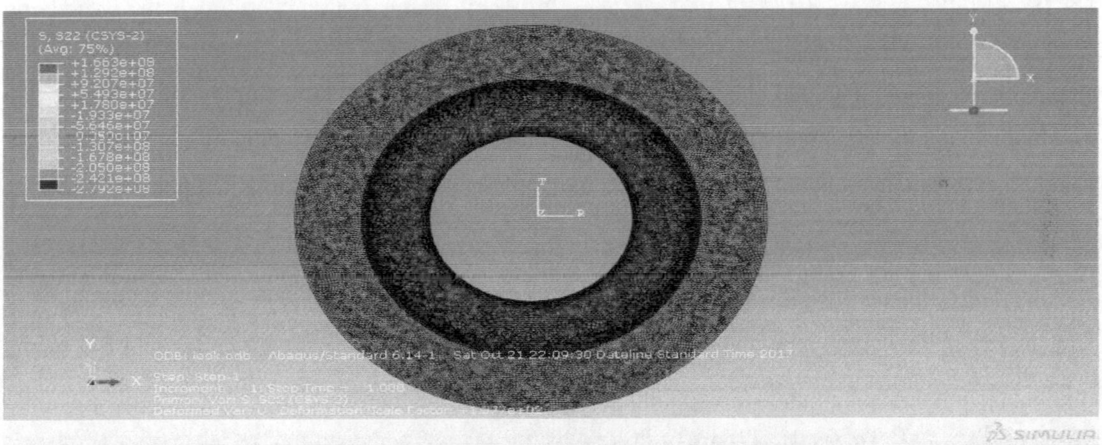


**Figure 7.** Variation of hoop stress along the radius subjected to pressure and temperature.

From these figures it is clear that maximum hoop stress occur at the interface. As hoop stress is responsible for the failure of the cylinder, extra material should be added at the interface to prevent failure. Extra material may be provided by using adhesive near the interface region. Stress continuity is a must for bimetallic joint otherwise joint will break. It is clear that using compound cylinder increases the pressure carrying capacity of the cylinder and extra material should be provided at the interface region to prevent failure.

IOP Conf. Series: Materials Science and Engineering **438** (2018) 012036 doi:10.1088/1757-899X/438/1/012036

## 5. Conclusion

In this work the displacement and stress distribution along radius subjected to internal and external pressure is investigated. The boundary conditions like pressure load and temperature are changed to observe the stress and displacement variation. Analytical solutions are presented. The results obtained in this analysis leads to the following conclusions:

a) In order to ensure the dependability of the cylinder, pressure and temperature gradient should be considered during design operations.
b) The tangential stress component is more responsive to temperature than the pressure.
c) The value of stresses are higher in the interface section than the other spaces.
d) Extra material should be added in the interface to prevent failure.
e) The internal pressure effect is more important when they are made from similar materials than dissimilar materials.

## 6. References

[1] Choudhury Susanta, Roy H., 2010 "Stress analysis of a thick walled cylinder". B Tech thesis, National Institute of Technology India.

[2] Q. Zhang, Z. W. Wang, C. Y. Tang, D. P. Hu, P. Q. Liu and L. Z. Xia, 2012., "Analytical solution of the thermo-mechanical stresses in a multilayered composite pressure vessel considering the influence of the closed ends", International Journal of Pressure Vessels and Piping, Vol. 98, PP.102-110.

[3] K. Vedeld, H. A. Sollund and J. Hellesland, 2015,"Closed analytical expressions for stress distributions in two-layer cylinders and their application to offshore lined and clad pipes", Journal of Offshore Mechanics and Arctic Engineering, Vol. 137, Isuue 2 PP. 021702-021702-9.

[4] H. A. Sollund, K. Vedeld and J. Hellesland, 2014, "Efficient analytical solution for heated and pressurized multi-layer cylinder Ocean Engineering", Vol.92, PP.285-295.

[5] E. David, 2005. "An overview of advanced materials for hydrogen storage", Journal of Material Processing Technology, Vol.162–163 PP.169-177.

[6] W. R. D. Wilson and W. J. Skelton, 1967, "Design of bi-metallic high pressure cylinders", Proc. Inst. Mech. Eng., Vol.3, PP.1-10.

[7] Bahoum Kaoutar, Dinay, 2017, "Stress analysis of compound cylinders subjected to thermo-mechanical loads". J. Mechanical Science and Technology, Vol.31, Issue 4, PP. 1805-1811.

[8] Patil S.A., Miraje, 2011, "Minimization of material volume of three layer compound cylinder having same materials subjected to internal pressure". Int. J. Engineering Science and Technology Vol.3, NO.8, PP.26-40.

[9] Lohar Hariram, 2013, "Steess analysis and brust pressure determination of two layered compound cylinder". IJEST Vol 5.

[10] Vasko Tajana , Franjo., 2014 "Stress and deformation analysis of compound cylinders in elastic plastic range" , PhD Thesis, University of Osijek.

IOP Conf. Series: Materials Science and Engineering **438** (2018) 012037 doi:10.1088/1757-899X/438/1/012037

# Quantification and Qualification of Silica Sand Extracted from Padma River Sand

**A.S.M.M. Rafi, U.F. Tasnim and M.S. Rahman**

Department of Glass and Ceramic Engineering, Rajshahi University of Engineering and Technology, Rajshahi 6204, Bangladesh

**Abstract.** The sands of Bangladesh's rivers are potentially and economically important hosts for silica sand resources. The work is based on extraction of silica sand from sands of the Padma River and quantification and qualification of extracted silica sand as glass sand. Characterization tests carried out on sand samples collected from sand bars of the Padma River in north-western Bangladesh revealed that the silica content of the sand is 76.4%. The extraction of silica sand from head sand is done by physical separation process. A chemical analysis of extracted sand is done by XRF analysis and the result shows that extracted sand contains 85.9% silica. The qualification analysis is done by glass making and characterizing the glass sample.

## *1. Introduction*

Silica sand is one of the most abundant minerals that can be found in diverse ways such as in quartz crystals, huge forming hills, quartz sand or silica sand, sand stone and so forth. It is quartz that over time, through the work of water and wind, has been broken down into tiny granules. Quartz is the most common mineral in the crust of the Earth comprising an estimated 35% of all rocks [1].

Global demand for industrial silica sand is forecast to advance 5.5 percent per year to 291 million metric tons in 2018, with a value of $12.5 billion. Several industries applied silica as a major component of production. In general, silica sand demand can be segmented into five major markets: glass, hydraulic fracturing, foundries, building products, and chemicals, as well as other smaller markets [2].

Padma is one of the major river in Bangladesh. The Padma River in central Bangladesh is approximately 100 km long and flows in a south-east direction from the confluence of the Jamuna (or Brahmaputra) and the Ganges to join the upper Meghna River, below which point it is known as the Meghna. River sands contain high amount of silica which can be used in many sectors such as glass making sectors, the photovoltaic applications required a high purity, construction etc. The Ganges-Padma is the major hydrodynamic system that formed one of the world's largest delta complex covering a major portion of the country and also a greater part of West Bengal in India [3].

The main objective of this work is extraction of silica sand from Padma river sand. Then the quantification and qualification analysis is done from the extracted silica sand. That is why, the objectives of this work are - extracting of silica sand from Padma river sand, quantifying of extracted silica sand and comparing with raw sand and checking the qualification of extracted silica sand.

## *2. Methodology*

The silica sand was collected from the bank of the Padma River at the Talaimari balughat where the sands were gathered for construction uses. The collection point co-ordination is latitude-24.359169 and longitude-88.625522.

For physical separation 5 kg raw sand samples were dried in the sun to remove moisture. Separation of heavy, medium and light minerals was carried out depending on specific gravity or density of the minerals using laboratory shaking table. After the gravity separation and drying, the light fractions of Padma River sand was run into the high intensity roll magnetic separator to separate the magnetic and nonmagnetic fractions. The samples were separated at 60 rpm (magnetic fraction) and 140 rpm (paramagnetic fraction) speed by High Intensity Rolling Magnetic separator (HIRMS). The nonmagnetic part was separated again by Induced Roller Magnetic Separation (IRMS) for high precision. The remaining light fraction was separated into ferromagnetic, paramagnetic and nonmagnetic fractions with an Induced Roll Magnetic Separator (IRMS) run at 0.3 A (2000 Gauss) and 3.0 A (20 000 Gauss).

The fractions separated by the Induced Roll Magnetic Separator (IRMS) were then processed by using an electrostatic plate separator (ESPS) operating at 25 kV and a feed rate of 20 rpm to separate the conductive and nonconductive minerals. The nonconductive sand particles contain high amount of silica.

Raw sand sample and extracted silica sand sample (nonconductive fraction of ESPS) were sieved for 15 minutes on Fritsch Vibratory Sieve Shaker using mesh 500μm, 250μm, 125μm and pan for determining the grain size distribution. Nonconductive particles of 250μm, 125μm size were mainly used as silica sand. 125μm size nonconductive sand was used for glass making.

Chemical analysis of Padma River sand and extracted silica sand was determined by X-ray fluorescence (XRF) Spectrometric method using Rigaku ZSX Primus XRF machine equipped with an end window 4 KW Rh-anode X-ray tube operated at 40 KV-60 mA for heavy elements and 30 KV-100 mA for light elements. By comparing the X-ray fluorescence (XRF) Spectrometric result of raw sand and extracted silica sand, the increased amount of silica content can be measured.

The qualification analysis of extracted sand as glass sand was analyzed by making the glass from the sand. Glass samples were made from extracted silica sand and laboratory grade silica sand (99.99%). For making low melting point glass $Na_2CO_3$, $K_2CO_3$ and $Li_2B_4O_7$ were used as fluxing agent. To make a firm glass object three batches were taken from each of the two different amount of batches. These batches are shown in table 1.

**Table 1.** Batch composition

| Sample name | Silica $SiO_2$ (gm.) | Sodium Carbonate $Na_2CO_3$ (gm.) | Potassium Carbonate $K_2CO_3$ (gm.) | Lithium Tetra Borate $Li_2B_4O_7$ (gm.) | Batch amount(gm.) |
|---|---|---|---|---|---|
| (a) | 3.1 | 2.73 | 1.18 | - | 7 |
| (b) | 3.1 | 2.73 | 1.18 | - | 7 |
| (c) | 3.1 | 2.73 | 1.18 | - | 7 |
| (d) | 4 | - | - | 4 | 8 |
| (e) | 4 | - | - | 4 | 8 |
| (f) | 2 | - | - | 6 | 8 |

Sample (a), (c), (d) and (f) were produced from extracted silica sand and sample (b) and (e) were produced from laboratory quartz. Moreover sample (f) contains more lithium tetra borate than the silica. In this work, sample (a) and (b) was melted in the muffle furnace and other samples were produced by the fusion bead.

The properties of prepared glasses were tested by Micro hardness tester to perform the Vickers test and measuring glass softening point of glass samples. Then the properties of the sample of extracted silica sand was compared with the properties of the sample of laboratory grade silica sand.

### *3. Experimental Results and Discussion*

#### *3.1 Physical Separation*

Physical separation was started with 5 kg raw sand. From gravity separation table the raw sand was separated into three portion light fraction, medium fraction and heavy fraction. Among them light fraction occupy 92% of total weight. After that this light fraction was fed into the high intensity roll magnetic separator, where it was separated into magnetic and nonmagnetic fraction. Here, magnetic fraction was 65% and nonmagnetic fraction was 35% of the total charge. It is observed that the silicon oxide ($SiO_2$) content is significantly degraded in the magnetic fraction and upgraded in nonmagnetic fraction [4]. It is also observed that the silicon dioxide ($SiO_2$) content in nonmagnetic fraction of Padma River sand is high which can be potential for glass production. This separation was done to remove iron bearing magnetic minerals from silica sand.

Further processing was continued with the nonmagnetic fraction. Induced Roller Magnetic separator separated the nonmagnetic part into paramagnetic fraction and nonmagnetic fraction. The paramagnetic fraction contains 12% of the total nonmagnetic fraction which was separated by HIRMS and the rest 88% is nonmagnetic. After this separation, the nonmagnetic part of the sample is more likely to have higher silica content. The paramagnetic portion (garnet, ilmenite) was separated from the sample. The ESPS separate the nonmagnetic fraction of IRMS into conductive part and nonconductive part. The result shows that the conductive fraction contains 22.04% of the total nonmagnetic fraction and the rest 77.96% is nonconductive. This separation was done to remove feldspar and other aluminosilicates mineral form silica sand. As silica sand is non-conductor it is easily separated by this process. ESPS separated the fine conductors from coarse nonconductor rich streams. The nonconductive portion obtained from the ESPS is mostly the silica.

So by the physical separation process the amount of extracted silica sand is 1108.4 gm. from 5 kg of raw sand of Padma River.

#### ***3.2 Characterization Analysis***

The raw sand analysis was started with 1.5 kg sand. From this analysis it was found that 72% of the raw sand is in between 500 to 125μm range.

After that, the grain size analysis of extracted sand was started with 1108.4 gm. of silica sand. After analyzing the extracted sand it was found that about 98% sand is in between 500 to 125μm range. According to ASTM standard [1] it is clear that the appropriate size range for silica sand is above 125μm to below 500μm.

The grain size distribution of the silica sand is improved by the physical separation process. This purification process increases the amount of silica at the standard range about 26.53% [6]. Results of the characterization analysis are listed in table 2.

**Table 2.** Results of sieve analysis of raw sand and extracted sand.

| Size of the sieve (μm) | Raw sand Amount of minerals (%) | Extracted sand Amount of minerals (%) |
|---|---|---|
| 500 | 17.00 | 0.49 |
| 250 | 33.67 | 46.91 |
| 125 | 38.25 | 51.25 |
| Pan | 11.33 | 1.35 |

#### ***3.3 Elemental Analysis***

Raw sand XRF analysis shows that it has relatively high silica content about 76%. So sand can be processed to improve the silica quality and degraded the impurities content. Sand of 125 μm size was taken for the analysis. By the analysis it is determined that the raw sand contains other than silica 10.62% $Al_2O_3$, 3.19% $K_2O$, 2.55% CaO, 3.39% $Fe_2O_3$ in an amount that is noticeable and have other minerals such as MnO, $Y_2O_3$, $Cr_2O_3$, NiO, $ZrO_2$, BaO, $Rb_2O$ etc.

After the physical separation was done. The sand of 125 μm size was taken to do the XRF analysis. The result shows that the silica content of the extracted sand is 85.85%. Some minerals like $ZnO_2$, $Y_2O_3$ and BaO were vanished by the physical separation process.

By the physical separation process silica content increases about 10%. Other impurities such as iron oxide, alumina, phosphorus penta oxide, manganese oxide, chromium oxide etc. are decreased in a large amount. The main impurity which has the most adverse effect on glass strength and color i.e., iron oxide decreased about 50% of its previous weight. Chemical leaching process can be involved for further purification. In table 3, the summarization of the elemental analysis is shown.

**Table 3.** Compositional analysis result of raw sand and extracted silica sand by XRF analysis.

| Name of the materials | Raw sand Elemental amount (%) | Extracted sand Elemental amount (%) |
|---|---|---|
| $Na_2O$ | 1.7107 | 1.8261 |
| MgO | 1.2113 | 0.0775 |
| $Al_2O_3$ | 10.6174 | 6.8625 |
| $SiO_2$ | 76.3709 | 85.8465 |
| $P_2O_5$ | 0.0887 | 0.02 |
| $SO_3$ | 0.0081 | 0.0105 |
| Cl | 0.0162 | 0.0216 |
| $K_2O$ | 3.1859 | 2.101 |
| CaO | 2.5473 | 0.9145 |
| $TiO_2$ | 0.3089 | 0.0792 |
| $Cr_2O_3$ | 0.3603 | 0.5566 |
| MnO | 0.0566 | 0.0201 |
| $Fe_2O_3$ | 3.398 | 1.6167 |
| NiO | 0.0082 | 0.0177 |
| ZnO | 0.0039 | Nil |
| $Rb_2O$ | 0.014 | 0.0075 |
| SrO | 0.0134 | 0.0122 |
| $Y_2O_3$ | 0.0095 | Nil |
| $ZrO_2$ | 0.0106 | 0.0098 |
| BaO | 0.0599 | Nil |

### *3.4 Qualification Analysis as Glass Sand*

*3.4.1 Glass Making Analysis.* The qualification analysis of extracted sand as glass sand was analyzed by making the glass from the sand. Low melting glass was tried to be produced. For glass preparation six attempts were taken. Among these six samples only two samples were succeed to produce glassy phase sample (a) and sample (b) (from table 1). In fusion bead ingredients were melted quickly. So, those samples didn't have enough time to go through refining process. Samples from fusion bead were not produced in proper glassy phase.

Due to high carbonate content in batch composition of sample (a) there was huge boiling. So the pouring was not successful and sample (a) was broken. Produced samples are shown in figure 1.

(a) (b)

Figure 1. Glass samples produced from batch composition (a) and (b).

Sample (a) is whitish and sample (b) is greenish in color. Both of them have some unmelted particles trapped inside the glass because of refining process.

*3.4.2 Micro Hardness.* Micro hardness test was done to the produced sample (a) and sample (b). Results of these tests are shown in table 4.

**Table 4.** Results of micro hardness test.

| Name of the sample | Micro hardness (HV) |
|---|---|
| Sample (a) | 712 |
| Sample (b) | 373 |

*3.4.3 Glass Softening Point.* After the micro hardness testing both this two sample go through softening point testing. The softening point of the samples are shown in table 5.

**Table 5.** Results of softening point test.

| Name of the sample | Softening point (̊c) |
|---|---|
| Sample (a) | 603 |
| Sample (b) | 603 |

## 4. Conclusion

As a commodity, silica sand is a low priced product but a primary ingredient for a diversity of products. On the whole, the consumption of silica sand can be measured as indicator to the global economy's trends and circumstances. In our developing country, it is the basic requirements for economic solvency to find out indigenous silica sources. This study is about extraction of silica sand from Padma river sand and analyze its' qualification. Following conclusions are made from this study.

- 98% silica sand is in the 125 to 500 μm size range.
- The XRF analysis result revealed that the silica sand content in the extracted sand is about 86%.
- Further purification should be done by chemical leaching to produce glass sand.

## 5. References

[1] IsiakaOlajideOdewale, Characteriztion of Unwana Beach Silica Sand and Its Industrial Applications, *International Journal of Science Innovations and Discoveries,* IJSID, 2013, 3(1), 93-100.

[2] Mohammad Mansour, Develop A Strategic Forecast of Silica Sand Based on Supply Chain Decomposition, *International Journal of Engineering (IJE)*, Volume (9): Issue (1): 2015, 9-27.

[3] http://en.banglapedia.org/index.php?title=Padma_River
https://en.wikipedia.org/wiki/Padma_River

[4] M. A. Rahman, Valuable heavy minerals from the Brahmaputra River sands of Northern Bangladesh, *Applied Earth Science,* Transactions of the Institutions of Mining and Metallurgy: Section B, Volume 125, 2016- Issue 3: Mineral Sands, 174-188.

[6] http://engb.eriez.com/resources/content/engb/documents/pdfs/Sand_Processing Techpaper.pdf

***Acknowledgments***

The authors are grateful for a research grant in the course of this investigation to the authority of RUET as well as Department of Glass and Ceramic Engineering, Rajshahi University of Engineering and Technology (RUET), Rajshahi-6204, Bangladesh.

The help of Institute of Mining, Mineralogy & Metallurgy (IMMM), Bangladesh Council of Scientific & Industrial Research (BCSIR), Joypurhat related to the thesis work were greatly acknowledged.

# AUTHOR INDEX

**Institute of Physics**
Dirac House, Temple Back
Bristol BS1 6BE UK

ISSN: 1757-8981
ISBN 978-1-5108-7656-9
9 781510 876569

CURRAN ASSOCIATES INC.
proceedings.com